Lecture Notes in Computer Scie

Commenced Publication in 1973
Founding and Former Series Editors:
Gerhard Goos, Juris Hartmanis, and Jan van Leeuwen

Lecture Notes in Computer Science

François Bry   Jan Małuszyński (Eds.)

Semantic Techniques for the Web

The REWERSE Perspective

 Springer

Volume Editors

François Bry
Institute for Informatics, University of Munic 2
Oettingenstr. 67, 80538 München, Germany
E-mail: bry@lmu.de

Jan Małuszyński
Department of Computer and Information Science, Linköping University
581 83 Linköping, Sweden
E-mail: janma@ida.liu.se

Library of Congress Control Number: 2009934695

CR Subject Classification (1998): H.4, H.3, C.2, H.5, J.1, K.4, K.6, I.2.11

LNCS Sublibrary: SL 3 – Information Systems and Application, incl. Internet/Web and HCI

ISSN   0302-9743
ISBN-10   3-642-04580-4 Springer Berlin Heidelberg New York
ISBN-13   978-3-642-04580-6 Springer Berlin Heidelberg New York

springer.com

© Springer-Verlag Berlin Heidelberg 2009
Printed in Germany

Typesetting: Camera-ready by author, data conversion by Scientific Publishing Services, Chennai, India
Printed on acid-free paper   SPIN: 12720248   06/3180   5 4 3 2 1 0

Preface

This volume in the LNCS series *State-of-the-Art Surveys* gives an overview of the main results achieved by the Network of Excellence REWERSE on "Reasoning on the Web" (http://rewerse.net). REWERSE was funded by the European Commission and Switzerland within the "6th Framework Programme" (FP6), in the period of March 1, 2004 through February 29, 2008.

In the beginning of 2004, the World Wide Web Consortium (W3C) was still working on the Web Ontology Language (OWL) and accepted its definition as a W3C Recommendation in April 2004. This work was part of the Semantic Web initiative aiming at the development of knowledge-based Web technologies. There was a consensus that rules will also play an important role in the Semantic Web. This issue was not yet on the W3C agenda but became the focus of REWERSE research. The objective of REWERSE was to define and to implement a coherent family of Web rule languages with well-defined semantics allowing reasoning on the Web. Another line of research focused on the development of support technologies for Semantic Web languages, such as component models or verbalization in controlled English. An important objective was also to demonstrate the usefulness of the Semantic Web techniques in real-life applications, particularly in bioinformatics. Members of REWERSE also participated in the work of the W3C Rule Interchange Format (RIF) Working Group, which was created in late 2005. Its goal is to develop a format for the interchange of rules in rule-based systems on the Semantic Web.

The research results of REWERSE appeared in over 400, mostly peer reviewed, publications in leading journals, international conferences and in the public deliverables accessible at http://www.rewerse.net/deliverables.html. The objective of this volume is to give a coherent perspective of the main topics and results of REWERSE. The material is organized in eight chapters. Each of the chapters addresses one of the main topics of REWERSE. The topics are highly relevant for the Semantic Web research.

Each chapter:

- Includes a state-of-the-art survey on the addressed topic; however, the level of details is limited and the interested reader is referred to the respective REWERSE publications
- Presents in a clear and cohesive way REWERSE contributions on the topics of the chapter, with indication of the development after REWERSE end, whenever applicable
- Provides an extensive bibliography and pointers to further literature.

Chapter 1 addresses the issue of combining rules and ontologies, thus the issue of integrating the ontology layer and the rule layer of the Semantic Web. The presentation of the state of the art is done using a classification of the approaches introduced in REWERSE. In the REWERSE work, presented in more detail, the integration is achieved by extending rules with queries to OWL ontologies. Three languages of this kind considered in REWERSE are dl-programs, HEX, and HD-rules. Their implementations are based on the hybrid reasoning principle: they re-use reasoners of the underlying rules languages and of OWL.

Chapter 2 discusses the issue of adapting query languages to the different Web formats instead of defining a specialized query language, for each format. The REWERSE contribution is the first truly versatile Web query language, Xcerpt, which enables access to multiple Web formats. Xcerpt can nevertheless be implemented as efficiently as a specialized language such as XQuery that is restricted to tree-shaped XML data. This result was achieved by developing a formal foundation for any Web query language, called CIQLog. The semantics of XQuery, XPath, Xcerpt, and SPARQL (as well as most other Web query languages) can be expressed conveniently in CIQLog. For CIQLog queries, a query algebra, called CIQCAG, was defined that specifies an evaluation method for CIQLog queries and thus queries in any of the above languages. CIQCAG significantly advances the current frontier of highly scalable (linear time) tree queries by defining a new class of data graphs (a proper superclass of trees and queries) that can be evaluated as efficiently as tree data. Furthermore, CIQCAG scales to arbitrary shapes of trees and data, yet provides for each restricted class the same or better complexity than all previous approaches.

Chapter 3 addresses the issue of evolution and reactivity in the Semantic Web. This concerns the vision of an active Web, where data sources evolve autonomously and react to events. In 2004, when the REWERSE project started, this topic was not yet explored in the context of the Semantic Web.

The REWERSE contribution presented in this chapter is a general framework for reactive Event-Condition-Action rules in the Semantic Web over heterogeneous component languages and the concrete homogeneous language XChange developed in this framework. The Xcerpt language discussed in Chapter 2 was re-used in this development as the query language for the conditions of the ECA rules.

Chapter 4 discusses rule-based policy representations and reasoning. Policies are used for protecting security and privacy in the context of the Semantic Web. The chapter makes a synthetic survey of the state of the art, comparing 12 relevant policy languages according to 10 criteria. The REWERSE work resulted in the Protune framework, presented in the chapter in more detail.

Chapter 5 discusses the question of suitable component models for Semantic Web languages. Such languages include ontology languages, data and metadata query languages, and Web-service workflow languages. As learned from years of experience in the development of complex software systems, languages deployed to practice arguably need to support some form of component-based development. Components enable higher software quality, better understanding and reusability of already developed artifacts. Any component approach contains an underlying component model, a description detailing what are valid components and how they can interact. With the plethora of languages developed for the Semantic Web, the question of their underlying component models was first addressed in REWERSE. The chapter presents a language-driven component model specification approach proposed by REWERSE. In this approach, a component model can be (automatically) generated in a given base language (actually, its specification, e.g., its grammar). As a consequence one can provide components for different languages and simplify the development of software artifacts used on the Semantic Web. The techniques presented in this chapter are illustrated by their application to the language Xcerpt of Chapter 2.

Chapter 6 addresses the issue of natural language support for the users of Semantic Web languages. Such languages have a very technical focus and fail to provide good usability for users with no background in formal methods. The chapter argues that controlled natural languages like Attempto Controlled English (ACE) can solve this problem and discusses the work done in REWERSE on this topic. ACE is a subset of English that can be translated into various logic-based languages, among them the Semantic Web standards OWL and SWRL. It is accompanied by a set of tools, namely, the parser APE, the Attempto Reasoner RACE, the ACE View ontology and rule editor, the semantic wiki AceWiki, and the Protune policy framework discussed in Chapter 4. The applications cover a wide range of Semantic Web scenarios, which shows how broadly ACE can be applied.

Chapter 7 addresses the issue of searching for relevant information on the Web, in particular the search of the biomedical literature. It gives a synthetic comparison of existing biomedical search engines and presents in more detail the semantic search engine GoPubMed, which was developed by REWERSE members. GoPubMed uses the background knowledge of existing ontologies to index the biomedical literature. The chapter discusses how the semantic search can contribute to overcome the limits of classic keywords-based search paradigms.

Chapter 8 discusses the use of ontologies and standards for information integration in bioinformatics. It describes properties of the different types of data sources, ontological knowledge and standards that are available on the Web. Moreover, it discusses how this knowledge can be used to support integrated access to multiple biological data sources, and presents an integration approach that combines the identified ontological knowledge and standards with traditional information integration techniques. The work in REWERSE that has been done on ontology-based data source integration, ontology alignment and integration using standards is discussed in more detail.

The authors and the editors gratefully acknowledge the contributions of the external reviewers. They helped improve the original submissions. The authors and the editors are also grateful to Uta Schwertel for her help in the initial phase of the publication process.

March 2009   François Bry
                                                                         Jan Małuszyński

List of Reviewers

Uwe Aßmann
José Júlio Alferes
Pedro Barahona
Johan Boye
Claudiu Duma
Michael Eckert
Thomas Eiter
Tim Furche

Pascal Hitzler
Patrick Lambrix
Jeff Z. Pan
Axel Polleres
Riccardo Rosati
Nahid Shahmehri
Paolo Traverso
Sanja Vranes

Table of Contents

5 Component Models for Semantic Web Languages 233
Jakob Henriksson and Uwe Aßmann

6 Controlled English for Reasoning on the Semantic Web 276
Juri Luca De Coi, Norbert E. Fuchs, Kaarel Kaljurand, and Tobias Kuhn

Chapter 1
Hybrid Reasoning with Rules and Ontologies

Włodzimierz Drabent[1,2], Thomas Eiter[3], Giovambattista Ianni[3,4],
Thomas Krennwallner[3], Thomas Lukasiewicz[5,3], and Jan Małuszyński[2]

[1] Institute of Computer Science, Polish Academy of Sciences,
ul. Ordona 21, PL – 01-237 Warszawa, Poland
[2] Department of Computer and Information Science, Linköping University,
SE – 581 83 Linköping, Sweden
[3] Institut für Informationssysteme, Technische Universität Wien
Favoritenstraße 9-11, A-1040 Vienna, Austria
[4] Dipartimento di Matematica, Università della Calabria
P.te P. Bucci, Cubo 30B, I-87036 Rende, Italy
[5] Computing Laboratory, University of Oxford
Wolfson Building, Parks Road, Oxford OX1 3QD, UK

Abstract. The purpose of this chapter is to report on work that has been done in the REWERSE project concerning hybrid reasoning with rules and ontologies. Two major streams of work have been pursued within REWERSE. They start from the predominant semantics of non-monotonic rules in logic programming. The one stream was an extension of non-monotonic logic programs under answer set semantics, with query interfaces to external knowledge sources. The other stream, in the spirit of the $\mathcal{AL}$-log approach of enhanced deductive databases, was an extension of Datalog (with the well-founded semantics, which is predominant in the database area). The former stream led to so-called non-monotonic dl-programs and HEX-programs, and the latter stream to hybrid well-founded semantics. Further variants and derivations of the formalisms (like a well-founded semantics for dl-programs, respecting probabilistic knowledge, priorities, etc.) have been conceived.

1.1 Introduction

The purpose of this chapter is to report on the work that has been done in REWERSE on hybrid reasoning with rules and ontologies. The importance of rules and ontologies for Web applications is reflected by the World Wide Web Consortium's[1] (W3C) proposal of the layered architecture of the Semantic Web, including the ontology layer and the rule layer. The ontology layer of the Semantic Web was quite developed already at the REWERSE start in 2004. In the same year, W3C adopted the Web Ontology Language (OWL) recommendation [32]. On the other hand, the rule layer was a topic addressed by many researchers but was not yet official subject of W3C activities.

---

[1] http://www.w3.org/

F. Bry and J. Maluszynski (Eds.): Semantic Techniques for the Web, LNCS 5500, pp. 1–49, 2009.

Integration of the rule layer with the ontology layer is necessary for rule-based applications using ontologies, like data integration applications. It can be achieved by combining existing ontology languages with existing rule languages, or by defining new languages, expressive enough to define ontologies, rules and their interaction. An important issue in combination of ontology languages and rule languages based on logics is the semantics of the combined language, as a foundation for development of sound reasoners. The REWERSE work reported in this chapter focused on *hybrid* reasoning, where the reasoner of the combined language reuses the existing reasoners of the component ontology language and rule language.

Motivated by the need for hybrid reasoning with rules and ontologies, two major streams of work have been pursued within REWERSE. They start from the predominant semantics of non-monotonic rules in logic programming. The one stream was an extension of non-monotonic logic programs under answer set semantics, with query interfaces to external knowledge sources. The other stream, in the spirit of the $\mathcal{AL}$-log [33] approach of enhanced deductive databases, was an extension of Datalog (with the well-founded semantics, which is predominant in the database area). The former stream lead to so-called non-monotonic dl-programs and HEX-programs, and the latter stream to hybrid well-founded semantics. Further variants and derivations of the formalisms (like a well-founded semantics for dl-programs, respecting probabilistic knowledge, priorities, etc.) have been conceived.

To put the REWERSE work in a broader perspective, the chapter begins with a concise introduction to the *Resource Description Framework* (RDF) layer, which sets the standard for the data model for the Semantic Web, to the RDF Schema, seen as a simple ontology language, and to OWL. We then discuss rule languages considered in integration proposals and present a classification of the major approaches to integration which uses the terminology of [4,81]. The remaining part of the chapter surveys the REWERSE work on hybrid integration of rules and ontologies.

1.2   Overview of Approaches

This section gives a brief survey of the approaches to combine or integrate reasoning with rules and ontologies on the Web. It starts with a brief introduction to the underlying formalisms of the Semantic Web, followed by discussion on the rule languages considered in integration proposals. Finally, a classification of the integration proposals is presented. For a more comprehensive survey, the interested reader is referred to [40].

1.2.1   RDF and RDF Schema

The Resource Description Framework (RDF) defines the data model for the Semantic Web as labeled, directed graphs. An RDF dataset (that is, an *RDF graph*) can be viewed as a set of the edges of such a graph, commonly represented by *triples* (or *statements*) of the form:

$$Subject\ Predicate\ Object$$

where

- the edge links *Subject*, which is a *resource* identified by a URI or a *blank node*, to *Object*, which is either another resource, a blank node, a *datatype literal*, or an *XML literal*;
- *Predicate*, in RDF terminology referred to as *property*, is the edge label.

The next example, originating from [40], illustrates the main concepts of RDF.

Example 1. Take a scenario in which three persons named Alice, Bob, and Charles, have certain relationships among each other: Alice knows both Bob and Charles, Bob just knows Charles, and Charles knows nobody.

For encoding the information that "a person called Bob knows a person called Charles" we need a vocabulary including concepts like "person" and "name". We can adopt the so-called FOAF (friend-of-a-friend) RDF vocabulary [84]. Then the statement can be given by the following RDF triples:

```
_:b rdf:type foaf:Person, _:b foaf:name "Bob", _:b foaf:knows _:c,
_:c rdf:type foaf:Person, and _:c foaf:name "Charles",
```

where the qualified names like `foaf:Person` are shortcuts for full URIs like `http://xmlns.com/foaf/0.1/Person`, making usage of *namespace prefixes* from XML, for ease of legibility. For instance, the triple

```
_:b foaf:name "Bob"
```

expresses that "someone has the name Bob." `_:b` is a blank node and can be seen as an anonymous identifier. In fact, the name for a blank node is meaningful only in the context of a given RDF graph; conceptually, blank node names can be uniformly substituted inside an RDF graph without changing the meaning of the encoded knowledge.

RDF information can be represented in different formats. One of the most common is the RDF/XML syntax.[2]. The much simpler Turtle[3] representation is adopted in SPARQL, the W3C standard language for querying RDF data. The information of the example can be encoded in Turtle as follows:

```
@prefix rdf: <http://www.w3.org/1999/02/22-rdf-syntax-ns#>.
@prefix foaf: <http://xmlns.com/foaf/0.1/>.
_:a rdf:type foaf:Person .
_:a foaf:name "Alice" .
_:a foaf:knows _:b .
_:a foaf:knows _:c .
_:b rdf:type foaf:Person .
_:b foaf:name "Bob" .
_:b foaf:knows _:c .
_:c rdf:type foaf:Person .
_:c foaf:name "Charles" .
```

---

[2] http://www.w3.org/TR/rdf-syntax-grammar/
[3] http://www.w3.org/TeamSubmission/turtle/

```
@prefix rdfs:  <http://www.w3.org/2000/01/rdf-schema#> .
@prefix rdf:   <http://www.w3.org/1999/02/22-rdf-syntax-ns#> .
@prefix foaf:  <http://xmlns.com/foaf/0.1/> .
<http://www.mat.unical.it/~ianni/foaf.rdf>
        a foaf:PersonalProfileDocument.
<http://www.mat.unical.it/~ianni/foaf.rdf> foaf:maker _:me .
<http://www.mat.unical.it/~ianni/foaf.rdf> foaf:primaryTopic _:me .
_:me a foaf:Person .
_:me foaf:name "Giovambattista Ianni" .
_:me foaf:homepage <http://www.gibbi.com> .
_:me foaf:phone <tel:+39-0984-496430> .
_:me foaf:knows [ a foaf:Person ;
        foaf:name "Axel Polleres" ;
        rdfs:seeAlso <http://www.polleres.net/foaf.rdf>].
_:me foaf:knows [ a foaf:Person ;
        foaf:name "Wolfgang Faber" ;
        rdfs:seeAlso <http://www.kr.tuwien.ac.at/staff/faber/foaf.rdf>].
_:me foaf:knows [ a foaf:Person ;
        foaf:name "Francesco Calimeri" ;
        rdfs:seeAlso <http://www.mat.unical.it/kali/foaf.rdf>].
_:me foaf:knows [ a foaf:Person .
        foaf:name "Roman Schindlauer" .
        rdfs:seeAlso <http://www.kr.tuwien.ac.at/staff/roman/foaf.rdf>].
```

Fig. 1. Giovambattista Ianni's personal FOAF file

A Turtle shortcut notation like

```
_:a rdf:type foaf:Person ;
    foaf:name "Alice" ;
    foaf:knows _:b ;
    foaf:knows _:c .
```

is a condensed version of the first four triples stated before.

Other common notations for RDF are N-Triples[4] and Notation 3[5].

Figure 1 shows some information about one of the authors of this article extracted from RDF data that are available on the Web. RDF defines a special property rdf:type,[6] abbreviated in Turtle syntax by the "a" letter. It allows the specification of "IS-A" relations, such as, for instance,

```
<http://www.mat.unical.it/~ianni/foaf.rdf> a foaf:PersonalProfileDocument.
```

in Figure 1 links the resource <http://www.mat.unical.it/~ianni/foaf.rdf> to the resource foaf:PersonalProfileDocument via rdf:type.

---

[4] http://www.w3.org/2001/sw/RDFCore/ntriples/

[5] http://www.w3.org/DesignIssues/Notation3.html

[6] short for the full URI http://www.w3.org/1999/02/22-rdf-syntax-ns#type

Types supported for RDF property values are URIs, or the two basic types, viz. rdf:Literal and rdf:XMLLiteral. Under the latter, a basic set of XML schema datatypes are supported.

The RDF Schema (RDFS) is a semantic extension of basic RDF. By giving special meaning to the properties rdfs:subClassOf and rdfs:subPropertyOf, to rdfs:domain and rdfs:range, as well as to several types (like rdfs:Class, rdfs:Resource, rdfs:Literal, rdfs:Datatype, etc.), RDFS allows to express simple taxonomies and hierarchies among properties and resources, as well as domain and range restrictions for properties.

The semantics of RDFS can be approximated by axioms in FOL, see e.g. [82]. Such a formalization can be used as a basis for RDFS reasoning, where the truth of a given triple t in a given RDF graph G under the RDFS semantics is decided.

1.2.2   The Web Ontology Language OWL

The next layer in the Semantic Web stack serves to formally define domain models as shared conceptualizations, called ontologies [55]. The Web Ontology Language OWL [32] is used to specify such domain models. The W3C document defines three languages OWL Lite, OWL DL, and OWL Full, with increasing expressive power. The first two are syntactic variants of expressive but decidable description logics (DLs) [8]. In particular, OWL DL coincides with with $\mathcal{SHOIN}(\mathbf{D})$ at the cost of imposing several restrictions on the usage of RDFS. These restrictions (e.g., disallowing that a resource is used both as a class and an instance) are lifted in OWL Full which combines the description logic flavor of OWL DL and the syntactic freedom of RDFS. For in-depth discussion of OWL Full, we refer the interested reader to the language specification [32].

Table 1. Mapping OWL DL Property axioms to DL and FOL

OWL property axioms as RDF triples	DL syntax	FOL short representation
$\langle P$ rdfs:domain $C\rangle$	$\top \sqsubseteq \forall P^-.C$	$\forall x, y. P(x, y) \supset C(x)$
$\langle P$ rdfs:range $C\rangle$	$\top \sqsubseteq \forall P.C$	$\forall x, y. P(x, y) \supset C(y)$
$\langle P$ owl:inverseOf $P_0\rangle$	$P \equiv P_0^-$	$\forall x, y. P(x, y) \equiv P_0(y, x)$
$\langle P$ rdf:type owl:SymmetricProperty $\rangle$	$P \equiv P^-$	$\forall x, y. P(x, y) \equiv P(y, x)$
$\langle P$ rdf:type owl:FunctionalProperty $\rangle$	$\top \sqsubseteq \leqslant 1P$	$\forall x, y, z. P(x, y) \wedge P(x, z) \supset y = z$
$\langle P$ rdf:type owl:InverseFunctionalProperty $\rangle$	$\top \sqsubseteq \leqslant 1P^-$	$\forall x, y, z. P(x, y) \wedge P(z, y) \supset x = z$
$\langle P$ rdf:type owl:TransitiveProperty $\rangle$	$P^+ \sqsubseteq P$	$\forall x, y, z. P(x, y) \wedge P(y, z) \supset P(x, z)$

While RDFS itself may already be viewed as a simple ontology language, OWL adds several features beyond RDFS' simple capabilities to define hierarchies (rdfs:subPropertyOf, rdfs:subClassOf) among properties and classes.[7] In particular, OWL allows to specify transitive, symmetric, functional, inverse, and inverse functional properties. Table 1 shows how OWL DL property axioms can

---

[7] As conventional in the literature, we use "concept" as a synonym for "class", and "role" as a synonym for "property."

Table 2. Mapping of OWL DL Complex Class Descriptions to DL and FOL

OWL complex class descriptions*	DL syntax	FOL short representation
owl:Thing	$\top$	$x = x$
owl:Nothing	$\bot$	$\neg x = x$
owl:intersectionOf $(C_1 \ldots C_n)$	$C_1 \sqcap \cdots \sqcap C_n$	$C_1(x) \wedge \cdots \wedge C_n(x)$
owl:unionOf $(C_1 \ldots C_n)$	$C_1 \sqcup \cdots \sqcup C_n$	$C_1(x) \vee \cdots \vee C_n(x)$
owl:complementOf (C)	$\neg C$	$\neg C(x)$
owl:oneOf $(o_1 \ldots o_n)$	$\{o_1, \ldots, o_n\}$	$x = o_1 \vee \cdots \vee x = o_n$
owl:restriction $(P$ owl:someValuesFrom $(C))$	$\exists P.C$	$\exists y.P(x,y) \wedge C(y)$
owl:restriction $(P$ owl:allValuesFrom $(C))$	$\forall P.C$	$\forall y.P(x,y) \supset C(y)$
owl:restriction $(P$ owl:value $(o))$	$\exists P.\{o\}$	$P(x,o)$
owl:restriction $(P$ owl:minCardinality $(n))$	$\geqslant nP$	$\exists y_1 \ldots y_n. \bigwedge\limits_{k=1}^{n} P(x,y_k) \wedge \bigwedge\limits_{i<j} y_i \neq y_j$
owl:restriction $(P$ owl:maxCardinality $(n))$	$\leqslant nP$	$\forall y_1 \ldots y_{n+1}. \bigwedge\limits_{k=1}^{n+1} P(x,y_k) \supset \bigvee\limits_{i<j} y_i = y_j$

*For reasons of legibility, we use a variant of the OWL abstract syntax [91] in this table.

be expressed in DL notation and in FOL. RDF triples $S\ P\ O$ are represented here as $P(S,O)$, since in description logics (and thus in OWL DL), predicate names and resources are assumed to be disjoint.

Moreover, OWL allows the specifications of complex class descriptions to be used in rdfs:subClassOf statements. Complex descriptions may involve class definitions in terms of union or intersection of other classes, as well as restrictions on properties. Table 2 gives an overview of the expressive possibilities of OWL for class descriptions and its semantic correspondences with description logics and first-order logics.[8] Such class descriptions can be related to each other using rdfs:subClassOf, owl:equivalentClass, and owl:disjointWith keywords, which allow us to express description logic axioms of the form $C_1 \sqsubseteq C_2$, $C_1 \equiv C_2$, and $C_1 \sqcap C_2 \sqsubseteq \bot$, respectively, in OWL.

Finally, OWL allows to express explicit equality or inequality relations between individuals by means of the owl:sameAs and owl:differentFrom properties.

For details on the description logic notions used in Tables 1 and 2, we refer the interested reader to, e.g., [8].

The next, more expressive, iteration of OWL (version 2)[9] is developed by W3C. According to the proposal OWL2 will be based on the decidable description logic $\mathcal{SROIQ}$ [60]. It will support additional features such as acyclic composition of properties, qualified number restrictions, and possibility to declare symmetry, reflexivity, or disjointness for properties.

Example 2 (Ontologies in Description Logics). A simple ontology about publications available online at http://asptut.gibbi.com/sandbox/reviewers.rdf includes OWL statements which can be represented by the following DL axioms:

---

[8] We use a simplified notion for the first-order logic translation here—actually, the translation needs to be applied recursively for any complex DL term. For a formal specification of the correspondence between DL expressions and first-order logic, cf. [8].

[9] http://www.w3.org/TR/owl2-syntax/

$$\exists ex{:}title.\top \sqsubseteq ex{:}Paper \tag{1}$$

$$\exists ex{:}title^{-}.\top \sqsubseteq xsd{:}string \tag{2}$$

$$ex{:}isAuthorOf^{-} \equiv dc{:}creator \tag{3}$$

$$ex{:}Publication \equiv ex{:}Paper \sqcap \exists ex{:}publishedIn.\top \tag{4}$$

$$\top \sqsubseteq\ \leqslant 1\ ex{:}publishedIn^{-} \tag{5}$$

$$ex{:}Senior \equiv foaf{:}Person \sqcap\ \geqslant 10\ ex{:}isAuthorOf \sqcap \tag{6}$$
$$\exists ex{:}isAuthorOf.ex{:}Publication$$

The axioms express the following information: *ex:title* is a datatype property on *ex:Papers* that takes strings as values (axioms (1) and (2)). Furthermore, the property *ex:isAuthorOf* is the inverse of the property *dc:creator* (axiom (3)). Next, the ontology defines in (4) a class *ex:Publication* which consists of all the papers which have been published, and in (5), we state that *ex:publishedIn* to be an inverse functional property (i.e., every paper is published in at most one venue). An *ex:Senior* researcher (6) is defined as a person who has at least ten papers, some of which are published.

1.2.3   Rule Languages for Integration

The rule languages considered in integration proposals are usually extensions of *Datalog*. Generally, rules have a form of "if" statements, where the predecessor, called the body of the rule, is a Boolean condition and the successor, called the head, specifies a conclusion to be drawn if the condition is satisfied.

In Datalog, the condition of a rule is a conjunction of zero or more atomic formulae of the form $p(t_1, \ldots, t_m)$ where p is an m-ary predicate symbol and $t_1, \ldots, t_m$ are terms which are constant symbols or variables.[10] The head of a rule is an atomic formula (atom). For example, the rule

$$auntOf(X, Y) \leftarrow parentOf(Z, Y), sisterOf(X, Z)$$

states that X is an aunt of Y if Z is a parent of Y and X is this parent's sister. The semantics of Datalog associates with every set of rules (rulebase) its least Herbrand model (see, e.g., [90]), where each ground (i.e., variable-free) atom is associated with a truth value true or false. The least Herbrand model is represented as the set of all atoms assigned to true. These are all the ground atoms which follow from the rules interpreted as implications in FOL. For example, the least Herbrand model of the rulebase consisting of the rule above and of the facts $parentOf(tom, john)$, $sisterOf(mary, tom)$ includes the formula $auntOf(mary, john)$. On the other hand, $auntOf(mary, tom)$ does not follow in this rulebase. Datalog with negation uses this to conclude $\neg auntOf(mary, tom)$. Datalog rulebases constitute a subclass of logic programs. The latter use FOL

---

[10] In logic programming, atomic formulae may in addition include terms built with n-ary function symbols.

terms, not necessarily restricted to constants and variables. Proposals for integration of rules and ontologies are mostly based on the following extensions of Datalog (which apply also to logic programs):

- **Datalog with negation-as-failure**, where the body may additionally include negation-as-failure (NAF) literals of the form *not a* where a is an atom. Intuitively, a NAF literal *not a* is considered true if it does not follow from the program that a is true. For example, $happy(john)$ can be concluded from the rulebase

$$happy(X) \leftarrow healthy(X), not\ hungry(X)$$
$$healthy(john) \leftarrow$$

  Two commonly accepted formalizations of this intuition are the stable model semantics and the well-founded semantics (see the survey [11]), which are introduced in more detail in Section 1.3. These semantics differ in their view of a belief state as a single classical model in which each atomic fact is either true and false, versus a three-valued model in which each fact is either true, false or *unknown*.

- **Extended Datalog.** This extension (see extended logic programs in [11]) makes it possible to state explicitly negative knowledge. This is achieved by allowing negative literals of the form $-p$, where "$-$" is called the *strong negation* connective, in the heads of rules as well as in the bodies. In addition NAF literals are also allowed in the bodies. For example the rule

$$-healthy(X) \leftarrow hasFever(X)$$

  allows to draw an explicit negative conclusion.

- **Rulebases with priorities.** Datalog rulebases employing strong negation may be inconsistent, i.e., may allow to draw contradictory conclusions. For example, the rules

$$fly(X) \leftarrow bird(X)$$
$$bird(Y) \leftarrow penguin(Y)$$
$$-fly(X) \leftarrow penguin(X)$$
$$penguin(tweety) \leftarrow$$

  allow to conclude $fly(tweety)$ and $-fly(tweety)$. In Defeasible Logic [2] and in Courteous Logic Programs [54], the user is allowed to specify a priority relation on rules of the rulebase to resolve contradictions in the derived conclusions.

- **Disjunctive Datalog** (see Disjunctive Logic Programs in [11]) admits disjunction of atoms in rule heads, and conjunction of atoms and NAF literals in the bodies, e.g.,

$$male(X) \lor female(X) \leftarrow person(X).$$

A commonly used semantics of Disjunctive Datalog rulebases is an extension of Answer Set Semantics.

The rule languages are supported by implementations which make it possible to query and/or to construct the models of rulebases.

1.2.4 Rule Interchange Format RIF

While there are already standard languages for ontologies viz. RDFS and OWL (which are becoming increasingly used), there is no standard for a rules language available yet.Many rules languages and systems have been proposed, and they offer varying features to reason over Semantic Web data. The *Rule Interchange Format* (RIF) working group of W3C is currently developing a standard exchange format for rules on the Web [13,12]. The *Rule Interchange Format Basic Logic Dialect* (RIF-BLD) [12] proposed by the group is basically a syntactic variant of Horn rules, which most available rule systems can process.

1.2.5 Approaches to Integration

Integration of a given rule language with a given ontology language is usually achieved by defining a common extension of both, to be called the integrated language. Alternatively, one can adopt an existing knowledge representation language expressive enough to represent rules and ontologies. As OWL is a standard ontology language the ontology languages considered in integration proposals are usually its subsets. The approaches can be classified by the degree of integration of rules and ontologies achieved in the integrated language (see e.g. [4,81]).

Heterogeneous Integration. In this approach, the distinction between rule predicates and ontology predicates is preserved in the integrated language. Integration of rules and ontologies is achieved by allowing ontology predicates in the rules of the integrated language. Assume for example that an ontology classifies courses as project courses and lecture courses.

$$Project \sqcup Lecture = Course$$

It also includes assertions like *Lecture*($cs05$), *Project*($cs21$) or *Course*($cs32$) (e.g. for courses including lectures and projects). The assertions indicate offered courses. A person is considered a student if he/she is enrolled in an offered lecture or project. This can be expressed by the following rules, using the ontology predicates

$$student(X) \leftarrow enrolled(X, Y), Lecture(Y)$$
$$student(X) \leftarrow enrolled(X, Y), Project(Y)$$

In addition, the rulebase includes enrollment facts, e.g., *enrolled*($joe, cs32$). The extended language allows thus to define ontologies using the constructs of the ontology language and the rulebases with rules referring to the ontologies. An extended rulebase together with an ontology is called a hybrid knowledge base. In heterogeneous approaches, implementations are often based on the hybrid reasoning principle, where a reasoner of the ontology language is interfaced with a reasoner of the rule language to reason in the integrated language.

Two kinds of heterogeneous approaches can be distinguished:

- **Loose coupling.** In this approach, the body of a rule may contain queries to the ontology. A ground set of rules with ontology queries can be reduced to a set of rules without ontology predicates. If the answer to a ground ontology query is positive the query is removed from the rule, otherwise the rule is removed from the set. The semantics of knowledge bases with loose coupling is based on this idea.

  With loose coupling applied to the example above, it cannot be concluded that Joe is a student. This is because neither $Lecture(cs32)$ nor $Project(cs32)$ can be derived from the ontology.

  Examples of loose coupling include:

  - **dl-programs** [46,43,47,42] combining (disjunctive) Datalog with negation under answer set semantics with OWL DL. So-called dl-queries, querying the ontology, are allowed in rule bodies. They may also refer to a variant of the ontology, where the set of its assertions is modified by the dl-query. This enables bi-directional flow of information between rules and ontologies. This work was partly supported by REWERSE and is discussed in more detail in Section 1.3.2.
  - **HEX-programs** [44] extending logic programs under the answer set semantics with support for higher/order and external atoms. This work was partly supported by REWERSE and is discussed in more detail in Section 1.3.3.
  - **TRIPLE** [101] a rule language with the syntax inspired by F-logic which admits queries to the ontology in rule bodies.
  - **SWI Prolog**[11] a logic programming system with a Semantic Web library which makes it possible to invoke RDF Schema and OWL reasoners from Prolog programs.

- **Tight integration.** In this approach, a semantics for the integrated language is given which defines models of hybrid knowledge bases by referring to the semantics of the original rule language and to the FOL models of the ontology. For example, tight integration of Datalog (without negation) with a Description Logic can be achieved within FOL by interpreting Datalog rules as implications. In this semantics, $student(joe)$ is a logical consequence of the example hybrid knowledge base. As $Course(cs32)$ is an assertion of the ontology, it follows by the axiom $Project \sqcup Lecture = Course$ that in any FOL model of the ontology $Project(cs32)$ or $Lecture(cs32)$ is true. As $enrolled(joe, cs32)$ is true in every model so the premises of at least one of the implications

$$student(joe) \leftarrow enrolled(joe, cs32), Lecture(cs32)$$
$$student(joe) \leftarrow enrolled(joe, cs32), Project(cs32)$$

  must be true in any model. Hence $student(joe)$ is concluded.

---

[11] http://www.swi-prolog.org/

Examples of tight integration include:

- $\mathcal{AL}$-**log** [33] and **CARIN** [68], classical works on integrating Datalog with a family of Description Logics under the FOL semantics.
- $\mathcal{DL}$+**log** [97] and its predecessor *r-hybrid knowledge bases* [96] integrating Disjunctive Datalog under Answer Set Semantics with OWL DL. For each FOL model of the ontology, the rules of the knowledge base are reduced to rules of Disjunctive Datalog, with stable models defined by the Answer Set Semantics. Similar to $\mathcal{DL}$+log is the approach of [57]. The guarded hybrid (g-hybrid) knowledge bases introduced therein integrate so-called *guarded programs* with ontologies in a particular DL close to OWL DL.
- **Hybrid Rules** [39] integrating logic programs under the well-founded semantics with OWL DL. For each FOL model of the ontology, the rules of the knowledge base are reduced to a logic program with the model defined by the well-founded semantics. This work was done within REW-ERSE and is reported in more details in Section 1.3.4.
- **Tightly Coupled dl-Programs** [73] combine disjunctive logic programs under the answer set semantics with description logics. They are based on a well-balanced interface between disjunctive logic programs and description logics, which guarantees the decidability of the resulting formalism without assuming syntactic restrictions. They faithfully extend both disjunctive programs and description logics. We refer to [73] for a detailed comparison to the above loosely coupled dl-programs.

The theoretical foundations developed by studying integration of ontologies with variants of Datalog provide a basis for further extensions. This includes dealing with uncertain and inconsistent knowledge, and using integrated Datalog-based languages as condition languages for ECA-rules.

Homogeneous Integration. The integrated language makes no distinction between rule predicates and ontology predicates. It includes the original rule language and the original ontology language as sublanguages. The integration is to be *faithful* in the sense that the sublanguages should have the same semantics as the respective original languages. Homogeneous integration is difficult to achieve since usually ontology languages are based on FOL and rule languages often support non-monotonic reasoning. An interesting related question is if existing proposals for heterogeneous integration can be embedded into more expressive logical languages.

Examples of homogeneous integration include:

- **DLP (Description Logic Programs)** [53], a language obtained by intersection of a Description Logic with Datalog rules interpreted as FOL implications. DLP has a limited expressive power, but a DLP ontology can be compiled into rules and easily integrated into a rulebase of a more expressive rule language. For example **Sweet Rules**[12] combine DLP and Datalog with

---

[12] http://sweetrules.projects.semwebcentral.org/

strong negation and priorities. The technique of compiling ontologies to rules is also used in DR-Prolog [3] based on Defeasible Logic [2].

- **SWRL (Semantic Web Rule Language)**[13] extending OWL DL with rules interpreted as FOL implications. Thus SWRL is based on FOL and does not offer nonmonotonic features, such as negation-as-failure. SWRL is undecidable. More recent works define decidable subsets of SWRL: Description Logic Rules [66] and ELP [67].

- **F-logic** [62] extending classical predicate calculus with the concepts of objects, classes, and types. It is expressive enough to represent ontologies, rules and their combinations [61].

- **Hybrid MKNF Knowledge Bases** [85,87] take Lifschitz's bimodal *Logic of Minimal Knowledge and Negation as Failure (MKNF)* [69] as a basis of faithful integration of Description Logic with Disjunctive Datalog. In addition, more recent results define the well-founded semantics for a subclass of Hybrid MKNF KBs [63,64].

- **Extended RDF Ontologies** [1] is an extension of RDF graphs with rules which admits NAF and strong negation. A stable model semantics defined for ERDF extends the semantics of RDF Schema. It is based on the partial logic of [56] and supports both closed-world and open-world reasoning.

The issue of embedding existing heterogeneous approaches into unifying logics was addressed by several authors. In particular, [31] shows how the Quantified Equilibrium Logic can be used for embedding heterogeneous approaches, like $\mathcal{DL}$+log and g-hybrid knowledge bases. The first-order autoepistemic logic [65] is considered as a unifying framework for integration of rules and ontologies in [29,30]. The latter paper shows how dl-programs, r-hybrid knowledge bases and hybrid MKNF knowledge bases can be embedded in this logic.

1.3   Hybrid Rules and Ontologies in REWERSE

In this section, we give a brief exposition of work that has been done in REW-ERSE regarding the combination of rules and ontologies. In fact, this problem has been approached in different ways, aiming at the support of different semantics and operability of the combination.

The main achievements are combinations for the two standard semantics of non-monotonic logic programs to date that were already mentioned in Section 1.2.3, viz. the *stable model semantics* [50] (which is called *answer set semantics* [52] in the version where strong negation is supported), and the *well-founded semantics* [103].

The *stable model semantics* [51] associates with each rulebase some (possibly zero) two-valued Herbrand models called *stable models* (or *answer sets*). Intuitively, a model is stable, if it can be recreated by applying the rules of the program starting from facts, where negation-as-failure in rule bodies is evaluated

---

[13] http://www.w3.org/Submission/SWRL/

with respect to that model. Formally, stable models may be defined by using the famous Gelfond-Lifschitz reduct [51].

The *well-founded semantics* [103] instead associates with a rulebase a unique (three-valued) Herbrand model, called the *well-founded* model of P, in which each ground atom is assigned one of three logical values true, false or unknown. Intuitively, the facts of a program should be true, and the ground atoms which are not instances of the head of any rule should be false. This information can be used to reason which other atoms must be true and which must be false in any Herbrand model. Such a reasoning gives in the limit the well-founded model, where the truth values of some atoms may still be undefined.

The properties and relationships between stable and well-founded semantics are well-understood and explored, and we do not embark on this issue here but refer to the literature, cf. [10]. We mention, though, that for a large class of programs relevant in practice (so-called stratified programs [11]), the two semantics coincide.

However, the different nature of the two semantics, and the available methods and algorithms for program evaluation in them is important with respect to possible combinations with ontologies. Indeed, the stable model semantics as a multiple-models semantics has to cope with several possible outcomes (that is, with nondeterminism in the evaluation), while the well-founded semantics as a canonical model semantics is determined; this makes it also more amenable to use proof-oriented methods for evaluation. In line with this, well-founded semantics engines (e.g., XSB) may be top-town oriented, while stable model engines, by current technology, are very much bottom up oriented (e.g., DLV and Smodels).

Within REWERSE, combinations of rules and ontologies have been developed that fall into the heterogeneous integration class described in Section 1.2.5. More in detail, *non-monotonic dl-programs* [46,43] and the more general HEX-*programs* [44] have been developed in order to have a loose coupling of OWL ontologies with nonmonotonic logic programs under the answer set semantics, while *Hybrid Rules (HD-rules)* have been developed in order to tightly couple OWL ontologies with nonmonotonic logic programs under the well-founded semantics.

The combinations faithfully extend the underlying logic programming seman- tics, and prototypes have been implemented that build on existing standard reasoning engines for logic programs and OWL ontologies. In fact, they were the first implementations of this kind, giving REWERSE a lead in the realization of expressive non-monotonic combinations of rules and ontologies. An application within REWERSE was a tool for computing credentials from rule-based policy specifications, based on the engine for HEX-programs.

In the following subsections, we briefly present the two streams of work that have been carried out by the groups in Linköping and Vienna, respectively. For space reasons, we must confine to the essential aspects and conveying the flavor; more details are available in the background publications.

1.3.1   Extensions of Expressive Non-monotonic Logic Programs by DL-Programs and HEX-Programs

The first stream of work for combining rules and ontologies in REWERSE was directed towards the stable models and answer set semantics, and led to two formalisms: dl-programs and HEX-programs.

The development of dl-programs was motivated by providing an extension to ordinary logic programs that allows one to couple a logic programming engine and description logic reasoner in a meaningful way. However, apart from the usual software engineering problems in coupling heterogeneous systems, the real challenge consisted in a smooth semantic integration, given that logic programs and OWL ontologies are based on rather different semantic grounds which are difficult to bridge (cf. [40]). To overcome this problem, as described in Section 1.2.5 non-monotonic dl-programs foster a loose integration, which takes an interfacing view where the logic program rules and the OWL ontologies can exchange information in terms of extensional data through so called *description-logic atoms* (dl-atoms), which may appear in the logic program. In a nutshell, such atoms can update and query an ontology, i.e., information can flow in both directions of the integrated knowledge bases.

This concept appeared to be quite fruitful and allows an easy definition of the semantics of dl-programs, by generalizing the stable model resp. answer set semantics of ordinary logic programs in a natural way. Furthermore, abstraction of description logic atoms to generic *external atoms* (which is somewhat related to the notion of generalized quantifiers in logic) opened the door to combine ordinary logic programs not only with ontologies, but with (in principle) any kind of external software via an interface at the extensional level. In particular, this facilitates to access and combine data and information in different formats (e.g., in OWL and RDF simultaneously), and to "out-source" parts of computations from the logic program to external functions, which can use tailored and problem-specific methods; the rules in the logic program then serve the role to generate different scenarios (e.g., by making guesses) and constrain solution candidates, for which the results of different computations might be suitably combined.

It turned out that such capabilities were useful for a problem of credential computation in rule-based policy specifications, and that a prototype for this task could be easily built on top of a prototype implementation of HEX-programs.

1.3.2   DL-Programs

Description logic programs (dl-programs), which had been introduced in [46], are a novel type of hybrid knowledge bases combining description logics and logic programs. They form another contribution to the attempt in finding an appropriate formalisms for combined rules and ontologies for the Semantic Web.

Roughly speaking, dl-programs consist of a normal logic program P and a description logic knowledge base (DL-KB) L. The logic program P might contain special devices called dl-atoms. Those dl-atoms may occur in the body of a rule and involve queries to L. Moreover, dl-atoms can specify an input to L before querying it, thus in dl-programs a bidirectional data flow is possible between the description logic component and the logic program.

The way dl-programs interface DL-KBs allows them to act as loosely coupled formalism. This feature brings the advantage of reusing existing logic programming and DL system in order to build an implementation of dl-programs.

In the following, we provide the syntax of dl-programs and an overview of the semantics. An in-detail treatise is given in [43].

Syntax of DL-Programs. Informally, a dl-program $KB = (L, P)$ consists of a description logic knowledge base L and a generalized normal program P, which may contain queries to L. Roughly, such a query asks whether a specific description logic axiom is entailed by L or not.

We first define dl-queries and dl-atoms, which are used to express queries to the description logic knowledge base L. A *dl-query* $Q(\mathbf{t})$ is either

- a concept inclusion axiom F or its negation $\neg F$, or
- of the forms $C(t)$ or $\neg C(t)$, where C is a concept and t is a term, or
- of the forms $R(t_1, t_2)$ or $\neg R(t_1, t_2)$, where R is a role and t_1, t_2 are terms.

A *dl-atom* has the form

$$\mathrm{DL}[S_1 \, op_1 p_1, \ldots, S_m \, op_m \, p_m; Q](\mathbf{t}), \qquad m \geq 0, \tag{7}$$

where each S_i is either a concept or a role, $op_i \in \{\uplus, \cup\!\!\!\!-, \cap\!\!\!\!-\}$, p_i is a unary resp. binary predicate symbol, and $Q(\mathbf{t})$ is a dl-query. We call $p_1, \ldots, p_m$ its *input predicate symbols*. Intuitively, $op_i = \uplus$ (resp., $op_i = \cup\!\!\!\!-$) increases S_i (resp., $\neg S_i$) by the extension of p_i, while $op_i = \cap\!\!\!\!-$ constrains S_i to p_i.

A *classical literal* (or simply literal) l is an atom p or a negated atom $-p$ with a rule predicate symbol (hence not a predicate symbol of L). A *dl-rule* r has the form

$$a \leftarrow b_1, \ldots, b_n, \mathrm{not} \, b_{n+1}, \ldots, \mathrm{not} \, b_m, \tag{8}$$

where a is a literal and any literal $b_1, \ldots, b_m$ may be a dl-atom. We define $H(r) = a$ and $B(r) = B^+(r) \cup B^-(r)$, where $B^+(r) = \{b_1, \ldots, b_n\}$ and $B^-(r) = \{b_{n+1}, \ldots, b_m\}$. If $B(r) = \emptyset$ and $H(r) \neq \emptyset$, then r is a *fact*. A *dl-program* $KB = (L, P)$ consists of a description logic knowledge base L and a finite set of dl-rules P.

The next example will illustrate main ideas behind the notion of dl-program.

Example 3. An existing network must be extended by new nodes (Fig. 2). The knowledge base L_N contains information about existing nodes $(n_1, \ldots, n_5)$ and their interconnections as well as a definition of "overloaded" nodes (concept *HighTrafficNode*), which are nodes with more than three connections:

$$\geq 1 \; wired \sqsubseteq Node; \quad \top \sqsubseteq \forall wired.Node; \quad wired = wired^-;$$
$$\geq 4 \; wired \sqsubseteq HighTrafficNode; \quad n_1 \neq n_2 \neq n_3 \neq n_4 \neq n_5;$$
$$Node(n_1); \quad Node(n_2); \quad Node(n_3); \quad Node(n_4); \quad Node(n_5);$$
$$wired(n_1, n_2); \quad wired(n_2, n_3); \quad wired(n_2, n_4);$$
$$wired(n_2, n_5); \quad wired(n_3, n_4); \quad wired(n_3, n_5).$$

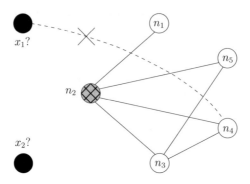

Fig. 2. Hightraffic network

In L_N, only n_2 is an overloaded node, and is highlighted in Fig. 2 with a criss-cross pattern.

To evaluate possible combinations of connecting the new nodes, the following program P_N is specified:

$$newnode(x_1). \tag{9}$$
$$newnode(x_2). \tag{10}$$
$$overloaded(X) \leftarrow \mathrm{DL}[wired \uplus connect; HighTrafficNode](X). \tag{11}$$
$$connect(X,Y) \leftarrow newnode(X), \mathrm{DL}[Node](Y), \tag{12}$$
$$\qquad \mathrm{not}\ overloaded(Y), \mathrm{not}\ excl(X,Y).$$
$$excl(X,Y) \leftarrow connect(X,Z), \mathrm{DL}[Node](Y), Y \neq Z. \tag{13}$$
$$excl(X,Y) \leftarrow connect(Z,Y), newnode(Z), newnode(X), Z \neq X. \tag{14}$$
$$excl(x_1,n_4). \tag{15}$$

Rules (9)–(10) define the new nodes to be added. Rule (11) imports knowledge about overloaded nodes in the existing network, taking new connections already into account. Rule (12) connects a new node to an existing one, provided the latter is not overloaded and the connection is not to be disallowed, which is specified by Rule (13) (there must not be more than one connection for each new node) and Rule (14) (two new nodes cannot be connected to the same existing one). Rule (15) states a specific condition: Node x_1 must not be connected with n_4.

Semantics of DL-Programs. Two different semantics have been defined for dl-programs, the (strong) answer-set semantics [46] and the well-founded semantics [47]. The former extends the notion of Gelfond-Lifschitz reduct incorporating the presence of dl-atoms: dl-programs can have, in general, multiple answer sets. The latter extends the well-founded semantics of [103] to dl-programs. The well-founded semantics is based on an appropriate notion of greatest unfounded set which embraces the presence of dl-atoms, and assigns a single three-valued model to every logic program.

More formally, given a consistent set I of classical literals (using the constants in P and L), I *satisfies* (i) a classical ground literal l, denoted $I \models_L l$, iff $l \in I$, and (ii) a dl-atom $a = \mathrm{DL}[\lambda; Q](\boldsymbol{c})$ with input list $\lambda = S_1 op_1 p_1, \ldots, S_m op_m p_m$, denoted $I \models_L a$, iff $L \cup \lambda(I) \models Q(\boldsymbol{c})$, where $\lambda(I) = \bigcup_{i=1}^m A_i(I)$ and

- $A_i(I) = \{S_i(\boldsymbol{d}) \mid p_i(\boldsymbol{d}) \in I\}$, for $op_i = \uplus$;
- $A_i(I) = \{\neg S_i(\boldsymbol{d}) \mid p_i(\boldsymbol{d}) \in I\}$, for $op_i = \cup\!\!\!-$;
- $A_i(I) = \{\neg S_i(\boldsymbol{d}) \mid p_i(\boldsymbol{d}) \in I$ does not hold$\}$, for $op_i = \cap\!\!\!-$.

Given a ground dl-rule r, we define (i) $I \models_L B(r)$ iff $I \models_L l$ for all $l \in B^+(r)$ and $I \not\models_L l$ for all $l \in B^-(r)$, and (ii) $I \models_L r$ iff $I \models_L H(r)$ whenever $I \models_L B(r)$. We say that I is a *model* of $KB = (L, P)$, or I *satisfies* KB, denoted $I \models KB$, iff $I \models_L r$ for all r in the grounding of P, $ground(P)$.

Strong answer sets can then be defined as follows. The *Gelfond-Lifschitz transform* of a dl-program $KB = (L, P)$ relative to consistent set I of ground literals for P is the dl-program $KB^I = (L, P^I)$, where P^I is obtained from $ground(P)$ by (i) deleting every rule r with $I \models_L l$ for some $l \in B^-(r)$ (ii) deleting all literals not b_i from all remaining rules. Assuming that all dl-atoms a that occur in P^I are monotone (i.e., $I \models_L a$ implies $I' \models_L a$, for all consistent sets $I \subseteq I'$ of ground literals for P), I is a *strong answer set* of KB iff it is a minimal model (w.r.t. set inclusion) of KB^I. For more details, see [43].

Example 4. As specified by the strong answer set semantics of dl-programs, the program (L_N, P_N) in Example 3 has four strong answer sets (we show only atoms with predicate *connect*): $M_1 = \{connect(x_1, n_1), connect(x_2, n_4), \ldots\}$, $M_2 = \{connect(x_1, n_1), connect(x_2, n_5), \ldots\}$, $M_3 = \{connect(x_1, n_5), connect(x_2, n_1), \ldots\}$, and $M_4 = \{connect(x_1, n_5), connect(x_2, n_4), \ldots\}$. Note that the ground dl-atom $\mathrm{DL}[wired \uplus connect; HighTrafficNode](n_2)$ from rule (3) is true in any partial interpretation of P_N. According to the proposed well-founded semantics for dl-programs in [47], the unique well-founded model of (L_N, P_N) contains thus $overloaded(n_2)$.

Features and Properties of DL-Programs. The strong answer set semantics of dl-programs is nonmonotonic, and generalizes the stable semantics of ordinary logic programs. In particular, satisfiable positive dl-programs (programs without default negation and $\cap\!\!\!-$ operator) have a least model semantics, and satisfiable stratified dl-programs have a unique minimal model which is iteratively described by a finite sequence of least models. Similarly, the well-founded semantics for dl-programs is a generalization of the well-founded semantics for ordinary logic programs. The two generalized semantics preserve some of the relationships that the answer set semantics and the well-founded semantics for normal programs have. In particular, given a knowledge base $KB = (L, P)$, the well-founded model of KB is contained in the set of cautious consequences of KB under strong answer set semantics [47,42]. Also, the two notions coincide in case stratified programs are considered.

The computational complexity of the formalism does not dramatically increase for dl-programs compared to normal logic programs: under strong answer set semantics, deciding satisfiability of general dl-programs over $\mathcal{SHIF}(\mathbf{D})$ DL-KBs is NEXP-complete, and $\mathrm{P^{NEXP}}$-complete if the DL-KB is in $\mathcal{SHOIN}(\mathbf{D})$. dl-programs have been generalized to a framework for incorporation of arbitrary knowledge sources other than description logic bases (see Section 1.3.3).

Applications. The bidirectional flow of knowledge between a description logic base and a logic program component enables a variety of possibilities. A major application for dl-programs is nonmonotonic reasoning on top of monotonic systems. We will present two flavors: *default logic* [95] and *closed world assumption* (CWA) [94]. Both reasoning applications can be implemented in dl-programs to support nonmonotonic reasoning for description logics.

We will give an example on how to implement default reasoning on top of ontologies. Since description logics are fragments of first-order logic, Reiter's default logic over description logics can be realized in dl-programs (cf. also terminological default logics [9]).

Let $\Delta = \langle L, D \rangle$ be a default theory, where

$$
L = \left\{ \begin{array}{l} redWine \sqsubseteq \neg whiteWine, lambrusco \sqsubseteq sparklingWine \sqcap redWine, \\ sparklingWine(veuveCliquot), lambrusco(lambrusco\_di\_modena) \end{array} \right\}
$$

and

$$
D = \left\{ \frac{sparklingWine(X) : whiteWine(X)}{whiteWine(X)}, \frac{whiteWine(X) : servedCold(X)}{servedCold(X)} \right\} .
$$

The embedding of Δ into a dl-program $KB^{df} = (L, P)$ is demonstrated next. Let P be the program

$$in_{whiteWine}(X) \leftarrow \mathrm{not}\ out_{whiteWine}(X) \tag{16}$$

$$out_{whiteWine}(X) \leftarrow \mathrm{not}\ in_{whiteWine}(X) \tag{17}$$

$$in_{servedCold}(X) \leftarrow \mathrm{not}\ out_{servedCold}(X) \tag{18}$$

$$out_{servedCold}(X) \leftarrow \mathrm{not}\ in_{servedCold}(X) \tag{19}$$

$$fail \leftarrow \mathrm{DL}[\lambda'; whiteWine](X), out_{whiteWine}(X), \mathrm{not}\ fail \tag{20}$$

$$fail \leftarrow \mathrm{DL}[\lambda'; servedCold](X), out_{servedCold}(X), \mathrm{not}\ fail \tag{21}$$

$$g_1(X) \leftarrow \mathrm{DL}[\lambda; sparklingWine](X), \mathrm{not}\ \mathrm{DL}[\lambda'; \neg whiteWine](X) \tag{22}$$

$$g_2(X) \leftarrow \mathrm{DL}[\lambda; whiteWine](X), \mathrm{not}\ \mathrm{DL}[\lambda'; \neg servedCold](X) \tag{23}$$

$$fail \leftarrow \mathrm{not}\ \mathrm{DL}[\lambda; whiteWine](X), in_{whiteWine}(X), \mathrm{not}\ fail \tag{24}$$

$$fail \leftarrow \mathrm{DL}[\lambda; whiteWine](X), out_{whiteWine}(X), \mathrm{not}\ fail \tag{25}$$

$$fail \leftarrow \mathrm{not}\ \mathrm{DL}[\lambda; servedCold](X), in_{servedCold}(X), \mathrm{not}\ fail \tag{26}$$

$$fail \leftarrow \mathrm{DL}[\lambda; servedCold](X), out_{servedCold}(X), \mathrm{not}\ fail \tag{27}$$

corresponding to the default rules in D, where λ and λ' are update lists of form $whiteWine \uplus g_1$, $servedCold \uplus g_2$ and $whiteWine \uplus in_{whiteWine}$, $servedCold \uplus in_{servedCold}$, respectively. Intuitively, the predicates $in_{whiteWine}$ and $in_{servedCold}$ encode that an individual is a member of concept $whiteWine$ and $servedCold$, resp. Similarly, $out_{whiteWine}$ and $out_{servedCold}$ are used to state that an individual is not a member of these concepts. The rules (16)–(19) guess an extension of those predicates, whereas (20) and (21) check whether the guessed model is compliant with the ontology L. Rules (22) and (23) are used to test the applicability of the defaults in D, and (24)–(27) then check whether the guess agrees with the semantics of default logic. In order to have models that agree with the conclusions of L, λ and λ' take over the task to communicate the current world view of P to the ontology L. Under strong answer set semantics, the above program has, as expected, among its cautious consequences the facts $in_{whiteWine}(veuveCliquot)$ and $out_{whiteWine}(lambrusco)$, which correctly denotes the fact that sparkling wines are white by default. Above encoding has been improved in [28], where various translations from default logic over description logic into cq-programs (see also Section 1.4.3) are given and further analyzed.

A second line of nonmonotonic reasoning is Reiter's CWA [95]. In this reasoning principle, we can infer the negative fact $\neg p(c)$ from a first-order theory T whenever we are unable to prove the positive fact $p(c)$ from T. The CWA of a theory T, denoted $CWA(T)$, is defined as the set of all literals $\{\neg p(c) \mid T \not\models p(c)\}$.

Take, for instance, the DL knowledge base

$$L = \{apple \sqsubseteq fruit, fruit(williams)\}$$

describing that apples are fruits, and that $williams$ is a particular fruit. The above knowledge base L leaves open whether $williams$ is an apple or not. Closing L by means of CWA enables us to deduce that $CWA(L) \models \neg apple(williams)$, i.e., under CWA we can infer that $williams$ is not an apple.

A particular encoding of above reasoning task in dl-programs is accomplished by the rule

$$\overline{apple}(X) \leftarrow not\, DL[apple](X) \;,$$

where $\overline{apple}$ is a fresh predicate. Given L, we can now infer $\overline{apple}(williams)$.

A well-known drawback of the CWA is that it faces inconsistency in case of disjunctive information. Let $L' = \{apple \sqcup pear(williams)\}$, i.e., $williams$ is an apple or a pear. Under CWA, we can infer that $williams$ is neither an apple nor a pear, which is inconsistent with our assertion in L'. The *extended closed-world assumption* (ECWA) is a refined version of CWA, which is able to treat cases like the one above in a reasonable manner. Full details on CWA and ECWA in dl-programs is given in [43].

1.3.3 HEX-Programs

HEX-programs [45] are declarative nonmonotonic logic programs with support for external knowledge and higher-order disjunctive rules. In spirit of dl-programs, they allow for a loose coupling between general external knowledge sources and declarative logic programs through the notion of external atoms, which take input from the logic program and exchange inferences with the external source. In addition, meta-reasoning tasks may be accomplished by means of higher-order atoms. HEX-programs are evaluated under a generalized answer-set semantics, thus are in principle capable of capturing many proposed extensions in answer-set programming.

Syntax of HEX-Programs. Let $\mathcal{C}$, $\mathcal{X}$, and $\mathcal{G}$ be mutually disjoint sets whose elements are called *constant names*, *variable names*, and *external predicate names*, respectively. Unless explicitly specified, elements from $\mathcal{X}$ (resp., $\mathcal{C}$) are denoted with first letter in upper case (resp., lower case), while elements from $\mathcal{G}$ are prefixed with the "&" symbol. We note that constant names serve both as individual and predicate names.

Elements from $\mathcal{C} \cup \mathcal{X}$ are called *terms*. A *higher-order atom* (or *atom*) is a tuple $(Y_0, Y_1, \ldots, Y_n)$, where $Y_0, \ldots, Y_n$ are terms; $n \geq 0$ is the *arity* of the atom. Intuitively, Y_0 is the predicate name, and we thus also use the more familiar notation $Y_0(Y_1, \ldots, Y_n)$. The atom is *ordinary*, if Y_0 is a constant.

For example, $(x, rdf\!:\!type, c)$, $node(X)$, and $D(a, b)$, are atoms; the first two are ordinary atoms.

An *external atom* is of the form

$$\&g[Y_1, \ldots, Y_n](X_1, \ldots, X_m) \; , \tag{28}$$

where $Y_1, \ldots, Y_n$ and $X_1, \ldots, X_m$ are two lists of terms (called *input* and *output* lists, respectively), and $\&g \in \mathcal{G}$ is an external predicate name. We assume that $\&g$ has fixed lengths $in(\&g) = n$ and $out(\&g) = m$ for input and output lists, respectively. Intuitively, an external atom provides a way for deciding the truth value of an output tuple depending on the extension of a set of input predicates: in this respect, an external predicate $\&g$ is equipped with a function $f_{\&g}$ evaluating to true for proper input values.

A *rule* r is of the form

$$\alpha_1 \vee \cdots \vee \alpha_k \leftarrow \beta_1, \ldots, \beta_m, \text{not } \beta_{m+1}, \ldots, \text{not } \beta_n \; , \tag{29}$$

where $m, k \geq 0$, $\alpha_1, \ldots, \alpha_k$ are atoms, and $\beta_1, \ldots, \beta_n$ are either atoms or external atoms. We define $H(r) = \{\alpha_1, \ldots, \alpha_k\}$ and $B(r) = B^+(r) \cup B^-(r)$, where $B^+(r) = \{\beta_1, \ldots, \beta_m\}$ and $B^-(r) = \{\beta_{m+1}, \ldots, \beta_n\}$. If $H(r) = \emptyset$ and $B(r) \neq \emptyset$, then r is a *constraint*, and if $B(r) = \emptyset$ and $H(r) \neq \emptyset$, then r is a *fact*; r is *ordinary*, if it contains only ordinary atoms. A HEX-*program* is a finite set P of rules. It is *ordinary*, if all rules are ordinary.

We next give an illustrative example.

Example 5 ([44]). Consider the following HEX-program P:

$$subRelation(brotherOf, relativeOf). \tag{30}$$

$$brotherOf(john, al). \tag{31}$$

$$relativeOf(john, joe). \tag{32}$$

$$brotherOf(al, mick). \tag{33}$$

$$invites(john, X) \lor skip(X) \leftarrow X \neq john, \&reach[relativeOf, john](X). \tag{34}$$

$$R(X, Y) \leftarrow subRelation(P, R), P(X, Y). \tag{35}$$

$$someInvited \leftarrow invites(john, X). \tag{36}$$

$$\leftarrow not\ someInvited. \tag{37}$$

$$\leftarrow \&degs[invites](Min, Max), Max > 2. \tag{38}$$

Informally, this program randomly selects a certain number of John's relatives for invitation. The first line states that *brotherOf* is a subrelation of *relativeOf*, and the next three lines give concrete facts. The disjunctive rule (34) chooses relatives, employing the external predicate *&reach*. This latter predicate takes in input a binary relation e and a node name n, returning the nodes reachable from n when traversing the graph described by e (see the following Example 7). Rule (35) axiomatizes subrelation inclusion exploiting higher-order atoms; that is, for those couples of binary predicates p, r for which it holds $subRelation(p, r)$, it must be that $r(x, y)$ holds whenever $p(x, y)$ is true.

The constraints (37) and (38) ensure that the number of invitees is between 1 and 2, using (for illustration) an external predicate *&degs* from a graph library. Such a predicate has a valuation function $f_{\&degs}$ where $f_{\&degs}(I, e, min, max)$ is true iff *min* and *max* are, respectively, the minimum and maximum vertex degree of the graph induced by the edges contained in the extension of predicate e in interpretation I.

Semantics of HEX-Programs. In the sequel, let P be a HEX-program. The *Herbrand base* of P, denoted HB_P, is the set of all possible ground versions of atoms and external atoms occurring in P obtained by replacing variables with constants from $\mathcal{C}$. The grounding of a rule r, $grnd(r)$, is defined accordingly, and the grounding of program P is given by $grnd(P) = \bigcup_{r \in P} grnd(r)$. Unless specified otherwise, $\mathcal{C}$, $\mathcal{X}$, and $\mathcal{G}$ are implicitly given by P.

Example 6 ([44]). Given $\mathcal{C} = \{edge, arc, a, b\}$, ground instances of $E(X, b)$ are for instance $edge(a, b)$, $arc(a, b)$, $a(edge, b)$, and $arc(arc, b)$; ground instances of $\&reach[edge, N](X)$ are all possible combinations where N and X are replaced by elements from $\mathcal{C}$, for instance $\&reach[edge, edge](a)$, $\&reach[edge, arc](b)$, $\&reach[edge, edge](edge)$, etc.

An *interpretation relative to* P is any subset $I \subseteq HB_P$ containing only atoms. We say that I is a *model* of atom $a \in HB_P$, denoted $I \models a$, if $a \in I$.

With every external predicate name $\&g \in \mathcal{G}$, we associate an $(n+m+1)$-ary Boolean function $f_{\&g}$ assigning each tuple $(I, y_1 \ldots, y_n, x_1, \ldots, x_m)$ either 0 or 1, where $n = in(\&g)$, $m = out(\&g)$, $I \subseteq HB_P$, and $x_i, y_j \in \mathcal{C}$. We say that $I \subseteq HB_P$ is a *model* of a ground external atom $a = \&g[y_1, \ldots, y_n](x_1, \ldots, x_m)$, denoted $I \models a$, if and only if $f_{\&g}(I, y_1, \ldots, y_n, x_1, \ldots, x_m) = 1$.

Example 7 ([44]). Let us associate with the external atom *&reach* a function $f_{\&reach}$ such that $f_{\&reach}(I, E, A, B) = 1$ iff B is reachable in the graph E from A. Let $I = \{e(b, c), e(c, d)\}$. Then, I is a model of $\&reach[e, b](d)$ since $f_{\&reach}(I, e, b, d) = 1$.

Note that in contrast to the semantics of higher-order atoms, which in essence reduces to first-order logic as customary (cf. [98]), the semantics of external atoms is in spirit of second order logic since it involves predicate extensions.

Considering example 5, as John's relatives are determined to be Al, Joe, and Mick, P has six answer sets, each of which contains one or two of the facts *invites(john, al)*, *invites(john, joe)*, and *invites(john, mick)*.

Let r be a ground rule. We define (i) $I \models H(r)$ iff there is some $a \in H(r)$ such that $I \models a$, (ii) $I \models B(r)$ iff $I \models a$ for all $a \in B^+(r)$ and $I \not\models a$ for all $a \in B^-(r)$, and (iii) $I \models r$ iff $I \models H(r)$ whenever $I \models B(r)$. We say that I is a *model* of a HEX-program P, denoted $I \models P$, iff $I \models r$ for all $r \in grnd(P)$. We call P *satisfiable*, if it has some model.

Given a HEX-program P, the *FLP-reduct* of P with respect to $I \subseteq HB_P$, denoted fP^I, is the set of all $r \in grnd(P)$ such that $I \models B(r)$. $I \subseteq HB_P$ is an *answer set of P* iff I is a minimal model of fP^I.

In principle, the truth value of an external atom depends on its input and output lists and on the entire model of the program. In practice, however, we can identify certain types of input terms that allow to restrict the input interpretation to specific relations. The Boolean function associated with the external atom $\&reach[edge, a](X)$ for instance will only consider the extension of the predicate *edge* and the constant value a for computing its result, and simply ignore everything else of the given input interpretation.

Features and Properties of HEX-Programs. As mentioned above, HEX-programs are a generalization of dl-programs, consisting indeed in a form of coupling of rules with arbitrary external computation sources, within a declarative logic-based setting. The higher-order features are similar to those of HiLog [26], i.e., the semantics of this high-order extension is still within first-order logic.

The semantics of HEX-programs conservatively extends ordinary answer-set programs, and it is easily extendable to support weak constraints [17]. External predicates can define other ASP features like aggregate functions [48]. Computational complexity of the language depends on external functions. The former is however not affected if external functions evaluate in polynomial time.

The dlvhex prototype,[14] an implementation of HEX-programs, is based on a flexible and modular architecture. The evaluation of the external atoms is realized by plugins, which are loaded at run-time. The pool of available external predicates can be easily customized by third-party developers.

---

[14] http://www.kr.tuwien.ac.at/research/systems/dlvhex/

Applications. HEX-programs have been applied in many applications in different contexts. Hoehndorf et al. [59] showed how to combine multiple biomedical upper ontologies by extending the first-order semantics of terminological knowledge with default logic. The corresponding prototype implementation of such kind of system is given by mapping the default rules to HEX-program. Fuzzy extensions of answer-set programs and their relationship to HEX-programs are given in [88,58]. The former maps fuzzy answer set programs to HEX-programs, whereas the latter defines a fuzzy semantics for HEX-programs and gives a translation to standard HEX-programs. In [89], the planning language $\mathcal{K}^c$ has been introduced which features external function calls in spirit of HEX-programs.

REWERSE has related applications, where a rule-based solution for solving *credential selection problems* (see below) was discussed. We also refer the reader to Chapter 3, which is devoted to reasoning about policies.

As stated in [16], selecting an "appropriate" set of credentials for satisfying trust negotiation tasks is an important problem. Since users typically want to disclose as little sensitive information as possible, an "appropriate" set of credentials is the least sensitive set of credentials needed to obtain a service. This minimization effort is referred to as the *credential selection problem* [16], which will be explained in the following.

In a nutshell, each participant in a rule-based policy specification environment expresses its policies by logic programs, and credentials provided by the requesting client are encoded by facts. The combination of the policies and a set of credentials should satisfy the given authorization request of the client.

More formally, a credential selection problem (CSEL) consists of

- a finite, stratified logic program P, representing the server's and client's policies,
- a goal G modeling the authorization requested by the client,
- a finite set of integrity constraints IC, representing forbidden combinations of credentials,
- a finite set of ground facts C, representing the portfolio of credentials and declarations of the client, and
- a *sensitivity aggregation function* $sen : 2^C \to \Sigma$, where Σ is a finite set (of *sensitivity values*) partially ordered by $\preceq$.

A solution for the credential selection problem is a set $S \subseteq C$ such that

1. $P \cup S \models G$,
2. $P \cup S \cup IC$ is consistent, and
3. $sen(S)$ is minimal among all S which satisfy 1. and 2.

Expressing this kind of credential selections is a valuable application for HEX-programs. The next example from [99] shows how to encode a CSEL instance in a HEX-program. The stratified program P is encoded in *Server policy*.

```
─────────────────────────────  Server policy  ─────────────────────────────
% if resource is public, no authentication is necessary
allow(download,Resource) :- public(Resource).

% user may download if she has a subscription and is authenticated
allow(download,Resource) :- authenticated(User),
                            hasSubscription(User,Subscription),
                            availableFor(Resource,Subscription).

% user may download if she has paid and is authenticated
allow(download,Resource) :- authenticated(User),
                            paid(User,Resource).

% user is authenticated, if she has a valid credential
authenticated(User) :- valid(Credential),
                       attr(Credential,name,User).

% a selected credential is valid, if its type is trusted
valid(Credential) :- selectedCred(Credential),
                     attr(Credential,type,T),
                     attr(Credential,issuer,CA),
                     isa(T,id),
                     trustedFor(CA,T).

% types that are ids, i.e., a hierarchy of identifiers
isa(id,id).
isa(ssn,id).
isa(passport,id).
isa(driving_license,id).
```

The goal G as fact *resource("paper01234.pdf")*, the set C of the client's credentials, and (implicitly by *credSens*) the set $\Sigma = \{1, 2, 4\}$ of sensitivity values is given below in *Client example*.

```
─────────────────────────────  Client example  ─────────────────────────────
hasSubscription("John Doe",law_basic).
hasSubscription("John Doe",computer_basic).

availableFor("paper01234.pdf",computer_basic).

% the client requests this goal G
resource("paper01234.pdf").

% credential authorities and their ID types
trustedFor("Open University",id).
trustedFor("Visa",id).
trustedFor("UK Government",ssn).

% next are three credentials and their
% associated properties and sensitivities
```

```
credential(cr01).
attr(cr01,type,id).
attr(cr01,name,"John Doe").
attr(cr01,issuer,"Open University").
credSens(cr01,1).

credential(cr02).
attr(cr02,type,ssn).
attr(cr02,name,"John Doe").
attr(cr02,issuer,"UK Government").
credSens(cr02,2).

credential(cr03).
attr(cr03,type,id).
attr(cr03,name,"John Doe").
attr(cr03,issuer,"Visa").
credSens(cr03,4).
```

The final part of the CSEL is given below, encoding the minimal $sen(S)$ of credentials $S \subseteq C$, which satisfies the program. The solution is given as ground facts with predicate $sens$. We make use of the external atom &$policy$, whose associated Boolean function $f_{\&policy}(I, p, n) = 1$ iff $n = \sum_{p(c,i) \in I} i$. That is, in a model of the program encoding our CSEL, $polSens(s)$ holds the sum s of sensitivity values i from all ground $sens(c, i)$ atoms.

—————————————————— **Optimization rules** ——————————————————

```
% open a search space
selectedCred(X) v -selectedCred(X)  :- credential(X).

sens(C,S)  :- selectedCred(C), credSens(C,S).

% remove models that don't accomplish the goal
:- not allow(download,R), resource(R).

% compute model sensitivity
polSens(S)  :- &policy[sens](S).

% select least sensitive model
:~ polSens(S).  [S:1]
```

The solution is given in the abridged answer set $\{sens(cr01,1), polSens(1), \ldots\}$, which specifies that credential $cr01$ is sufficient for achieving the goal G and is the least sensitive one among the possible subsets of the credentials C. Note that the program makes use of the weak constraint construct :~ `polSens(S)`. `[S:1]`, whose intuitive meaning is adding a cost s to answer sets in which `polSens(s)` holds, and then selecting optimal answer sets by minimizing the costs (for an in-detail account on weak constraints (in HEX-programs) we refer to [17,99]).

1.3.4   Extensions of Well-Founded Semantics by Hybrid Well-Founded Semantics

This section gives an introduction to the REWERSE work on tight integration presented in [39,38]. This work developed a framework for hybrid combination of normal logic programs under the well-founded semantics with various theories of the FOL. The hybrid programs defined in this way extend faithfully both normal programs and the underlying theories. The framework gives principles of implementation showing how a rule engine supporting the well-founded semantics of normal programs can be combined with a reasoner for the underlying theory to get a reasoner for the hybrid programs which is sound w.r.t. their declarative semantics. The implemented instance of the framework was the language of HD-rules [38,37], integrating Datalog with negation and OWL DL. The framework itself is not restricted to Datalog; compound terms are permitted in addition to constants.

To present this work we first illustrate on an example the well-founded semantics of normal programs. Then we discuss the syntax and the declarative semantics of the hybrid programs. Finally we explain the principles of the operational semantics.

The Well-Founded Semantics. The well-founded semantics associates with a normal logic program a unique *3-valued Herbrand model*, called its *well-founded model*. For the Herbrand base $\mathcal{H}$ of a program denote $\neg \mathcal{H} = \{ \neg a \mid a \in \mathcal{H} \}$. Then a *3-valued Herbrand interpretation* $\mathcal{I}$ of P is a subset of $\mathcal{H} \cup \neg \mathcal{H}$ such that for no ground atom A both A and $\neg A$ are in $\mathcal{I}$. Intuitively, the set $\mathcal{I}$ assigns the truth value **t** (true) to all its members. Thus A is false (has the truth value **f**) in $\mathcal{I}$ iff $\neg A \in \mathcal{I}$, and $\neg A$ is false in $\mathcal{I}$ iff $A \in \mathcal{I}$. If $A \notin \mathcal{I}$ and $\neg A \notin \mathcal{I}$ then the truth value of A (and that of $\neg A$) is **u** (undefined). The truth value of compound formulae is defined in a usual way. For instance the truth value of $F_1 \wedge F_2$ is **t** if the truth values of both F_1 and F_2 are **t**, it is **f** if the truth value of some of them is **f**, and it is **u** if some of them has the truth value **u** and none has **f**. The notation $\mathcal{I} \models_3 F$ will be used to denote that a formula F is true in a 3-valued interpretation $\mathcal{I}$.

We illustrate the notion of the well-founded model by some examples. Several (equivalent) formal definitions can be found elsewhere, see for instance [102,5,49].

Example 8. The well-founded model of program $\{ p \leftarrow p;\ q \leftarrow \neg p;\ r \leftarrow q, \neg r \}$ is $\{ \neg p, q \}$. Informally, the value of p is false independently from the values of q, r (as p is defined by a single rule $p \leftarrow p$). From $\neg p$ we derive q (by rule $q \leftarrow \neg p$). However neither r nor $\neg r$ can be derived. The program does not have stable models.

Example 9. A two person game consists in moving a token between vertices of a directed graph. Each move consists in traversing one edge from the actual position. Each of the players in order makes one move. The graph is described by a database of facts $m(X, Y)$ corresponding to the edges of the graph. A position

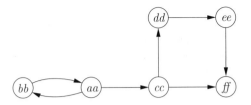

Fig. 3. The game graph

X is said to be a *winning position* X if there exists a move from X to a position Y which is a losing (non-winning) position:

$$w(X) \leftarrow m(X, Y), \neg w(Y)$$

Consider the graph in Fig. 3 and assume that it is encoded by the facts $m(bb, aa)$, $m(aa, bb), \ldots, m(ee, ff)$ of the program. Now ff is a losing position – there is no move from ff. This is reflected by the well-founded model of the program; in the model $w(ff)$ is false, as no program rule has a ground instance $w(ff) \leftarrow \ldots$ with a true body (as the program contains no fact of the form $m(ff, t)$). Thus ee is a winning position: $w(ee)$ is true, due to rule instance $w(ee) \leftarrow m(ee, ff), \neg w(ff)$. Similarly, $w(cc)$ is true and $w(dd)$ is false. However each of aa, bb is neither winning nor losing, from each of them the player has an option of moving to the other one. Literals $w(aa), w(bb)$ have value **u** in the well-founded model of the program. The model contains the following literals with the predicate symbol w: $w(cc), w(ee), \neg w(dd), \neg w(ff)$.

The program has two stable models, in each of them $w(aa)$ and $w(bb)$ have the opposite logical values (and the values of $w(cc), w(ee), w(dd), w(ff)$ are the same as those in the well-founded model).

The previous sections considered logic programs under the stable model semantics (or, more generally, answer set semantics). Both semantics coincide for a wide class of programs relevant in practice, including the stratified programs. The well-founded model of such a program is 2-valued, and it is its unique stable model. Usually the pragmatics of knowledge representation is different for the two semantics. With the answer set semantics, each stable model represents a solution to a problem. With the well-founded semantics, the solutions are represented by consequences (i.e. answers) of the program.

Hybrid Programs. Informally, a *hybrid program* consists of a set of axioms $\mathcal{T}$, called *external theory* and of a generalized normal program P, which may contain formulae of the language of $\mathcal{T}$, called *constraints*[15] in the bodies of the rules.

More precisely, one considers a first-order alphabet including, as usual, disjoint alphabets of predicate symbols $\mathcal{P}$, function symbols $\mathcal{F}$ (including a set of

---

[15] This term is used due to similarities with constraint logic programming [83].

constants) and variables $\mathcal{V}$. Following the heterogeneous approach to integration, it is assumed that $\mathcal{P}$ consists of two disjoint sets $\mathcal{P}_R$ (*rule predicates*) and $\mathcal{P}_C$ (*constraint predicates*). The atoms and the literals constructed with these predicates are called respectively *rule atoms* (*rule literals*) and *constraint atoms* (*constraint literals*). The bodies of the rules of normal programs over alphabets $\mathcal{P}_R$, $\mathcal{F}$, $\mathcal{V}$ may now be extended with constraints over alphabets $\mathcal{P}_C$, $\mathcal{F}$, $\mathcal{V}$. (It is also allowed that the set of function symbols of the external theory $\mathcal{T}$ is a subset of $\mathcal{F}$.) In a particular instance of the framework one has to be specific about the kind of formulae allowed as constraints of the rules.

A *hybrid rule* has the form

$$a \leftarrow c, b_1, \ldots, b_n, \tag{39}$$

where c is a constraint over $\mathcal{P}_C$, $\mathcal{F}$, $\mathcal{V}$ and $b_1, \ldots, b_n$ are rule literals. If constraints allow quantifiers, some variables may not be free in a rule. A safeness restriction on the syntax of rules introduced in [36] for discussing semantic issues is somewhat elaborate. A sufficient condition for a rule to be safe is that each its free variable has to appear in a positive rule literal.

A *hybrid program* is a pair $(P, \mathcal{T})$ where P is a set of hybrid rules and $\mathcal{T}$ is a set of axioms over $\mathcal{P}_C$, $\mathcal{F}$, $\mathcal{V}$. A hybrid program is said to be safe if all its rules are safe.

Remember that we deal with two kinds of negation: the classical negation of FOL and nonmonotonic negation of the rules. The former is applied to (formulae containing only) constraint predicates, and the latter only to (atoms with) rule predicates. So the same symbol $\neg$ can be used to denote both.

The following example [39] shows a safe hybrid program with constraints referring to an ontology.

Example 10. Consider a classification of geographical locations. For example the classification may concern the country (Finland (Fi), Norway (No), etc.), the continent (Europe (E), etc.), and possibly other categories. We specify a classification by axioms in a DL logic. The ontology provides, among others, the following information

- subclass relations (T-box axioms): e.g. ($Fi \sqsubseteq E$);
- classification of some given locations represented by constants (A-box axioms). For instance, assuming that the positions of Example 9 represent locations we may have: bb is a location in Finland ($Fi(bb)$), cc is a location in Europe ($E(cc)$).

Now the ontology will be used as an external theory for a program. We describe a variant of the game from Example 9, with the rules subject to additional restrictions (see Fig. 4). Assume that the positions of the graph represent geographical locations described by the ontology. The restrictions will be expressed as ontological constraints added in rule bodies. For instance let constraints be added to

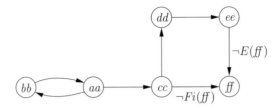

Fig. 4. The modified game graph

the facts $m(ee, ff)$ and $m(cc, ff)$:

$$w(X) \leftarrow m(X, Y), \neg w(Y)$$

$$
\begin{array}{ll}
m(bb, aa) & m(cc, ff) \leftarrow \neg Fi(ff) \\
m(aa, bb) & m(ee, ff) \leftarrow E(ff) \\
m(aa, cc) & \\
m(cc, dd) & \\
m(dd, ee) &
\end{array}
$$

Intuitively, this would mean that the move from ee to ff is allowed only if ff is in Europe and the move from cc to ff – only if ff is not in Finland. These restrictions may influence the outcome of the game: ff will still be a losing position but if the axioms of the ontology do not allow to conclude that ff is in Europe, we cannot conclude that ee is a winning position. However, we can conclude that if ff is not in Europe then it cannot be in Finland. Thus, at least one of the conditions $E(ff)$, $\neg Fi(ff)$ holds. Therefore cc is a winning position: If $E(ff)$ then, as in Example 9, ee is a winning position, dd is a losing one, hence cc is a winning position. On the other hand, if $\neg Fi(ff)$ the move from cc to ff is allowed in which case cc is a winning position.

An example employing non-nullary function symbols is given in [37].

Declarative Semantics. The declarative semantics of hybrid programs is defined as a generalization of the well-founded semantics of normal programs; it refers to the (2-valued) models of the external theory $\mathcal{T}$ of a hybrid program. Given a hybrid program $(P, \mathcal{T})$ we cannot define a unique well-founded model of P since we have to take into consideration the logical values of the constraints in the rules. However, for any given model M of $\mathcal{T}$ one can consider the well-founded model of the normal program P/M obtained by replacing the constraints in the rules by their logical values in M.

More precisely, let $ground(P)$ be the set of ground instances of the hybrid rules in P. Then P/M is the normal program obtained from $ground(P)$ by

– removing each rule constraint C which is true in M (i.e. $M \models C$),
– removing each rule whose constraint C is not true in M, (i.e. $M \not\models C$).

The well-founded model of P/M is called the *well-founded model of P based on M*.

A formula F (over $\mathcal{P}_R, \mathcal{F}, \mathcal{V}$) **holds** (is *true*) in the well-founded semantics of a hybrid program $(P, \mathcal{T})$ (denoted $(P, \mathcal{T}) \models_{wf} F$) iff $M \models_3 F$ for each well-founded model M of $(P, \mathcal{T})$.

We say that F is *false* in the well-founded semantics of $(P, \mathcal{T})$ if $(P, \mathcal{T}) \models_{wf} \neg F$, and that F is *undefined* if the logical value of F in each well-founded model of $(P, \mathcal{T})$ is **u**. Notice that there is a fourth case: if F does not have the same logical value in all well-founded models of P then F is neither true, nor false, nor undefined. Notice that the negation in the rule literals is nonmonotonic, and the negation in the constraints is that from the external theory, thus monotonic.

Example 11. For the hybrid program $(P, \mathcal{T})$ of Example 10 we have to consider models of the ontology $\mathcal{T}$. For every model M_0 of $\mathcal{T}$ such that $M_0 \models E(f\!f)$ the program P/M_0 includes the fact $m(ee, f\!f)$. The well-founded model of P/M_0 includes thus the literals $\neg w(f\!f), w(ee), \neg w(dd), w(cc)$ (independently of whether $M_0 \models Fi(f\!f)$).

On the other hand, for every model M_1 of the ontology such that $M_1 \models \neg Fi(f\!f)$ the program P/M_1 includes the fact $m(cc, f\!f)$. The well-founded model of P/M_1 includes thus the literals $\neg w(f\!f), w(cc)$ (independently of whether $M_1 \models E(f\!f)$).

Notice that each of the models of the ontology falls in one of the above discussed cases. Thus, $w(cc)$ and $\neg w(f\!f)$ hold in the well-founded semantics of the hybrid program, while $w(ee), \neg w(ee), w(dd)$ and $\neg w(dd)$ do not hold in it (provided that the logical value of $E(f\!f)$ is not the same in all the models of $\mathcal{T}$). The logical value of $w(aa)$ and that of $w(bb)$ is **u** in each well-founded model of the program. Thus $w(aa)$ and $w(bb)$ are undefined in the well-founded semantics of the program, and $w(dd)$ and $w(ee)$ are not (they are neither true, nor false, nor undefined).

Consider a case of hybrid rules without negative rule literals. So the non-monotonic negation does not occur. Such rules can be seen as implications of FOL and treated as axioms added to $\mathcal{T}$. For such case the well-founded semantics and the logical consequence $\models$ of FOL are similar. They are not equivalent, as the well-founded semantics deals only with Herbrand models of the rules. However they coincide in the following sense. (1) For any ground rule atom A if $(P, \mathcal{T}) \models_{wf} A$ then $P \cup \mathcal{T} \models A$. The reverse implication does not hold[16]. (2) Assume that only such interpretation domains are considered in which each element is a value of a ground term, and the values of distinct terms are distinct. Then A is true in all models of $P \cup \mathcal{T}$ iff $(P, \mathcal{T}) \models_{wf} A$, for any rule atom A.

As the well-founded semantics of normal programs is undecidable, so is the well-founded semantics of hybrid programs. It is however decidable for Datalog hybrid programs with decidable external theories.

The declarative semantics of hybrid programs is based on Herbrand models of the rules. Thus it treats distinct terms as having distinct values. The syntactic

---

[16] As a counterexample take $P = \{ p \leftarrow q(x), r(x); \; r(x) \leftarrow \}$ and $\mathcal{T} = \{ \exists x.q(x) \}$. $P \cup \mathcal{T} \models p$ but $(P, \mathcal{T}) \not\models_{wf} p$, as there exist models of $\mathcal{T}$ in which each ground atom $q(t)$ is false.

equality of the well-founded semantics may be different from the equality of the external theory. This may lead to strange consequences. For instance consider a hybrid program $(P, \mathcal{T})$, where $P = \{\, p(a)\,\}$. Both $p(a)$ and $\neg p(b)$ hold in the well-founded semantics of $(P, \mathcal{T})$, even if $\mathcal{T}$ implies that $a = b$. One may avoid such anomalies by requiring that – speaking informally – terms which are equal according to $\mathcal{T}$ are treated in the same way by the rules of P. For more details the reader is referred to [36]. To avoid technical difficulties let us require that, in what follows, the external theory satisfies the axioms of the free equality theory (CET, Clark equality theory [27]). Thus ground terms have the same values (in a model of $\mathcal{T}$) iff they are syntactically equal.

Operational Semantics. In this section we present the operational semantics [39,36] of hybrid programs. The semantics is a basis for implementation; a prototype implementation has been described in [38,37]. Our presentation is informal. For a precise description the user is referred to [39,36].

Like in logic programming, the task of a computation is to find instances of a given goal formula G which are true in the well-founded semantics of a given program. Similarly to logic programming, the operational semantics is defined in terms of search trees. It is based on the idea of constructive negation presented in [34,35]. In that work the only constraint predicate was the equality and the constraint theory was the free equality theory (CET) [27].

The operational semantics is similar to SLDNF- and SLS-resolution [70,93], extended by handling constraints originating from the hybrid rules. For an input goal a derivation tree is constructed; its nodes are goals. Whenever a negative literal is selected in some node, a subsidiary derivation tree is constructed. In logic programming an answer to a goal is a binding for goal variables. In hybrid programs an answer is a constraint satisfiable in a given theory $\mathcal{T}$. Thus, to develop an implementation it is necessary to have a constraint solver for $\mathcal{T}$. However, the constraint solver is only used as a black box deciding the satisfiability of a given constraint.

A *hybrid goal* (shortly: *goal*) has the form

$$c, b_1, \ldots, b_m$$

where $m \geq 0$, each b_i is a rule literal and c is a constraint, called the constraint of the goal. The definition of safeness and the sufficient condition for safeness applies also to hybrid goals.

The computation is controlled by a selection function which selects a rule literal in a goal. If the selected literal is positive the goal is resolved, as usual, by matching the selected rule literal with the head of a renamed variant of a hybrid rule of the program. However, the unification is replaced by adding a constraint to the derived goal. More precisely, consider a goal $G = c, \overline{L}, b, \overline{L'}$ and a rule $r = h \leftarrow c', \overline{K}$, such that no variable occurs both in G and r. The goal

$$G' \;=\; b{=}h, c, c', \overline{L}, \overline{K}, \overline{L'}$$

is said to be **derived** from G by r, with the selected atom b, if the constraint $b{=}h, c, c'$ is satisfiable. As usual, several rules of the program may match the selected atom and give rise to different computations, visualized by a tree with nodes labeled by goals and edges representing the derivation steps. A goal c with no rule literals is called *successful* (and ends a *successful branch* of the tree).

Consider such a tree with root G, with no negative literal selected. Let us denote by $c|_G$ the constraint $\exists \ldots c$, where the quantified variables are those variables of c that do not occur (free) in G. Then by an *answer* of the tree we mean any constraint $(c_1 \vee \ldots \vee c_n)|_G$, where $c_1, \ldots, c_n$ are some of the successful leaves of the tree. Every answer, speaking informally, implies G in the well-founded semantics of the program. (A precise formulation is given later on.) The most general answer is the disjunction of all the successful leaves. If the Herbrand universe is infinite, the set of (the constraints of) successful leaves may be infinite and the most general answers may not exist.

The negation of the most general answer is a *negative answer* of the tree; it implies $\neg G$ in the well-founded semantics of the program. Less general negative answers may be obtained without constructing the whole tree. By a *cross-section* we mean a set F of tree nodes such that each successful branch has a node in F. If $c_1, \ldots, c_n$ are the constraints of the goals of a cross-section then $\neg((c_1 \vee \ldots \vee c_n)|_G)$ is a negative answer. We skip here a definition of negative answers corresponding to infinite cross-sections. Each negative answer implies that the root G is false.

In the general case negative literals may be selected in the tree, and we have to deal with three logical values $\mathbf{t}$, $\mathbf{u}$, $\mathbf{f}$. Due to this we introduce two kinds of trees, *t-trees* and *tu-trees*. A t-tree tells when its root G is $\mathbf{t}$ (in the well-founded semantics of the program). Speaking informally, each answer of the t-tree implies G. A tu-tree tells when its root G is $\mathbf{t}$ or $\mathbf{u}$; G being $\mathbf{t}$ or $\mathbf{u}$ implies some answer of the tree. Thus each negative answer of the tu-tree implies $\neg G$. We are interested in the answers of t-trees and the negative answers of tu-trees.

The two kinds of trees differ by the treatment of negative selected literals. In a t-tree, when a negative literal $\neg b$ is selected in a goal $G' = c, \overline{L}, \neg b, \overline{L'}$ then a subsidiary tu-tree for c, b is constructed, and some its negative answer d is obtained. The literal $\neg b$ is replaced by d. If the resulting constraint c, d is satisfiable then the obtained goal $G'' = c, d, \overline{L}, \overline{L'}$ is the (only) child of G' in the t-tree. Otherwise (c, d unsatisfiable) G is a leaf. (An informal justification is that d implies that $\neg b$ is $\mathbf{t}$.) One may avoid constructing the subsidiary tu-tree; then G' does not have a child (as $d = \neg c$ is a trivial negative answer of any tu-tree for c, b, and $c, \neg c$ is unsatisfiable).

In a tu-tree, when a negative literal $\neg b$ is selected in a goal $G' = c, \overline{L}, \neg b, \overline{L'}$ then a subsidiary t-tree for c, b is constructed, and some its answer c' is obtained. The literal $\neg b$ is replaced by the negation $d = \neg c'$ of c'. If c, d is satisfiable then the obtained goal $G'' = c, d, \overline{L}, \overline{L'}$ is the (only) child of G' in the tu-tree. Otherwise G' is a leaf. (An informal justification is that c' implies $\neg b$ being $\mathbf{f}$; hence $\neg b$ being $\mathbf{t}$ or $\mathbf{u}$ implies d.) One may avoid constructing the subsidiary tu-tree; then G' has $G'' = c, \overline{L}, \overline{L'}$ as its child (as $c' = \textbf{false}$ is a trivial answer of any t-tree).

To avoid circularity (e.g. a t-tree for p refers to a tu-tree for q, and the latter tree refers to the former) *ranks* are assigned to the trees, similarly as it is done in the definitions of SLDNF- and SLS-resolutions [70,93].

We now illustrate the operational semantics of hybrid programs by an example. In the example we apply certain simplifications to the tree nodes.[17] The same example without the simplifications is presented in [39,36].

Example 12. Consider the hybrid program of Example 10. A query $w(cc)$ can be answered by constructing the following trees: a t-tree for $w(cc)$, a tu-tree for $w(dd)$, a t-tree for $w(ee)$, and a tu-tree for $w(ff)$; of ranks 3, 2, 1, 0 respectively. In the goals with more than one rule literals, the selected one is underscored.

$$
\begin{array}{cc}
w(cc) & w(dd) \\
| & | \\
\underline{m(cc,Y)}, \neg w(Y) & \underline{m(dd,Y')}, \neg w(Y') \\
\diagup \quad \diagdown & | \\
\neg Fi(ff), \neg w(ff) \quad \neg w(dd) & \neg w(ee) \\
| \qquad\qquad | & | \\
\neg Fi(ff) \qquad E(ff) & \neg E(ff)
\end{array}
$$

$$
\begin{array}{cc}
w(ee) & \\
| & \\
\underline{m(ee,Y'')}, \neg w(Y'') & w(ff) \\
| & | \\
E(ff), \neg w(ff) & \underline{m(ff,Y')}, \neg w(Y') \\
| & \\
E(ff) &
\end{array}
$$

The empty cross-section of the tu-tree for $w(ff)$ provides a negative answer **true**, the t-tree for $w(ee)$ has an answer $E(ff)$, the tu-tree for $w(dd)$ has a negative answer $E(ff)$ (the cross-section consisting of the leaf), and the t-tree for $w(cc)$ has an answer $\neg Fi(ff) \vee E(ff)$ (which in $\mathcal{T}$ is equivalent to **true**).

An implementation of hybrid programs based on the described ideas is presented in [38,37]. The operational semantics makes it possible to employ an existing constraint solver (e.g. a description logic reasoner) and treat it as a black box. Also, construction of t-trees and tu-trees can be implemented on top of a Prolog system with the well-founded semantics. Thus the costs of implementation is rather low. We also mention that – similarly as in CLP – it is not necessary to check satisfiability of the constraint for each tree node. The answers (negative answers) of trees obtained in such way are logically equivalent to those described

---

[17] Any constraint may be replaced by an equivalent one. In any node $C, \overline{L}$ of a t- or tu- tree for G, the constraint C of the node can be replaced by $C|_{G,\overline{L}}$. Instead of referring to a lower rank tree for C, A, a tree for $(C|_A), A$ can be used. Also, a goal may be replaced by a logically equivalent one (e.g. $X=a, p(X)$ by $X=a, p(a)$). These modifications do not change the (negative) answers of the trees. This is rather obvious in the particular example; we omit a formal justification for a general case.

above. This provides an opportunity to decrease the interaction with the constraint solver, and to improve efficiency. The prototype of [38,37] invokes the solver once, after having constructed the main tree.

For safe hybrid programs the operational semantics is sound w.r.t. to the well-founded semantics. More precisely if $(P, \mathcal{T})$ is a safe hybrid program and $G = c_0, \overline{L}$ a goal then for any substitution θ

1. if c is an answer of a t-tree for $(P, \mathcal{T})$ and G, and $\mathcal{T} \models c\theta$ then $(P, \mathcal{T}) \models_{\mathrm{wf}} \overline{L}\theta$;
2. if c is a negative answer of a tu-tree for $(P, \mathcal{T})$ and G, and $\mathcal{T} \models c\theta$ then $(P, \mathcal{T}) \models_{\mathrm{wf}} \neg \overline{L}\theta$.

The safeness condition may be abandoned, if additional restrictions are imposed on the existential quantifier used in constraints see [36] for details). This is related to the fact that constraints are interpreted on arbitrary domains, without assuming that each element of a domain is represented by ground term, while the well-founded semantics defines a Herbrand model.

In the general case, the operational semantics is not complete. The reason is that only finite constraint formulae are used as (negative) answers. However the method is complete in the case of Datalog, with safe rules and goals. More precisely, assume that the Herbrand universe is finite. Consider a safe program $(P, \mathcal{T})$ and a safe goal $G = c_0, \overline{L}$. For any grounding substitution θ for the variables of G such that $c_0\theta$ is satisfiable

1. if $(P, \mathcal{T}) \models_{\mathrm{wf}} \overline{L}\theta$ then there exists a t-tree (of a finite rank) for G with an answer c such that $\mathcal{T} \models c\theta$;
2. if $(P, \mathcal{T}) \models_{\mathrm{wf}} \neg \overline{L}\theta$ then there exists a tu-tree (of a finite rank) for G with a negative answer c such that $\mathcal{T} \models c\theta$.

A stronger result, which includes independence from the selection rule, also holds.

1.4 Variants and Extensions of the Basic Formalisms

In this section, we discuss some variants and extensions of the above basic formalisms, which have been crafted in order to make them more versatile or to overcome some restrictions. More specifically, we summarize extensions of the basic formalisms that allow for handling uncertainty and vagueness. We also describe an extension of loosely coupled dl-programs by (unions of) conjunctive queries as dl-atoms and disjunctions in rule heads (called *cq-programs*).

From a more general perspective, during the recent years, handling uncertainty and vagueness has started to play an important role in Semantic Web research. A recent forum for approaches to uncertainty reasoning in the Semantic Web is the annual *Workshop on Uncertainty Reasoning for the Semantic Web (URSW)*. There also exists a W3C Incubator Group on *Uncertainty Reasoning for the World Wide Web*. The research focuses especially on probabilistic and fuzzy extensions of description logics, ontology languages, and formalisms integrating rules and ontologies. Note that probabilistic formalisms allow to encode ambiguous information, such as "John is a student with the probability 0.7 and a teacher with the probability 0.3", while fuzzy approaches allow to encode vague or imprecise information, such as "John is tall with the degree of truth 0.7". Formalisms for dealing with uncertainty and vagueness are especially applied in ontology mapping, data integration, information retrieval, and database querying. Vagueness and imprecision also abound in multimedia information processing and retrieval, and are an important aspect of natural language interfaces to the Web.

We first consider extensions of dl-programs by probabilistic uncertainty, and we then discuss fuzzy extensions. We finally focus on cq-programs.

1.4.1 Probabilistic DL-Programs

We now summarize the main ideas behind loosely and tightly coupled probabilistic dl-programs, introduced in [71,74,75,19] and [18,22,20,21], respectively. For further details on the syntax and semantics of these programs, their background, and their semantic and computational properties, we refer to the above works.

Loosely coupled probabilistic dl-programs [71,74,75] are a combination of loosely coupled dl-programs under the answer set and the well-founded semantics with probabilistic uncertainty as in Bayesian networks. Roughly, they consist of a loosely coupled dl-program (L, P) under different "total choices" B (they are the full joint instantiations of a set of random variables, and they serve as pairwise exclusive and exhaustive possible worlds), and a probability distribution μ over the set of total choices B. One then obtains a probability distribution over Herbrand models, since every total choice B along with the loosely coupled dl-program produces a set of Herbrand models of which the probabilities sum up to $\mu(B)$. As in the classical case, the answer set semantics of loosely coupled probabilistic dl-programs is a refinement of the well-founded semantics of loosely coupled probabilistic dl-programs. Consistency checking and tight query processing (i.e., computing the entailed tight interval for the probability of a conditional or unconditional event) for in such probabilistic dl-programs under the answer set semantics can be reduced to consistency checking and query processing in loosely coupled dl-programs under the answer set semantics, while tight query processing under the well-founded semantics can be done in an anytime fashion by reduction to loosely coupled dl-programs under the well-founded semantics. For suitably restricted description logic components, the latter can be done in polynomial time in the data complexity. Query processing in the special case of stratified loosely coupled probabilistic dl-programs can be reduced to computing the canonical model of stratified loosely coupled dl-programs. Loosely coupled probabilistic dl-programs can especially be used for (database-oriented) probabilistic data integration in the Semantic Web, where probabilistic uncertainty is used to handle inconsistencies between different data sources [19].

Example 13. A university database may use a loosely coupled dl-program (L, P) to encode ontological and rule-based knowledge about students and exams. A probabilistic dl-program $KB = (L, P', C, \mu)$ then additionally allows for encoding probabilistic knowledge. For example, the following two probabilistic rules in P' along with a probability distribution on a set of random variables may express that if two master (resp., bachelor) students have given the same exam, then there is a probability of 0.9 (resp., 0.7) that they are friends:

$$
\begin{aligned}
friends(X, Y) \;\leftarrow\; & given\_same\_exam(X, Y), DL[master\_student(X)], \\
& DL[master\_student(Y)], \; choice_m \; ; \\
friends(X, Y) \;\leftarrow\; & given\_same\_exam(X, Y), DL[bachelor\_student(X)], \\
& DL[bachelor\_student(Y)], \; choice_b \; .
\end{aligned}
$$

Here, we assume the set $C = \{\{choice_m, not\_choice_m\}, \{choice_b, not\_choice_b\}\}$ of values of two random variables and the probability distribution μ on all their four joint instantiations, given by μ: $choice_m, not\_choice_m, choice_b, not\_choice_b \mapsto$ $0.9, 0.1, 0.7, 0.3$ under probabilistic independence. For example, $choice_m, choice_b$ is associated with the probability $0.9 \times 0.7 = 0.63$. Asking about the entailed tight interval for the probability that $john$ and $bill$ are friends can then be expressed by a probabilistic query of the form $\exists(friends(john, bill))[R, S]$, whose answer depends on the available concrete knowledge about $john$ and $bill$ (whether they have given the same exams, and are both master or bachelor students).

Tightly coupled probabilistic dl-programs [18,22] are a tight combination of disjunctive logic programs under the answer set semantics with description logics and Bayesian probabilities. They are a logic-based representation formalism that naturally fits into the landscape of Semantic Web languages. Tightly coupled probabilistic dl-programs can especially be used for representing mappings between ontologies [20,21], which are a common way of approaching the semantic heterogeneity problem on the Semantic Web. In this application, they allow in particular for resolving inconsistencies and for merging mappings from different matchers based on the level of confidence assigned to different rules (see below). Furthermore, tightly coupled probabilistic description logic programs also provide a natural integration of ontologies, action languages, and Bayesian probabilities towards Web Services. Consistency checking and query processing in tightly coupled probabilistic dl-programs can be reduced to consistency checking and cautious/brave reasoning, respectively, in tightly coupled disjunctive dl-programs. Under certain restrictions, these problems have a polynomial data complexity.

Example 14. The two correspondences between two ontologies O_1 and O_2 that (i) an element of *Collection* in O_1 is an element of *Book* in O_2 with the probability 0.62, and (ii) an element of *Proceedings* in O_1 is an element of *Proceedings* in O_2 with the probability 0.73 (found by the matching system hmatch) can be expressed by the following two probabilistic rules:

$$
\begin{aligned}
O_2 : Book(X) &\leftarrow O_1 : Collection(X) \wedge hmatch_1; \\
O_2 : Proceedings(X) &\leftarrow O_1 : Proceedings(X) \wedge hmatch_2.
\end{aligned}
$$

Here, we assume the set $\mathcal{C} = \{\{hmatch_i, not\_hmatch_i\} \mid i \in \{1, 2\}\}$ of values of random variables and the probability distribution μ on all joint instantiations of these variables, given by μ: $hmatch_1, not\_hmatch_1, hmatch_2, not\_hmatch_2 \mapsto$ $0.62, 0.38, 0.73, 0.27$ under probabilistic independence.

Similarly, two other correspondences between O_1 and O_2 (found by the matching system falcon) are expressed by the following two probabilistic rules:

$$O_2: InCollection(X) \leftarrow O_1: Collection(X) \wedge falcon_1;$$
$$O_2: Proceedings(X) \leftarrow O_1: Proceedings(X) \wedge falcon_2,$$

where we assume the set $\mathcal{C}' = \{\{falcon_i, not\_falcon_i\} \mid i \in \{1,2\}\}$ of values of random variables and the probability distribution μ' on all joint instantiations of these variables, given by μ': $falcon_1, not\_falcon_1, falcon_2, not\_falcon_2 \mapsto$ $0.94, 0.06, 0.96, 0.04$ under probabilistic independence.

Using the trust probabilities 0.55 and 0.45 for hmatch and falcon, respectively, for resolving inconsistencies between rules, we can now define a merged mapping set that consists of the following probabilistic rules:

$$O_2: Book(X) \leftarrow O_1: Collection(X) \wedge hmatch_1 \wedge sel\_hmatch_1;$$
$$O_2: InCollection(X) \leftarrow O_1: Collection(X) \wedge falcon_1 \wedge sel\_falcon_1;$$
$$O_2: Proceedings(X) \leftarrow O_1: Proceedings(X) \wedge hmatch_2;$$
$$O_2: Proceedings(X) \leftarrow O_1: Proceedings(X) \wedge falcon_2,$$

Here, we assume the set $\mathcal{C}''$ of values of random variables and the probability distribution μ'' on all joint instantiations of these variables, which are obtained from $\mathcal{C} \cup \mathcal{C}'$ and $\mu \cdot \mu'$ (which is defined as $(\mu \cdot \mu')(B\,B') = \mu(B) \cdot \mu'(B')$, for all joint instantiations B of $\mathcal{C}$ and B' of $\mathcal{C}'$), respectively, by adding the values $\{sel\_hmatch_1, sel\_falcon_1\}$ of a new random variable along with the probabilities $sel\_hmatch_1, sel\_falcon_1 \mapsto 0.55, 0.45$ under probabilistic independence, for resolving the inconsistency between the first two rules.

1.4.2   Fuzzy DL-Programs

We next briefly describe loosely and tightly coupled fuzzy dl-programs, which have been introduced in [72,76] and [78,80], respectively, and extended by probabilities in [77] and by a top-k retrieval technique in [79], respectively. All these fuzzy dl-programs have natural special cases where query processing can be done in polynomial time in the data complexity. For further details on their syntax and semantics, background, and properties, we refer to the above works.

Towards dealing with vagueness and imprecision in the reasoning layers of the Semantic Web, loosely coupled (normal) fuzzy dl-programs under the answer set semantics [72,76] are a generalization of normal dl-programs under the answer set semantics by fuzzy vagueness and imprecision in both the description logic and the logic program component. This is the first approach to fuzzy dl-programs that may contain default negations in rule bodies. Query processing in such fuzzy dl-programs can be done by reduction to normal dl-programs under the answer set semantics. In the special cases of positive and stratified loosely coupled fuzzy dl-programs, the answer set semantics coincides with a canonical least model and an iterative least model semantics, respectively, and has a characterization in terms of a fixpoint and an iterative fixpoint semantics, respectively.

Example 15. Consider the fuzzy DL knowledge base L of a car shopping Web site, which defines especially (i) the fuzzy concepts of sports cars (*SportsCar*), "at most 22 000 €" (*LeqAbout22000*), and "around 150 horse power" (*Around150HP*), (ii) the attributes of the price and of the horse power of a car (*hasInvoice* and *hasHP*, respectively), and (iii) the properties of some concrete cars (such as a *MazdaMX5Miata* and a *MitsubishiES*). Then, a loosely coupled fuzzy dl-program $KB = (L, P)$ is given by the set of fuzzy dl-rules P, which contains only the following fuzzy dl-rule encoding the request of a buyer (asking for a sports car costing at most 22 000 € and having around 150 horse power), where $\otimes$ may be the conjunction strategy of, e.g., Gödel Logic (that is, $x \otimes y = \min(x, y)$ for all $x, y \in [0, 1]$, used to evaluate the logical connectives $\wedge$ and $\leftarrow$ on truth values):

$$query(x) \leftarrow_{\otimes} DL[SportsCar](x) \wedge_{\otimes} DL[\exists hasInvoice.LeqAbout22000](x) \wedge_{\otimes}$$
$$DL[\exists hasHP.Around150HP](x) \geq 1 \,.$$

The above fuzzy dl-program $KB = (L, P)$ is positive, and has a minimal model M_{KB}, which defines the degree to which some concrete cars in the DL knowledge base L match the buyer's request, for example,

$$M_{KB}(query(MazdaMX5Miata)) = 0.36 \,, \quad M_{KB}(query(MitsubishiES)) = 0.32 \,.$$

That is, the *MazdaMX5Miata* is ranked top with the degree 0.36, while the *MitsubishiES* is ranked second with the degree 0.32.

Towards an infrastructure for additionally handling uncertainty in the reasoning layers of the Semantic Web, probabilistic fuzzy dl-programs [77] combine fuzzy description logics, fuzzy logic programs (with stratified default-negation), and probabilistic uncertainty in a uniform framework for the Semantic Web. Intuitively, they allow for defining several rankings on ground atoms using fuzzy vagueness, and then for merging these rankings using probabilistic uncertainty (by associating with each ranking a probabilistic weight and building the weighted sum of all rankings). Such programs also give rise to important concepts dealing with both probabilistic uncertainty and fuzzy vagueness, such as the expected truth value of a crisp sentence and the probability of a vague sentence.

Example 16. A loosely coupled probabilistic fuzzy dl-program is given by a suitable fuzzy DL knowledge base L and the following set of fuzzy dl-rules P, modeling some query reformulation / retrieval steps using ontology mapping rules:

$$query(x) \leftarrow_{\otimes} SportyCar(x) \wedge_{\otimes} hasPrice(x, y_1) \wedge_{\otimes} hasPower(x, y_2) \wedge_{\otimes}$$
$$DL[LeqAbout22000](y_1) \wedge_{\otimes} DL[Around150HP](y_2) \geq 1, \quad (40)$$
$$SportyCar(x) \leftarrow_{\otimes} DL[SportsCar](x) \wedge_{\otimes} sc_{pos} \geq 0.9 \,, \quad (41)$$
$$hasPrice(x, y) \leftarrow_{\otimes} DL[hasInvoice](x, y) \wedge_{\otimes} hi_{pos} \geq 0.8 \,, \quad (42)$$
$$hasPower(x, y) \leftarrow_{\otimes} DL[hasHP](x, y) \wedge_{\otimes} hhp_{pos} \geq 0.8 \,, \quad (43)$$

where we assume the set $C = \{\{sc_{pos}, sc_{neg}\}, \{hi_{pos}, hi_{neg}\}, \{hhp_{pos}, hhp_{neg}\}\}$ of values of random variables and the probability distribution μ on all joint instantiations of these variables, given by μ: $sc_{pos}, sc_{neg}, hi_{pos}, hi_{neg}, hhp_{pos},$ $hhp_{neg} \mapsto 0.91, 0.09, 0.78, 0.22, 0.83, 0.17$ under probabilistic independence. Rule (40) is the buyer's request, but in a "different" terminology than the one of the car selling site. Rules (41)–(43) are so-called ontology alignment mapping rules. For example, rule (41) states that the predicate "SportyCar" of the buyer's terminology refers to the concept "SportsCar" of the selected site with probability 0.91.

The following may be some tight consequences of the above probabilistic fuzzy dl-program (where for ground atoms q, we use $(\mathbf{E}[q])[L, U]$ to denote that the expected truth value of q lies in the interval $[L, U]$):

$$(\mathbf{E}[query(MazdaMX5Miata)])[0.21, 0.21], \quad (\mathbf{E}[query(MitsubishiES)])[0.19, 0.19].$$

That is, the *MazdaMX5Miata* is ranked first with the degree 0.21, while the *MitsubishiES* is ranked second with the degree 0.19.

Tightly coupled fuzzy dl-programs under the answer set semantics [78,80] are a tight integration of fuzzy disjunctive logic programs under the answer set semantics with fuzzy description logics. From a different perspective, they are a generalization of tightly coupled disjunctive dl-programs by fuzzy vagueness in both the description logic and the logic program component. This is the first approach to fuzzy dl-programs that may contain disjunctions in rule heads. Query processing in such programs can essentially be done by a reduction to tightly coupled disjunctive dl-programs. A closely related work [79] explores the problem of evaluating ranked top-k queries. It shows in particular how to compute the top-k answers in data-complexity tractable tightly coupled fuzzy dl-programs.

Example 17. A tightly coupled fuzzy dl-program $KB = (L, P)$ is given by a suitable fuzzy DL knowledge base L and the set of fuzzy rules P, which contains only the following fuzzy rule (where $x \otimes y = \min(x, y)$):

$$query(x) \leftarrow_\otimes SportyCar(x) \wedge_\otimes hasInvoice(x, y_1) \wedge_\otimes hasHorsePower(x, y_2) \wedge_\otimes$$
$$LeqAbout22000(y_1) \wedge_\otimes Around150(y_2) \geqslant 1.$$

Informally, *query* collects all sports cars, and ranks them according to whether they cost at most around 22 000 € and have around 150 HP. Another fuzzy rule involving also a negation in its body and a disjunction in its head is given as follows (where $\ominus x = 1 - x$ and $x \oplus y = \max(x, y)$):

$$Small(x) \vee_\oplus Old(x) \leftarrow_\otimes Car(x) \wedge_\otimes hasInvoice(x, y) \wedge_\otimes$$
$$not_\ominus GeqAbout15000(y) \geqslant 0.7.$$

This rule says that a car costing at most around 15 000 € is either small or old. Notice here that *Small* and *Old* may be two concepts in the fuzzy DL knowledge base L. That is, the tightly coupled approach to fuzzy dl-programs under the answer set semantics also allows for using the rules in P to express relationships between the concepts and roles in L. This is not possible in the loosely coupled approach to fuzzy dl-programs under the answer set semantics in [72,76], since the dl-queries there can only occur in rule bodies, but not in rule heads.

1.4.3 CQ-Programs

An extension for dl-programs are cq-programs [41], which allow for expressing *(union of) conjunctive queries* (U)CQ over description logics in the dl-atoms, and disjunctions in the head of the rules.

This approach for hybrid reasoning with rules and ontologies is following the loose coupling approach, i.e., it is a heterogeneous integration that differentiates between logic programming predicates and description logic concept and roles. cq-programs benefit of some of the advantages of the loose coupling approach, such as the possibility of immediate integration of existing solvers for the implementation of the language. Also, the clear separation of the involved components enables the possibility of designing a modular architecture, as may be imagined.

In contrast with dl-programs, the cq-program combination is tighter in a sense that it allows to existentially quantify over unknown individuals that are implicit in a DL knowledge base.

Example 18. Consider the following simplified version of a scenario in [86].

$$L = \left\{ \begin{array}{l} hates(Cain, Abel), hates(Romulus, Remus), \\ father(Cain, Adam), father(Abel, Adam), \\ father \sqsubseteq parent, \\ \exists father.\exists father^-.\{Remus\}(Romulus) \end{array} \right\}$$

$$P = \{BadChild(X) \leftarrow \mathrm{DL}[parent](X, Z), \mathrm{DL}[parent](Y, Z), \mathrm{DL}[hates](X, Y)\}$$

Apart from the explicit facts, L states that each *father* is also a *parent* and that *Romulus* and *Remus* have a common father. The single rule in P specifies that an individual hating a sibling is a *BadChild*. From this dl-program, *BadChild(Cain)* can be concluded, but not *BadChild(Romulus)*.

Instead of P, let us use

$$P' = \{BadChild(X) \leftarrow \mathrm{DL}[parent(X, Z), parent(Y, Z), hates(X, Y)](X, Y)\},$$

where the body of the rule is a CQ $\{parent(X, Z), parent(Y, Z), hates(X, Y)\}$ to L with distinguished variables X and Y. We then obtain the desired result; that is, we can derive the fact *BadChild(Romulus)*.

The semantics of the cq-programs is in spirit of dl-programs, and mainly differs in the generalized entailment notion for cq-atoms, which extend that of dl-atoms. Informally, a cq-atom α is in form $\mathrm{DL}[\lambda; q](\boldsymbol{X})$, where q can be a union of conjunctive queries with output variables $\boldsymbol{X}$, while λ represents a list of modifiers for the description logic base L at hand, with the same meaning given in dl-programs. The CQ-extension adds additional expressiveness to dl-programs, as is evident by results that show an increase in complexity from NEXP to 2-EXP for the description logic $\mathcal{SHIF}(\mathbf{D})$.

A further plus of this extension is that it opens the floodgates for exploiting optimizations in dl-programs, via a technique able to produce rewritten programs where the computational burden can be shifted to and from one of the two reasoners at hand. For instance, conjunctions of atoms can be computed, whenever semantically equivalent, on the description logic base side instead that on the logic program side. In [41], several forms of optimizing rewriting rules have been defined to rewrite DL-queries in rule bodies to more efficient ones. Experimental results comparing unoptimized to rewritten programs show a substantial performance improvement.

1.5   Conclusion

In this chapter, we have briefly shown work that has been done in REWERSE on the issue of combining rules and ontologies. To this end, we have first given an overview of different combination approaches, which have been systematically grouped into a classification that takes different degree of integration and of rules and ontologies into account.

We have then presented the two streams of genuine approaches which have been pursed in REWERSE by the groups in Vienna and Linköping, respectively, to give meaningful and expressive combinations that faithfully generalize the stable models and the well-founded semantics of logic programs, respectively, leading to nonmonotonic combinations of rules and ontologies whose prototype implementations reflected the state of the art in this area. Furthermore, several extensions to these approaches have been briefly discussed, which address needs such as handling probabilistic information, fuzzy values, or more expressive queries to ontologies than simple instance checks or consistency tests.

While the work on combinations of rules and ontologies in REWERSE has broken new ground and was fruitfully taken up by other groups within REWERSE but also outside (in particular, HEX-programs and dlvhex have found applications in various contexts), its impact on the development of the rules layer of the Semantic Web, and in particular to emerging standards, has yet to materialize. The reason is that, different from ontologies, the standardization of rules that is targeted by the RIF working group of the W3C (see Section 1.2.4) is a formidable challenge, given that there are very many notions of rules and their semantics; this is one of the reasons that, at the time of this writing, merely a compromise for a core rule dialect (RIF-BLD) is what has been achieved so far; features such as negation (even stratified one) have been targeted in more comprehensive packages, but not realized so far. We expect that stable models and the well-founded semantics will be the premier semantics reflected in a RIF standard for non-monotonic negation that is beyond stratified negation in logic programs, and that the ideas and concepts which have been developed in the REWERSE streams will impact on the definition of possible interfacing between rules and ontologies in the emerging standards.

At present, the issue of combining rules and ontologies for the Semantic Web is not regarded to be satisfactorily solved; a number of different approaches have been made so far, but they all have some features that do not suggest them to be regarded the ultimate solution to the problem; let alone that perhaps there is no single, "universal" such solution, but a range of different solutions which cater different features and needs that have to be fulfilled in different contexts.

This already manifests in different types of rules; in REWERSE, the focus was on logic-based rules, but other rule types such as production rules are equally important and require different treatment; in fact, an integration of production rules and ontologies with bidirectional information flow is an interesting subject for future work, in which the operational and logical semantics of the rules and the ontology, respectively, have to be bridged. The OntoRule project will within the "7th Framework Program" (FP7) of the EU Commission target business rules and policies, and will to a great deal be based on a lower layer that integrates production rules and ontologies, aside with logic programming rules. By way of this project, results of REWERSE will migrate more towards practical exploitation and into commercial rules engines.

In order to make expressive combinations of rules and ontologies available for deployment to applications, a number of research tasks remain to be pursued.

Currently, we lack extended case studies and large scale examples beyond the toy examples that have been considered in the seminal papers that introduced the approaches. Such case studies might provide helpful insight and give some guidance in the development of a "gold standard" for rules plus ontologies. At the least, required constructs in the language, be they just syntactic sugar or really increasing the expressiveness of a formalism, should be identifiable in this way. The trouble is, however, to single out a set of representative cases, which is by no means trivial. A benchmark suite would be very valuable and, if carefully composed, undoubtedly an important step forward.

Another issue are complex data structures, and realizations of the combinations beyond the Datalog fragment. Indeed, in practice one needs to handle complex data that are aggregations of other data, such as records, lists, sets, etc. Such data structures can be modeled in many logic programming systems using function symbols, and support in terms of explicit syntax is offered. They can also be modeled in the hybrid well-founded semantics (Section 1.3.4) and its implementations; this approach deals with function symbols of arbitrary arities. However, in the current solvers for stable model and answer set semantics, function symbols are largely banned because they are a well-known source of undecidability, even in rather plain settings; only more recently, work on decidable classes and prototype implementations of stable models semantics with function symbols has been carried out (cf. [15,14,24,100] and references therein), and function symbols also increasingly attract attention as a modeling construct. The DLV-Complex system [23,24] aims at providing functions symbols in a decidable setting, giving support to lists and sets along with libraries for their manipulations. It remains to see how logic programs in this setting can be combined with ontologies; semantically, the gap between rules and ontologies widens by the use of such function symbols, and decidability issues has to be reconsidered.

An obvious task is the development of better algorithms and efficient implementations. The current prototype implementations serve more as proofs of concept and experimental testbeds, but are not largely optimized. There is a lot of room for improvement, even though the optimization methods are expected to be tailored to a particular semantics and implementation setting. The intertwining of a rules and an ontology engine, as done in the prototype implementations of dl-programs and HD-rules, imposes specific requirements that can not be easily transferred to other implementations. Developing an integrated engine that processes rules and ontologies en par is an interesting issue; whether a conversion of logic programs into ontology axioms or vice versa a mapping of ontologies into logic programming rules is a viable approach remains to be explored. This, however, may work well for fragments of combinations in which such conversions are easily possible.

In close connection to the previous issue are semantic and computational properties of combinations. There is clearly a trade-off between the expressiveness of a formalism on the one hand and its intrinsic complexity on the other. If we expect to have fast reasoning over knowledge bases with large extensional part, comprising millions (or even billions) of facts, then naturally the reasoning tasks per se must not have high intrinsic complexity. For this reason, it is important to have an understanding of the complexity characteristics of combinations, to know about fragments with tractable and low complexity (just polynomial time as such might not be sufficient for practical applications, if the data volume is large), and to respect such characteristics in implementations in a way that easy instances are solved with little effort while more computation time is spent on harder instances. Recent research on rules and conjunctive query answering over description logics from the lower expressiveness end like $\mathcal{EL}$ and $\mathcal{EL}++$ [6,7], or DL-Lite [25,92] may be here a starting point.

Finally, an important issue is also to combine knowledge sources beyond rules and ontologies. Indeed, a rule base and an ontology may be just two components in an information system that consists of many other components that are in different formats. And while throughout this chapter, the rules and the ontology have been considered as more or less integral parts of one description, this picture may no longer be valid if the components are independently conceived and autonomous, like they happen to be in a peer to peer system. In such a case, also the viewpoint of semantic combination should be rather different, and incorporating trust is an important requirement.

Acknowledgments

The work in this chapter was partially supported by the European Commission through the project REWERSE (IST-2003-506779), in which the main technical results were obtained, and the project Ontorule (FP7 231875), as well as by the Austrian Science Fund (FWF) grants P17212, P20840, P20841 and the Italian National Project Interlink II04CG8AGG. Thomas Lukasiewicz has been supported by the German Research Foundation (DFG) under the Heisenberg Programme.

References

1. Analyti, A., Antoniou, G., Damásio, C.V., Wagner, G.: Stable model theory for extended RDF ontologies. In: Gil, Y., Motta, E., Benjamins, V.R., Musen, M.A. (eds.) ISWC 2005. LNCS, vol. 3729, pp. 21–36. Springer, Heidelberg (2005)
2. Antoniou, G., Maher, M., Billington, D.: Defeasible Logic versus Logic Programming. Journal of Logic Programming 41(1), 45–57 (2000)

3. Antoniou, G., Bikakis, A.: DR-Prolog: A system for defeasible reasoning with rules and ontologies on the Semantic Web. IEEE Trans. Knowl. Data Eng. 19(2), 233–245 (2007)
4. Antoniou, G., Damásio, C.V., Grosof, B., Horrocks, I., Kifer, M., Maluszynski, J., Patel-Schneider, P.F.: Combining rules and ontologies. A survey. FP6 NoE REWERSE, Deliverable I3-D3, http://rewerse.net/deliverables/m12/i3-d3.pdf
5. Apt, K.R., Bol, R.N.: Logic programming and negation: A survey. J. Log. Program. 19(20), 9–71 (1994)
6. Baader, F., Brandt, S., Lutz, C.: Pushing the $\mathcal{EL}$ envelope. In: Proceedings of the Nineteenth International Joint Conference on Artificial Intelligence, IJCAI 2005, Edinburgh, UK. Morgan-Kaufmann Publishers, San Francisco (2005)
7. Baader, F., Brandt, S., Lutz, C.: Pushing the $\mathcal{EL}$ envelope further. In: Clark, K., Patel-Schneider, P.F. (eds.) Proceedings of the OWLED 2008 DC Workshop on OWL: Experiences and Directions (2008)
8. Baader, F., Calvanese, D., McGuinness, D.L., Nardi, D., Patel-Schneider, P.F. (eds.): The Description Logic Handbook: Theory, Implementation, and Applications, 2nd edn. Cambridge University Press, Cambridge (2007)
9. Baader, F., Hollunder, B.: Embedding Defaults into Terminological Knowledge Representation Formalisms. Journal of Automated Reasoning 14(1), 149–180 (1995)
10. Baral, C.: Knowledge Representation, Reasoning and Declarative Problem Solving. Cambridge University Press, Cambridge (2003)
11. Baral, C., Gelfond, M.: Logic Programming and Knowledge Representation. Journal of Logic Programming 19(20), 73–148 (1994)
12. Boley, H., Kifer, M. (eds.): RIF Basic Logic Dialect, W3C Working Draft (July 2008), http://www.w3.org/TR/2008/WD-rif-bld-20080730/
13. Boley, H., Kifer, M., Pătrânjan, P.L., Polleres, A.: Rule interchange on the web. In: Antoniou, G., Aßmann, U., Baroglio, C., Decker, S., Henze, N., Patranjan, P.-L., Tolksdorf, R. (eds.) Reasoning Web 2007. LNCS, vol. 4636, pp. 269–309. Springer, Heidelberg (2007)
14. Bonatti, P., Baselice, S.: Composing normal programs with function symbols. In: Garcia de la Banda, M., Pontelli, E. (eds.) ICLP 2008. LNCS, vol. 5366, pp. 425–439. Springer, Heidelberg (2008)
15. Bonatti, P.A.: Reasoning with infinite stable models. Artificial Intelligence 156(1), 75–111 (2004)
16. Bonatti, P.A., Eiter, T., Faella, M.: Advanced Policy Queries. REWERSE Deliverable I2-D6, Dipartimento di Scienze Fisiche - Sezione di Informatica, University of Naples "Federico II" (2006)
17. Buccafurri, F., Leone, N., Rullo, P.: Strong and Weak Constraints in Disjunctive Datalog. In: Dix, J., Furbach, U., Nerode, A. (eds.) LPNMR 1997. LNCS (LNAI), vol. 1265, pp. 2–17. Springer, Heidelberg (1997)
18. Calì, A., Lukasiewicz, T.: Tightly integrated probabilistic description logic programs for the Semantic Web. In: Dahl, V., Niemelä, I. (eds.) ICLP 2007. LNCS, vol. 4670, pp. 428–429. Springer, Heidelberg (2007)
19. Calì, A., Lukasiewicz, T.: An approach to probabilistic data integration for the Semantic Web. In: da Costa, P.C.G., d'Amato, C., Fanizzi, N., Laskey, K.B., Laskey, K.J., Lukasiewicz, T., Nickles, M., Pool, M. (eds.) URSW 2005-2007. LNCS, vol. 5327, pp. 52–65. Springer, Heidelberg (2008)

20. Calì, A., Lukasiewicz, T., Predoiu, L., Stuckenschmidt, H.: Rule-based approaches for representing probabilistic ontology mappings. In: da Costa, P.C.G., d'Amato, C., Fanizzi, N., Laskey, K.B., Laskey, K.J., Lukasiewicz, T., Nickles, M., Pool, M. (eds.) URSW 2005-2007. LNCS, vol. 5327, pp. 66–87. Springer, Heidelberg (2008)
21. Calì, A., Lukasiewicz, T., Predoiu, L., Stuckenschmidt, H.: Tightly integrated probabilistic description logic programs for representing ontology mappings. In: Hartmann, S., Kern-Isberner, G. (eds.) FoIKS 2008. LNCS, vol. 4932, pp. 178–198. Springer, Heidelberg (2008)
22. Calì, A., Lukasiewicz, T., Predoiu, L., Stuckenschmidt, H.: Tightly coupled probabilistic description logic programs for the Semantic Web. In: Journal on Data Semantics XII (2009)
23. Calimeri, F., Cozza, S., Ianni, G.: External sources of knowledge and value invention in logic programming. Annals of Mathematics and Artificial Intelligence 50(3-4), 333–361 (2007)
24. Calimeri, F., Cozza, S., Ianni, G., Leone, N.: Computable functions in ASP: Theory and implementation. In: de La Banda, M., Pontelli, E. (eds.) ICLP 2008. LNCS, vol. 5366, pp. 407–424. Springer, Heidelberg (2008)
25. Calvanese, D., De Giacomo, G., Lembo, D., Lenzerini, M., Rosati, R.: Tractable reasoning and efficient query answering in description logics: The DL-Lite family. Journal of Automated Reasoning 39(3), 385–429 (2007)
26. Chen, W., Kifer, M., Warren, D.S.: Hilog: A foundation for higher-order logic programming. Journal of Logic Programming 15(3), 187–230 (1993)
27. Clark, K.L.: Negation as failure. In: Gallaire, H., Minker, J. (eds.) Logic and Databases, pp. 293–322. Plenum Press (1978)
28. Dao-Tran, M., Eiter, T., Krennwallner, T.: Realizing Default Logic over Description Logic Knowledge Bases. In: Proceedings of the 10th European Conference on Symbolic and Quantitative Approaches to Reasoning with Uncertainty (ECSQARU 2009). Springer, Heidelberg (to appear, 2009)
29. de Bruijn, J., Eiter, T., Polleres, A., Tompits, H.: Embedding non-ground logic programs into autoepistemic logic for knowledge-base combination. In: Veloso, M.M. (ed.) IJCAI, pp. 304–309 (2007)
30. de Bruijn, J., Eiter, T., Tompits, H.: Embedding approaches to combining rules and ontologies into autoepistemic logic. In: Brewka, G., Lang, J. (eds.) Proceedings of the Eleventh International Conference on Principles of Knowledge Representation and Reasoning (KR 2008), pp. 485–495. AAAI Press, Menlo Park (2008)
31. de Bruijn, J., Pearce, D., Polleres, A., Valverde, A.: Quantified Equilibrium Logic and Hybrid Rules. In: Marchiori, M., Pan, J.Z., Marie, C.d.S. (eds.) RR 2007. LNCS, vol. 4524, pp. 58–72. Springer, Heidelberg (2007)
32. Dean, M., Schreiber, G., Bechhofer, S., van Harmelen, F., Hendler, J., Horrocks, I., McGuinness, D.L., Patel-Schneider, P.F., Stein, L.A.: OWL Web Ontology Language Reference, W3C Recommendation (February 2004)
33. Donini, F., Lenzerini, M., Nardi, D., Schaerf, A.: AL-Log: Integrating Datalog and description logics. Intelligent Information Systems 10(3), 227–252 (1998)
34. Drabent, W.: SLS-resolution without floundering. In: Pereira, L.M., Nerode, A. (eds.) Proc. 2nd International Workshop on Logic Programming and Non-Monotonic Reasoning, pp. 82–98. MIT Press, Cambridge (1993)
35. Drabent, W.: What is failure? An approach to constructive negation. Acta Informatica 32(1), 27–59 (1995)
36. Drabent, W., Maluszynski, J.: Hybrid Rules with Well-Founded Semantics. Preliminarily accepted to Knowledge and Information Systems (2009)

37. Drabent, W., Henriksson, J., Maluszynski, J.: HD-rules: A hybrid system interfacing Prolog with DL-reasoners. In: Proceedings of the ICLP 2007 Workshop on Applications of Logic Programming to the Web, Semantic Web and Semantic Web Services (ALPSWS 2007). CEUR Workshop Proceedings, vol. 287 (2007), http://www.ceur-ws.org/Vol-287

38. Drabent, W., Henriksson, J., Maluszynski, J.: Hybrid reasoning with rules and constraints under well-founded semantics. In: Marchiori, M., Pan, J.Z., de Sainte Marie, C. (eds.) RR 2007. LNCS, vol. 4524, pp. 348–357. Springer, Heidelberg (2007)

39. Drabent, W., Małuszyński, J.: Well-founded Semantics for Hybrid Rules. In: Marchiori, M., Pan, J.Z., de Sainte Marie, C. (eds.) RR 2007. LNCS, vol. 4524, pp. 1–15. Springer, Heidelberg (2007)

40. Eiter, T., Ianni, G., Krennwallner, T., Polleres, A.: Rules and Ontologies for the Semantic Web. In: Baroglio, C., Bonatti, P.A., Maluszynski, J., Marchiori, M., Polleres, A., Schaffert, S. (eds.) Reasoning Web 2008. LNCS, vol. 5224, pp. 1–53. Springer, Heidelberg (2008), Slides available at http://rease.semanticweb.org/

41. Eiter, T., Ianni, G., Krennwallner, T., Schindlauer, R.: Exploiting conjunctive queries in description logic programs. Annals of Mathematics and Artificial Intelligence (2009); published online 27 January 2009, doi:10.1007/s10472-009-9111-3. Also available as Tech. Rep. INFSYS RR-1843-08-02, Inst. of Information Systems, TU Vienna

42. Eiter, T., Ianni, G., Lukasiewicz, T., Schindlauer, R.: Well-founded semantics for description logic programs in the Semantic Web. Technical Report INFSYS RR-1843-09-01, Institut für Informationssysteme, Technische Universität Wien, A-1040 Vienna, Austria (March 2009)

43. Eiter, T., Ianni, G., Lukasiewicz, T., Schindlauer, R., Tompits, H.: Combining answer set programminag with description logics for the Semantic Web. Artificial Intelligence 172(12-13), 1495–1539 (2008)

44. Eiter, T., Ianni, G., Schindlauer, R., Tompits, H.: A Uniform Integration of Higher-Order Reasoning and External Evaluations in Answer Set Programming. In: International Joint Conference on Artificial Intelligence (IJCAI 2005), Edinburgh, UK, August 2005, pp. 90–96 (2005)

45. Eiter, T., Ianni, G., Schindlauer, R., Tompits, H.: Effective integration of declarative rules with external evaluations for semantic-web reasoning. In: Sure, Y., Domingue, J. (eds.) ESWC 2006. LNCS, vol. 4011, pp. 273–287. Springer, Heidelberg (2006)

46. Eiter, T., Lukasiewicz, T., Schindlauer, R., Tompits, H.: Combining answer set programming with description logics for the Semantic Web. In: Dubois, D., Welty, C., Williams, M.A. (eds.) Proceedings Ninth International Conference on Principles of Knowledge Representation and Reasoning (KR 2004), Whistler, British Columbia, Canada, June 2-5, pp. 141–151. Morgan Kaufmann, San Francisco (2004)

47. Eiter, T., Lukasiewicz, T., Schindlauer, R., Tompits, H.: Well-founded semantics for description logic programs in the Semantic Web. In: Antoniou, G., Boley, H. (eds.) RuleML 2004. LNCS, vol. 3323, pp. 81–97. Springer, Heidelberg (2004)

48. Faber, W., Pfeifer, G., Leone, N., Dell'Armi, T., Ielpa, G.: Design and Implementation of Aggregate Functions in the DLV System. Theory and Practice of Logic Programming 8(5-6), 545–580 (2008)

49. Ferrand, G., Deransart, P.: Proof method of partial correctness and weak completeness for normal logic programs. J. Log. Program. 17(2/3&4), 265–278 (1993)

50. Gelfond, M., Lifschitz, V.: The stable model semantics for logic programming. In: Kowalski, R.A., Bowen, K. (eds.) Proceedings of the Fifth International Conference on Logic Programming, pp. 1070–1080. MIT Press, Cambridge (1988)
51. Gelfond, M., Lifschitz, V.: The Stable Model Semantics for Logic Programming. In: Logic Programming: Proceedings Fifth Intl. Conference and Symposium, pp. 1070–1080. MIT Press, Cambridge (1988)
52. Gelfond, M., Lifschitz, V.: Classical Negation in Logic Programs and Disjunctive Databases. New Generation Computing 9, 365–385 (1991)
53. Grosof, B., Horrocks, I., Volz, R., Decker, S.: Description Logic Programs: Combining Logic Programs with Description Logic. In: Proceedings of 12th International Conference on the World Wide Web (2003)
54. Grosof, B.N.: Prioritized Conflict Handling for Logic Programs. In: Małuszyński, J. (ed.) Logic Programming, Proceedings of the 1997 International Symposium, pp. 197–211. MIT Press, Cambridge (1997)
55. Gruber, T.R.: A Translation Approach to Portable Ontology Specifications. Knowledge Acquisition 5(2), 199–220 (1993)
56. Herre, H., Jaspars, J., Wagner, G.: Partial logics with two kinds of negation as a foundation for knowledge-based reasoning. In: Gabbay, D., Wansing, H. (eds.) What is Negation?, pp. 121–159. Kluwer Academic Publishers, Dordrecht (1999)
57. Heymans, S., de Bruijn, J., Predoiu, L., Feier, C., Van Nieuwenborgh, D.: Guarded hybrid knowledge bases. TPLP 8(3), 411–429 (2008)
58. Heymans, S., Toma, I.: Ranking services using fuzzy hex-programs. In: Calvanese, D., Lausen, G. (eds.) RR 2008. LNCS, vol. 5341, pp. 181–196. Springer, Heidelberg (2008)
59. Hoehndorf, R., Loebe, F., Kelso, J., Herre, H.: Representing default knowledge in biomedical ontologies: Application to the integration of anatomy and phenotype ontologies. BMC Bioinformatics 8(1), 377 (2007)
60. Horrocks, I., Kutz, O., Sattler, U.: The even more irresistible $\mathcal{SROIQ}$. In: Proceedings of the 10th International Conference of Knowledge Representation and Reasoning (KR 2006), pp. 57–67 (2006)
61. Kifer, M.: Rules and Ontologies in F-Logic. In: Eisinger, N., Małuszyński, J. (eds.) Reasoning Web. LNCS, vol. 3564, pp. 22–34. Springer, Heidelberg (2005)
62. Kifer, M., Lausen, G., Wu, J.: Logical foundations of object-oriented and frame-based languages. J. ACM 42(4), 741–843 (1995)
63. Knorr, M., Alferes, J.J., Hitzler, P.: A Well-founded Semantics for Hybrid MKNF Knowledge Bases. In: DL 2007, 20th Int. Workshop on Description Logics, pp. 347–354. Bozen-Bolzano University Press (2007)
64. Knorr, M., Alferes, J.J., Hitzler, P.: A coherent well-founded model for hybrid mknf knowledge bases. In: Ghallab, M., Spyropoulos, C.D., Fakotakis, N., Avouris, N.M. (eds.) ECAI. Frontiers in Artificial Intelligence and Applications, vol. 178, pp. 99–103. IOS Press, Amsterdam (2008)
65. Konolige, K.: Quantification in autoepistemic logic. Fundam. Inform. 15(3-4), 275–300 (1991)
66. Krötzsch, M., Rudolph, S., Hitzler, P.: Description logic rules. In: Ghallab, M., Spyropoulos, C.D., Fakotakis, N., Avouris, N.M. (eds.) ECAI 2008 - 18th European Conference on Artificial Intelligence, Patras, Greece, July 21-25. Frontiers in Artificial Intelligence and Applications, vol. 178, pp. 80–84. IOS Press, Amsterdam (2008)
67. Krötzsch, M., Rudolph, S., Hitzler, P.: Elp: Tractable rules for OWL 2. In: Sheth, A.P., Staab, S., Dean, M., Paolucci, M., Maynard, D., Finin, T., Thirunarayan, K. (eds.) ISWC 2008. LNCS, vol. 5318, pp. 649–664. Springer, Heidelberg (2008)

68. Levy, A., Rousset, M.: CARIN: A representation language combining Horn rules and description logics. Artificial Intelligence 104(1-2), 165–209 (1998)
69. Lifschitz, V.: Nonmonotonic databases and epistemic queries. In: Proceedings IJCAI 1991, pp. 381–386 (1991)
70. Lloyd, J.W.: Foundations of logic programming, Second extended edn. Springer series in symbolic computation. Springer, New York (1987)
71. Lukasiewicz, T.: Probabilistic description logic programs. In: Godo, L. (ed.) ECSQARU 2005. LNCS, vol. 3571, pp. 737–749. Springer, Heidelberg (2005)
72. Lukasiewicz, T.: Fuzzy description logic programs under the answer set semantics for the Semantic Web. In: Proceedings RuleML 2006, pp. 89–96. IEEE Computer Society, Los Alamitos (2006)
73. Lukasiewicz, T.: A novel combination of answer set programming with description logics for the Semantic Web. In: Franconi, E., Kifer, M., May, W. (eds.) ESWC 2007. LNCS, vol. 4519, pp. 384–398. Springer, Heidelberg (2007)
74. Lukasiewicz, T.: Probabilistic description logic programs. Int. J. Approx. Reasoning 45(2), 288–307 (2007)
75. Lukasiewicz, T.: Tractable probabilistic description logic programs. In: Prade, H., Subrahmanian, V.S. (eds.) SUM 2007. LNCS, vol. 4772, pp. 143–156. Springer, Heidelberg (2007)
76. Lukasiewicz, T.: Fuzzy description logic programs under the answer set semantics for the Semantic Web. Fundam. Inform. 82(3), 289–310 (2008)
77. Lukasiewicz, T., Straccia, U.: Description logic programs under probabilistic uncertainty and fuzzy vagueness. In: Mellouli, K. (ed.) ECSQARU 2007. LNCS, vol. 4724, pp. 187–198. Springer, Heidelberg (2007)
78. Lukasiewicz, T., Straccia, U.: Tightly integrated fuzzy description logic programs under the answer set semantics for the Semantic Web. In: Marchiori, M., Pan, J.Z., Marie, C.d.S. (eds.) RR 2007. LNCS, vol. 4524, pp. 289–298. Springer, Heidelberg (2007)
79. Lukasiewicz, T., Straccia, U.: Top-k retrieval in description logic programs under vagueness for the Semantic Web. In: Prade, H., Subrahmanian, V.S. (eds.) SUM 2007. LNCS, vol. 4772, pp. 16–30. Springer, Heidelberg (2007)
80. Lukasiewicz, T., Straccia, U.: Tightly coupled fuzzy description logic programs under the answer set semantics for the Semantic Web. Int. J. Semantic Web Inf. Syst. 4(3), 68–89 (2008)
81. Małuszyński, J.: Integration of Rules Ontologies. In: Liu, L., Özsu, M.T. (eds.) Encyclopedia of Database Systems. Springer, Heidelberg (2009)
82. Marin, D.: A formalization of RDF. Technical Report TR/DCC-2006-8, TR Dept. Computer Science, Universidad de Chile (2006)
83. Marriott, K., Stuckey, P.J., Wallace, M.: Constraint logic programming. In: Handbook of Constraint Programming. Elsevier, Amsterdam (2006)
84. Miller, L., Brickley, D.: The Friend of a Friend (FOAF) Project (since 2000), http://www.foaf-project.org/
85. Motik, B., Rosati, R.: A faithful integration of description logics with logic programming. In: IJCAI 2007, Proceedings of the 20th International Joint Conference on Artificial Intelligence, pp. 477–482 (2007)
86. Motik, B., Sattler, U., Studer, R.: Query answering for OWL-DL with rules. J. Web Sem. 3(1), 41–60 (2005)
87. Motik, B., Rosati, R.: Closing semantic web ontologies. Technical report, University of Manchester (2006) (version March 7, 2007), http://web.comlab.ox.ac.uk/people/Boris.Motik/pubs/ mr06closing-report.pdf

88. Van Nieuwenborgh, D., De Cock, M., Vermeir, D.: Computing Fuzzy Answer Sets Using dlvhex. In: Dahl, V., Niemelä, I. (eds.) ICLP 2007. LNCS, vol. 4670, pp. 449–450. Springer, Heidelberg (2007)

89. Van Nieuwenborgh, D., Eiter, T., Vermeir, D.: Conditional Planning with External Functions. In: Baral, C., Brewka, G., Schlipf, J. (eds.) LPNMR 2007. LNCS, vol. 4483, pp. 214–227. Springer, Heidelberg (2007)

90. Nilsson, U., Małuszyński, J.: Logic, Programming and Prolog, 2nd edn. John Wiley and Sons, Chichester (1995), http://www.ida.liu.se/~ulfni/lpp/

91. Patel-Schneider, P.F., Hayes, P., Horrocks, I.: OWL Web Ontology Language Semantics and Abstract Syntax, W3C Recommendation (February 2004)

92. Poggi, A., Calvanese, D., De Giacomo, G., Lembo, D., Lenzerini, M., Rosati, R.: Linking data to ontologies. Journal of Data Semantics X, 133–173 (2008)

93. Przymusinski, T.C.: On the declarative and procedural semantics of logic programs. Journal of Automated Reasoning 5, 167–205 (1989)

94. Reiter, R.: On Closed World Data Bases. In: Gallaire, H., Minker, J. (eds.) Logic and Data Bases, pp. 55–76. Plenum Press, New York (1978)

95. Reiter, R.: A Logic for Default Reasoning. Artificial Intelligence 13(1-2), 81–132 (1980)

96. Rosati, R.: On the decidability and complexity of integrating ontologies and rules. Journal of Web Semantics 3(1), 61–73 (2005)

97. Rosati, R.: $\mathcal{DL}+log$: Tight integration of Description Logics and disjunctive Datalog. In: Doherty, P., Mylopoulos, J., Welty, C.A. (eds.) KR, pp. 68–78. AAAI Press, Menlo Park (2006)

98. Ross, K.A.: Modular stratification and magic sets for datalog programs with negation. J. ACM 41(6), 1216–1266 (1994)

99. Schindlauer, R.: Answer-Set Programming for the Semantic Web. PhD thesis, Vienna University of Technology, Austria (December 2006)

100. Simkus, M., Eiter, T.: FDNC: Decidable non-monotonic disjunctive logic programs with function symbols. In: Dershowitz, N., Voronkov, A. (eds.) LPAR 2007. LNCS, vol. 4790, pp. 514–530. Springer, Heidelberg (2007); Full paper to appear in ACM TOCL

101. Sintek, M., Decker, S.: TRIPLE - A Query, Inference, and Transformation Language for the Semantic Web. In: Horrocks, I., Hendler, J. (eds.) ISWC 2002. LNCS, vol. 2342, p. 364. Springer, Heidelberg (2002), http://triple.semanticweb.org/

102. van Gelder, A., Ross, K.A., Schlipf, J.S.: Unfounded Sets and Well-founded Semantics for General Logic Programs. In: Principles of Database Systems, pp. 221–230. ACM, New York (1988)

103. Van Gelder, A., Ross, K.A., Schlipf, J.S.: The Well-Founded Semantics for General Logic Programs. Journal of the ACM 38(3), 620–650 (1991)

Chapter 2
Four Lessons in Versatility
or How Query Languages Adapt to the Web

François Bry, Tim Furche, Benedikt Linse, Alexander Pohl,
Antonius Weinzierl, and Olga Yestekhina

Institute for Informatics, University of Munich,
Oettingenstraße 67, D-80538 München, Germany
http://www.pms.ifi.lmu.de/

Abstract. Exposing not only human-centered information, but machine-processable data on the Web is one of the commonalities of recent Web trends. It has enabled a new kind of applications and businesses where the data is used in ways not foreseen by the data providers. Yet this exposition has fractured the Web into islands of data, each in different Web formats: Some providers choose XML, others RDF, again others JSON or OWL, for their data, even in similar domains. This fracturing stifles innovation as application builders have to cope not only with one Web stack (e.g., XML technology) but with several ones, each of considerable complexity.

With Xcerpt we have developed a rule- and pattern based query language that aims to give shield application builders from much of this complexity: In a single query language XML and RDF data can be accessed, processed, combined, and re-published. Though the need for combined access to XML and RDF data has been recognized in previous work (including the W3C's GRDDL), our approach differs in four main aspects: (1) We provide a single language (rather than two separate or embedded languages), thus minimizing the conceptual overhead of dealing with disparate data formats. (2) Both the declarative (logic-based) and the operational semantics are unified in that they apply for querying XML and RDF in the same way. (3) We show that the resulting query language can be implemented reusing traditional database technology, if desirable. Nevertheless, we also give a unified evaluation approach based on interval labelings of graphs that is at least as fast as existing approaches for tree-shaped XML data, yet provides linear time and space querying also for many RDF graphs.

We believe that Web query languages are the right tool for declarative data access in Web applications and that Xcerpt is a significant step towards a more convenient, yet highly efficient data access in a "Web of Data".

2.1 Introduction

The one undeniable trend in the development of the Web has been a move from human-centered information to more machine-processable data. This trend is a part of most visions for the future of the Web, may they be called "Web 2.0", "Semantic Web", "Web of Data", "Linked Data". There is a reason that this trend underlies so many of the visions for a future Web: With machine-processable data, other agents than the owner or publisher of data can create novel applications, e.g., by using the data in a

F. Bry and J. Maluszynski (Eds.): Semantic Techniques for the Web, LNCS 5500, pp. 50–160, 2009.

context never envisioned by the data owner, by presenting it in different ways or media, or by enhancing or mixing it with other data.

Unfortunately, though machine-processable data is called for by many of these visions, they do not agree on the *data format*. For human-centered information, HTML has clearly dominated the Web. For machine-processable data, Web 2.0 APIs and publishers tend to use XML, JSON, or YAML, Semantic Web publishers RDF and/or OWL. This way, application designers are either impeded from using data published in, say, RDF, if they are used to data in, say, XML or they have to cope with not only one (already fairly complex) stack of Web technologies but several.

The need for a more integrated, easier access to Web data has been recognized: For instance, the W3C has proposed a means of accessing XML data as RDF (GRDDL [54]). Other approaches integrate existing RDF query languages into XML query languages (XSPARQL [7], [79]) or vice versa ([60], SPAT[1]. In this work, we present a different answer to this problem: a single, unified language, called Xcerpt, that can query both XML and RDF with the same ease. Previous approaches require the user to learn (a) an XML (usually XPath or XQuery), (b) an RDF query language (usually SPARQL), and (c) how concepts from RDF and XML are mapped to each other, if at all. In our approach, we first develop *a query language flexible enough to deal with most Web data* (in the spirit of, though with quite different focus and result than [138]). Then we only have to teach the user how to query RDF resp. XML with that query language, reusing as much of the data and query concepts between the two settings as possible. Not only does this reduce the learning curve for the user considerably, it also makes it easy to extend the approach with further Web formats such as JSON, YAML, or Topic Maps.

We introduce Xcerpt in Sections 2.3.1 and 2.3.2 after a brief recall of the basics of the two Web formats considered here, XML and RDF, in Section 2.2.

But defining a language for unified access to XML and RDF is just how the story begins. For the approach to be feasible, we require two more ingredients: 1. a simple semantics that is nevertheless versatile enough to cover the specifics of both XML and RDF. 2. an evaluation engine that is competitive to engines specialized to XML or RDF data only.

In Sections 2.4.1 to 2.4.4 we propose two different ways to define the (declarative) semantics of Xcerpt: The first uses a modified form of simulation to describe which queries match what data. It is flexible enough to deal with queries on XML and RDF data and can be defined very concisely. We show in Section 2.4.2 and 2.4.3 how to adapt the (well-founded) semantics of rule programs with negation to use simulation rather than term equality/instantiation.

This gives an easy, straightforward definition of the semantics of Xcerpt. However, the disadvantage is that required notion of simulation is not as well studied as term equality and not supported by existing database or rule technologies. Therefore, we show in Section 2.4.4 how Xcerpt can be translated into standard Datalog with negation and value invention ($\text{Datalog}^{\neg}_{new}$) which can be evaluated by most SQL-database engines and many rule engines. Not only do we show how to translate Xcerpt into $\text{Datalog}^{\neg}_{new}$, but we do the same for XPath, XQuery, and SPARQL, thus establishing a uniform formal foundation for all these languages (that we exploit in Section 2.5.1 for

---

[1] http://www.w3.org/2007/01/SPAT/

a unified evaluation engine for all these languages). Moreover, we use the translation to prove several complexity and expressiveness features. Most importantly, we show that full Xcerpt is unsurprisingly Turing-complete, that stratification does not limit the expressiveness of Xcerpt, and identify a decidable fragment (weakly-recursive Xcerpt).

For $Datalog^-_{new}$ and thus for Xcerpt, whether on XML or RDF data, we define a novel evaluation algorithm and indexing scheme in Section 2.5.1, thus turning to the second of the two missing ingredients, the competitive evaluation engine. We show how to extend tree labeling schemes (such as the pre/post-encoding [81]) to graph data in a novel way: Where previous such approaches [6,151,48,145] can not guarantee linear time and space evaluation of acyclic conjunctive queries on interesting super-classes of trees, our approach exhibits such a class: the continuous-image graphs. On this significant super-class of trees we can still maintain linear time and space evaluation. The basic idea of the approach is a generalized interval labeling together with (most importantly) a novel join algorithm for intermediary answers represented by intervals.

Together with the results from Section 2.4.4, we thus obtain a surprisingly large linear time and space fragment of Xcerpt, viz. (weakly-recursive) acyclic Xcerpt on continuous-image graphs, a novel super-class of trees. The same also applies to, e.g., SPARQL.

To complete the evaluation of Xcerpt, we not only need an efficient evaluation engine for Xcerpt queries, but for Xcerpt *rules*. Section 2.5.2 gives a first step towards such a rule engine for Xcerpt. It introduces *simulation unification* as an extended, more flexible form of unification that is adapted to Xcerpt's notion of simulation discussed above. Based on simulation unification, we show how subsumption can be exploited to define an efficient resolution with tabling for locally stratified Xcerpt programs.

To summarize, the theme of this chapter is the investigation of how to address the increasing number of diverse data formats being introduced on the Web. We suggest as a solution, Xcerpt,

1. *a versatile query language* that allows access to both XML and RDF in the same language, sharing concepts as much as possible (Section 2.3– 2.3.3). It is complemented by
2. *a versatile declarative semantics* based on a form of simulation adapted to Web data that is easy to understand, yet can be translated to standard database and rule technology, as can XPath, XQuery, and SPARQL (Section 2.4– 2.4.4). For that semantics (and thus for Xcerpt, XPath, XQuery, and SPARQL), we propose
3. *a versatile evaluation algorithm* that is able to provide the best-known complexity for acyclic conjunctive queries on tree-shaped XML data, manages to maintain that complexity for many RDF graphs, and yet can also operate on arbitrary graphs (Section 2.5.1). We extend that evaluation algorithm towards a full versatile rule-based query language for the Web like Xcerpt by illustrating how resolution with tabling can be adapted to use
4. *a versatile form of subsumption* based on simulation unification for determining where previously computed answers to a sub-query can be reused for further sub-queries (Section 2.5.2).

The structure of this chapter follows the four perspectives on addressing the rising amount of Web data formats: data, query language, semantics, and evaluation. While the parts on data and query are to some extent necessary for understanding the parts on semantics and evaluation, the latter two are fairly independent. It is not necessary to understand the details of the semantics for the evaluation or vice versa. We discuss querying XML and RDF in separate sections (Section 2.3.1 and 2.3.2) and refer to the sub-language of Xcerpt used for XML access as Xcerpt^{XML}, to that used for RDF access as Xcerpt^{RDF}. In both sections we highlight where specific concepts are needed and where the same concept can be used for both XML and RDF access.

Related Work. To keep the parts fairly self-contained and to avoid overly long preliminaries, we decided to address related work in each part separately.

In particular, Section 2.3.4 compares Xcerpt, in particular its features for accessing RDF, with SPARQL, the W3C proposal for querying RDF and a number of its extensions. Section 2.4.4 gives a brief comparison of the challenges when translating Xcerpt to Datalog$^-_{new}$, i.e., to existing database and rule technology, compared to XPath, XQuery or SPARQL. For the evaluation, we extensively compare our approach with existing labeling schemes for tree and graph data in Section 2.5.1. The basic principles of the evaluation algorithm are discussed in the context of related work in Section 2.5.1.

As pointed out there are a number of previous approaches to integrating XML and RDF access. These can be divided in two categories: Approaches such as GRDDL [57] use two separate query languages to first transform data from one format in the other and then to query only in the latter format. The advantage of this approach is that existing language engines can be used as is. The second kind of approaches is exemplified by XSPARQL [128] and [79]: Here one of the languages is embedded into the other, providing an interface between the two languages (of varying sophistication). The advantage is that we can now transfer results in both directions, the disadvantage is that new query engines or rather involved query translations are needed[2].

For both approaches there are two main shortcomings:

1. The user has to learn two different query languages that were designed entirely separate.
2. Since the language engines remain entirely separate, these approaches first transform *all data* (without respect to what is actually queried), then load all transformed data in the second query engine, only then it is filtered by the conditions of the queries in the target format. Thus, there is no chance for goal-driven query evaluation and even static propagation of query conditions from queries in the target format to the transformation queries are very hard due to the starkly varying semantics of the two languages involved.

Though Xcerpt and Xcerpt^{RDF} still require the use to learn some concepts specific to XML or RDF, they are design to share concepts where possible. Furthermore, we use a unified semantics thus allowing for full static cross-format optimization and a unified evaluation allowing for dynamic cross-format optimization.

Neither the above approaches not Xcerpt addresses integrating also queries on (OWL) ontologies, e.g., in the style of [62]. Though this is certainly an important issue, it is out

---

[2] In Section 2.4.4 we illustrate a first step towards a translation approach.

of the scope of this chapter. We believe that some of the discussed issues apply also in that context (in particular, the treatment of blank nodes in RDF in the semantics and evaluation of Xcerpt), but there are many more issues when considering even conjunctive queries on ontologies that would need addressing.

2.2 Versatile Data

2.2.1 Extensible Markup Language (XML)

XML [28] is, by now, *the* foremost data representation format for the Web and for semi-structured data in general. It has been adopted in a stupendous number of application domains, ranging from document markup (XHTML, Docbook [150]) over video annotation (MPEG 7 [110]) and music libraries (iTunes[3]) to preference files (Apple's property lists [9]), build scripts (Apache Ant[4]), and XSLT [95] stylesheets. XML is also frequently adopted for serialization of (semantically) richer data representation formats such as RDF or TopicMaps.

XML is a generic markup language for describing the structure of data. Unlike in HTML (HyperText Markup Language), the predominant markup language on the web, neither the tag set nor the semantics of XML are fixed. XML can thus be used to derive markup languages by specifying tags and structural relationships.

The following presentation of the information in XML documents is oriented along the XML Infoset [56] which describes the information content of an XML document. The XQuery data model [67] is, for the most parts, closely aligned with this view of XML documents.

Following the XPath and XQuery data model, we provide a *tree shaped* view of XML data. This deviates from the Infoset where valid ID/IDREF links are resolved and thus the data model is graph, rather than tree shaped. This view is adopted in some XML query languages such as Xcerpt [40] and Lorel [3], but most query languages follow XPath and XQuery and consider XML tree shaped.

XML in 500 Words. The core provision of XML is a syntax for representing hierarchical data. Data items are called elements in XML and enclosed in start and end *tags*, both carrying the same tag names or *labels*. <author>...</author> is an example of such an element. In the place of '...', we can write other elements or character data as *children* of that element. The following listing shows a small XML fragment that illustrates elements and element nesting:

```
  <bib xmlns:dc="http://purl.org/dc/elements/1.1/">
2   <article journal="Computer Journal" id="12">
      <dc:title>...Semantic Web...</dc:title>
4     <year>2005</year>
      <authors>
6       <author>
          <first>John</first> <last>Doe</last> </author>
```

---

[3] http://www.apple.com/itunes/
[4] http://ant.apache.org/

```
 8      <author>
          <first>Mary</first> <last>Smith</last> </author>
10    </authors>
    </article>
12  <article journal="Web Journal">
      <dc:title>...Web...</dc:title>
14    <year>2003</year>
      <authors>
16      <author>
          <first>Peter</first> <last>Jones</last> </author>
18      <author>
          <first>Sue</first> <last>Robinson</last> </author>
20    </authors>
    </article>
22 </bib>
```

In addition, we can observe *attributes* (name, value pairs associated with start tags) that are essentially like elements but may only contain character data, no other nested attributes or elements. Also, by definition, *element order* is significant, attribute order is not. For instance

```
<author><last>Doe</last><first>John</first></author>
```

represents different information than the `author` element in lines 6–9, but

```
<article id="12" journal="Computer Journal">...</article>
```

represents the same element information item as lines 2–15.

Figure 1 gives a graphical representation of the XML document that is referenced in preceding illustrations. When represented as a graph, an XML document without links is a labeled tree where each node in the tree corresponds to an element and its type. Edges connect nodes and their children, that is, elements and the elements nested in them, elements and their content and elements and their attributes. Since the visual distinction between the parent-child relationship can be made without edge labels and since attributes are not addressed or receive no special treatment in the research presented in this text, edges will not be labeled in the following figures.

Elements, attributes, and character data are XML's most common information types. In addition, XML documents may also contain *comments*, *processing instructions* (name-value pair with specific semantics that can be placed anywhere an element can be placed), *document level information* (such as the XML or the document type declarations), *entities*, and *notations*, which are essentially just other kinds of information containers.

On top of these information types, two additional facilities relevant to the information content of XML documents are introduced by subsequent specifications: Namespaces [26] and Base URIs [109]. Namespaces allow partitioning of element labels used in a document into different namespaces, identified by a URI. Thus, an element is no longer labeled with a single label but with a triple consisting of the *local name*, the *namespace prefix*, and the *namespace URI*. E.g., for the `dc:title` element in line 3, the local name is `title`, the namespace prefix is `dc`, and the namespace URI (called "name" in [56]) is `http://purl.org/dc/elements/1.1/`. The latter can be derived by

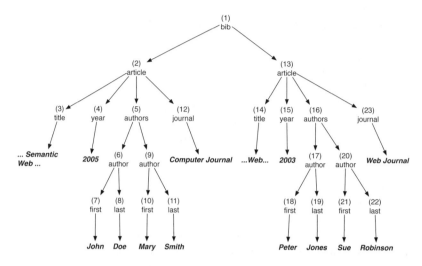

Fig. 1. Visual representation of sample XML document

looking for a *namespace declaration* for the prefix dc. Such a declaration is shown in line 1: xmlns:dc="http://... It associates the prefix dc with the given URI in the scope of the current element, i.e., for that element and all elements contained within unless there is another nested declaration for dc, in which case that declaration takes precedence. Thus, we can associate with each element a set of *in-scope namespaces*, i.e., of pairs namespace prefix and URI, that are valid in the scope of that element. Base URIs [109] are used to resolve relative URIs in an XML document. They are associated with elements using xml:base="http://... and, as namespaces, are inherited to contained elements unless a nested xml:base declaration takes precedence.

The above features of XML are covered by most query languages. Additionally some languages (most notably XQuery) also provide access to type information associated via DTD or XML Schema [66]. These features are mentioned below where appropriate but not discussed in detail here.

2.2.2 Resource Description Framework (RDF)

As the second preeminent data format on the Semantic Web, the *Resource Description Format (RDF)* [108,101,85] is emerging. RDF is, though much less common than XML, a widespread choice for interchanging (meta-) data together with descriptions of the schema and, in contrast to XML, a basic description of its semantics of that data.

Not to distract from the salient points of the discussion, we omit typed literals (and named graphs) from the following discussion.

RDF in 500 Words. RDF graphs contain simple statements about *resources* (which, in other contexts, are be called "entities", "objects", etc., i.e., elements of the domain that may partake in relations). Statements are triples consisting of subject, predicate, and object, all of which are resources. If we want to refer to a specific resource, we use (supposedly globally unique) URIs, if we want to refer to a resource for which we know that it exists and maybe some of its properties, we use *blank nodes* which play the role of existential quantifiers in logic. However, blank nodes may not occur in predicate position. Finally, for convenience, we can directly use *literal values* as objects.

RDF may be serialized in many formats (for a recent survey see [20]), such as RDF/XML [15], an XML dialect for representing RDF, or Turtle [13] which is also used in SPARQL. The following Turtle data represents roughly the same data as the XML document discussed in the previous section:

```
  @prefix dc: <http://purl.org/dc/elements/1.1/> .
2 @prefix dct: <http://purl.org/dc/terms/> .
  @prefix vcard: <http://www.w3.org/2001/vcard-rdf/3.0#> .
4 @prefix bib: <http://www.edutella.org/bibtex#> .
  @prefix ex: <http://example.org/libraries/#> .
6 ex:smith2005 a bib:Article ; dc:title "...Semantic Web..." ;
      dc:year "2005" ;
8     ex:isPartOf [ a bib:Journal ;
          bib:number "11"; bib:name "Computer Journal" ] ;
10    bib:author [ a rdf:Bag ;
          rdf:_1 [ a bib:Person ;
12            bib:last "Smith" ; bib:first "Mary" ] ;
          rdf:_2 [ a bib:Person ;
14            bib:first "John" ; bib:last "Doe" ] ] .
```

Following the definition of namespace prefixes used in the remainder of the Turtle document (omitting common RDF namespaces), each line contains one or more statements separated by colon or semi-colon. If separated by semi-colon, the subject of the previous statement is carried over. E.g., line 1 reads as ex:smith2005 is a (has rdf:type) bib:Article and has dc:title "...Semantic Web...". Lines 3–4 show a blank node: the article is part of some entity which we can not (or don't care to) identify by a unique URI but for which we give some properties: it is a bib:Journal, has bib:number "11", and bib:name "Computer Journal".

Figure 2 shows a visual representation of the above RDF data, where we distinguish literals (in square boxes) and classes, i.e., resources that can be used for classifying other resources, and thus can be the object of an rdf:type statement (in square boxes with rounded edges) from all other resources (in plain ellipses).

What sets RDF apart from XML and justifies its role as *the* data format for the *Semantic* Web is that RDF data comes with attached meaning, that allows us to infer additional knowledge beyond what is stated explicitly. Query languages are usually expected to behave consistent w.r.t. some form of RDF entailment (e.g., simple, full, or RDFS entailment), i.e., graphs equivalent under the respective entailment yield the same answers. Simply stated, rather than just consulting the actual RDF data for answering a query, we might also need to consider additional, inferred triples depending on the form of entailment chosen. E.g., when querying for resources of type bib:Publication we might also want to return bib:Articles if we have the additional information that bib:Article is a sub-class of bib:Publication. SPARQL, e.g., is designed to be agnostic of the particular entailment used: it can be used to query RDF data under any of the above mentioned entailment forms.

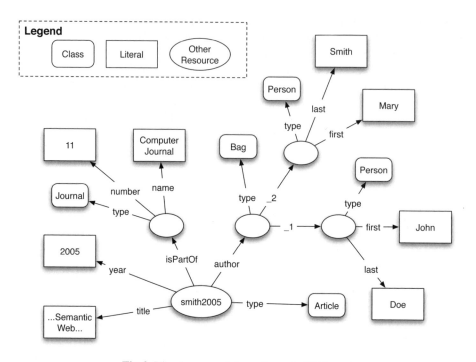

Fig. 2. Visual representation of sample RDF graph

In the following, we assume familiarity with the notion of RDF entailment, interpretation, model, as well as the RDFS semantics from [85].

2.3 Versatile Queries

With the rise of a plethora of different semi-structured Web formats, versatility [32] has become the central requirement for web query languages. Besides the well-known and ubiquitous formats HTML, XML and RDF, there are quite a lot of less familiar formats such as RDFa [5,4] for embedding RDF information in HTML pages, the microformats [98] geo, hCard, hCalendar, hResume, etc., the ISO-standard Topic Maps [74,123]. We call a web query language *format versatile*, if it can handle, merge or transform data in different formats within the same query program. The need for integrating data from different formats has been acknowledged by partial solutions such as GRDDL [57,149,72], hGRDDL [4] and XSPARQL [7]. All these solutions have in common that they try to solve the problem of web data integration by applying a mix of already established technologies such as XSLT transformations, DOM manipulations, and a combination of XML and RDF query languages such as XQuery and SPARQL. It is thus unsurprising that understanding these solutions requires a large background knowledge of the employed technologies, and that the methods are much more complicated than they could be if a format-versatile language particularly geared at integrating data from different web formats was employed.

Besides format versatility, we distinguish two other kinds of versatility: *schema* and *representational versatility*. A web query language is called *schema versatile*, if it can handle and intermediate between different schemata (i.e. schema heterogeneity) on the Web. Usage of different schemata for representing similar data is very common and well-studied in the field of data integration [146,106]. Since the Web is being enhanced with structured and semantically rich data, data integration on the Web [102] has also received considerable attention and has spurred the growth of ontology alignment [117,63,64] research. Schema heterogeneity on the Web is encountered whenever two ontologies describe the same kind of information on the Web, but employ different languages for this end.

Finally, *representational heterogeneity* is encountered in XML dialects such as RDF/XML, where the same information is represented differently due to the use of syntactic sugar notations – e.g. for rdf:type arcs or for the concise notation of literals, URIs or RDF containers. Moreover, representational heterogeneity is present in any XML dialect that does not enforce any order of the information that it provides, since for serialization an arbitrary order must be chosen. We call a language *representational versatile*, if it can query data agnostic of the representational variant chosen.

In this and the following sections, we show how the design of Xcerpt query terms, construct terms and rules has lead to a versatile language with respect to all three issues – format, schema and representation. This section starts out by looking at Xcerpt from an abstract point of view, its relationship to logic programming and the interface defined by Xcerpt terms. In Section 2.3.1, we introduce the sub-language of Xcerpt for querying, construction and transformation of XML, called Xcerpt^{XML}. It is introduced along the example of harvesting search results and microformat information of personal profile pages of a social network. In Section 2.3.2, Xcerpt's RDF querying capabilities, referred to as Xcerpt^{RDF}, are presented with special emphasis on treating RDF specifies such as containers, collections and reifications. Finally, in Section 2.3.3, we present a use-case on combining microformat information harvested with Xcerpt^{XML} and RDF data queried with Xcerpt^{RDF}, thus combining versatile querying in XML and RDF.

Xcerpt Terms from an Abstract Point of View: Simulation, Substitutions, and Application of Substitution Sets. Xcerpt is a rule and pattern based language inspired by logic programming, but with significantly richer querying capabilities that are necessitated by the semi-structured nature of data on the Web.

In contrast to Prolog unification, Xcerpt uses a more involved kind of unification called *simulation unification*[5] to extract bindings of logical variables from Web data.

While Prolog rules consist of possibly non-ground terms in the head and the body of a rule, Xcerpt distinguishes between *construct terms* and *query terms* to be used in the heads and the bodies of rules, respectively. This differentiation is necessary because the semi-structured nature of data on the Web requires expressive query constructs – such as descendant, subterm negation, optionality – only in the query part of a rule (i.e. in the query terms), and constructs for reassembling the data – such as grouping – only in the construction part (i.e. the construct terms). Additionally, Xcerpt offers data terms as an abstraction of XML (and thus also HTML) and RDF data. Xcerpt terms fulfill the

---

[5] The term *simulation* is derived from graph simulation as defined in [1].

following three properties: (i) any data term is also a query term, (ii) any data term is also a construct term, and (iii) the intersection between the set of construct terms and query terms is exactly the set of data terms, where some subterms may be substituted by variables.

Also Prolog differentiates between terms and ground terms and facts. In Prolog it holds that any ground term is a fact (i.e. data). In Xcerpt, however, a term may very well be ground, but still be only an incomplete description of data – i.e. a query. Xcerpt terms are formally – but, for the sake of brevity, not in their entirety – defined in Section 2.4.1.

The differences between Prolog Unification and Simulation unification can be briefly summarized as follows:

- *Non-Symmetry of simulation unification.* Whereas Prolog unification is a symmetric operation on two generally non-ground terms, Xcerpt simulation unfication is a non-symmetric relation having a query term as the first argument, and a construct term as the second.
- *Different types of variables.* While Prolog Unification only allows for one single type of variable that will bind to any type of term, Xcerpt differentiates between different types of variables. Obviously the types of variables also differ with the data format that is being queried (XML, RDF, Topic Maps, Microformats, etc). When querying XML data, Xcerpt distinguishes between term variables, that bind to an entire XML fragment and label variables, that bind to a qualified or local name only.[6]
- *Notations for querying incomplete data.* Due to the almost schemaless nature of data on the Web, Xcerpt terms must be able to incompletely specify or describe the data that is being searched for. These notations include optionality of subterms, subterms at arbitrary depth and negated subterms and are introduced in detail in Section 2.3.1.
- *Substitution sets instead of substitutions.* While in Prolog one can find a single most general unifier for two terms t_1 and t_2 up to variable renaming, this is not true for Xcerpt. Simulation unification between two Xcerpt terms xt_1 and xt_2 results in a set of substitutions (that may very well contain only a single substitution or none at all), which is due to the richer kind of simulation and the deeper structure of data found on the Web. Imagine, for example, a biological database in XML format on the Web that contains data about enzymes and chemical reactions they catalyze. Although the database may be contained in a single XML document, the query for all pairs of enzymes and catalyzed reactions should, obviously, return more than a single tuple.

Feature unification [94,93], i.e. unification between feature terms, has been investigated in linguistics to aid automatic translation of natural language texts. Feature terms are used as an abstract representation of text, and are similar to semi-structured expressions as far as they can be arbitrarily nested as XML documents, may contain nodes that are entirely represented by their properties (just as RDF blank nodes), and in that the

---

[6] Variables for term identifiers and for XML attributes are not considered in this survey for the sake of brevity.

order of subterms may or may not be relevant. In contrast to simulation unification, feature unification is symmetric, and feature terms do not provide constructs for specifying incompleteness in depth or different types of variables. Finally, feature unification does not return sets of variable bindings but serves to translate text from one natural language to another.

Matching or – in Xcerpt terminology – simulating queries with data is only one of two steps in the transformation of semi-structured data. Just as Prolog, but more consequently (because of aggregation), Xcerpt clearly separates extraction of data (the data is bound to variables within *rule bodies*) and construction of new data (reassembling the data by application of substitution sets to *rule heads*).[7] This separation contrasts with XML query languages such as XQuery and XSLT, in which querying and construction is *intertwined*. Construction of new data with rule based languages is achieved by applying a substitution to a term. As mentioned above, however, Xcerpt does not deal with ordinary substitutions, but with *substitution sets*, and moreover, it differentiates between different kinds of terms. Therefore, we must be more specific: Construction of new data in Xcerpt is achieved by applying *sets* of substitutions to *construct* terms. The step from single substitutions to substitution sets allows the introduction of grouping constructs and aggregations to rule-based web querying. In the absence of grouping and aggregation constructs, application of substitution sets does not result in a single Xcerpt term, but in a set of terms (which may very well be unary or even empty).

The above discussion of Xcerpt terms can be summarized by the following interface (written as a functional type signature) of an Xcerpt term:

$$simulates :: QueryTerm \rightarrow ConstructTerm \rightarrow Bool \quad (1)$$

$$simulation\_unify :: QueryTerm \rightarrow ConstructTerm \rightarrow SubstitutionSet \quad (2)$$

$$apply\_substitution\_set :: SubstitutionSet \rightarrow ConstructTerm \rightarrow [DataTerm] \quad (3)$$

The function *simulates* returns true for a query term q and a construct term t if and only if the substitution set returned for $simulation\_unify(q,t)$ is non-empty. In addition to three above mentioned functions, a function which decides the subsumption relationship between two Xcerpt query terms is required if an optimized tabling algorithm for backward chaining evaluation of a multi-rule program is to be used. For more information about the subsumption relationship between Xcerpt query terms see Section 2.5.2.

In Section 2.3.1, we informally introduce the XML processing capabilities of Xcerpt, Xcerpt[XML] terms, Xcerpt[XML] simulation unification and the application of substitution sets to Xcerpt[XML] terms. In Section 2.3.2 we do the same for Xcerpt[RDF].

2.3.1 XML Queries—Examples and Patterns

A large number of query languages for XML data have been proposed in the past. They range from navigational languages such as XSLT [96] XQuery [142], their common

---

[7] Queries against a single Prolog rule, such as the append rule, may indeed be used to achieve both: concatenation of lists and finding components of a list. Still, querying is performed by matching *rule bodies* with terms, and data construction by filling in bindings for variables in *rule heads*.

subset XPath [19], and Quilt [47] (the predecessor of XQuery) over pattern based languages such as XML-QL [58], UnQL [41] and Xcerpt to visual query languages such as visXcerpt [17], XQBE [12] and XML-GL [46]. For a comprehensive survey over XML query languages, their expressive power and language constructs, see [14], for a comparison of Lorel, XML-QL, XML-GL, XSL and XQL see [23].

In this section, we introduce the XML processing capabilities of Xcerpt, taking Web search results, personal profile pages from the LinkedIn social network and FOAF documents as a running example. With this data, the following task will be accomplished:

- We will extract links to LinkedIn profile pages from search results of the Google search engine. These search results are wrapped within deeply nested HTML which primarily serve presentation purposes, and snippets of text extracted from the indexed pages. By matching among others `class` and `id` attributes, only the relevant links will be extracted.
- From the profile pages relevant data of the curriculum vitae of the persons is identified and extracted by exploiting the microformat vocabularies `hresume`, `hcalendar` and `hcard` which are integrated into the HTML pages for semantic enrichment of the textual content.
- Finally, FOAF documents are queried to find additional information not present in the LinkedIn profile. Since FOAF is an RDF format that may be serialized in RDF/XML, we will discuss the syntactic XML structure of these documents and their correspondence to Xcerpt^{RDF} query terms in this section, but use Xcerpt^{RDF} to query their contents in Section 2.3.2.

Xcerpt^{XML} Data and Rules. This section introduces Xcerpt^{XML} data terms, that abstract from XML documents, ignoring XML specifities such as processing instructions, comments, entities and DTDs. Xcerpt^{XML} terms are introduced to allow a more concise representation of XML data that can be extended to form queries and construct patterns to be used in rules.

Rules are written in a similar fashion to Datalog or Prolog rules, and have the following general form:

```
CONSTRUCT <CONSTRUCTTERM> FROM <QUERY> END
```

Xcerpt queries are enclosed between the **FROM** and **END** keywords and are *matched* – in Xcerpt terminology *simulated* – with data. Due to Xcerpt's answer closedness (see Definition 1 for details), data may also be used as queries. To see how XML is represented as Xcerpt data, consider the FOAF document in Listing 1.1 and the corresponding Xcerpt data term in Listing 1.2.

FOAF is an acronym for "Friend-Of-A-Friend", which is a vocabulary for specifying relationships among people, their personal information such as adresses, education and contact information. FOAF is primarily an RDF vocabulary, and is therefore semantically richer than plain XML data, but most FOAF documents are serialized in RDF/XML. Therefore, FOAF documents serialized in RDF/XML can be queried or transformed *syntactically* (on the XML level) or *semantically* (on the RDF level). While this section deals with syntactic transformations of Web data, semantic queries, transformations and reasoning using Xcerpt^{RDF} are discussed in Section 2.3.2.

```
 <rdf:RDF xmlns:rdf="http://www.w3 ... rdf-syntax-ns#"
2        xmlns:rdfs="http://www.w3 ... rdf-schema#"
         xmlns:foaf="http://xmlns.com/foaf/0.1/"
4        xml:base="http://www.example.com/">
   <foaf:PersonalProfileDocument
       rdf:about="descriptions/Bill.foaf">
6    <foaf:maker rdf:resource="#me"/>
     <foaf:primaryTopic rdf:resource="#me"/>
8  </foaf:PersonalProfileDocument>
   <foaf:Person rdf:ID="me">
10   <foaf:givenname>Bill</foaf:givenname>
     <foaf:mbox_sha1sum>5e22c ... 35b9</foaf:mbox_sha1sum>
12   <foaf:depiction rdf:ID="images/bill.png"/>
     <foaf:knows>
14     <foaf:Person>
         <foaf:name>Hillary</foaf:name>
16       <foaf:mbox_sha1sum>1228 ... 2f5</foaf:mbox_sha1sum>
         <rdfs:seeAlso rdf:ID="descriptions/Hillary.foaf"/>
18     </foaf:Person>
     </foaf:knows>
20 </foaf:Person>
   </rdf:RDF>
```

Listing 1.1. A friend-of-a-friend document

```
1 declare namespace rdf "http://www.w3 ... rdf-syntax-ns#";
  declare namespace rdfs "http://www.w3 ... rdf-schema#";
3 declare namespace foaf "http://xmlns.com/foaf/0.1/"
  declare xml-base "http://www.example.com/"
5
  rdf:RDF [
7  foaf:PersonalProfileDocument
       (rdf:about="descriptions/Bill.foaf") [
     foaf:maker (rdf:resource="#me"),
9    foaf:primaryTopic (rdf:resource="#me") ],
   foaf:Person (rdf:ID="#me") [
11   foaf:givenname [ "Bill" ],
     foaf:mbox_sha1sum [ "5e22c ... 35b9" ],
13   foaf:depiction (rdf:ID="images/bill.png"),
     foaf:knows [
15     foaf:Person [
         foaf:name [ "Hillary" ],
17       foaf:mbox_sha1sum [ "1228 ... 2f5" ]
         rdfs:seeAlso (rdf:ID="descriptions/Hillary.foaf") ] ] ]
           ]
```

Listing 1.2. A friend-of-a-friend-document written as an Xcerpt data term

Listings 1.1 and 1.2 exhibit an overwhelming similarity. Therefore, we will only quickly discuss the points in which the data term representation deviates from the XML serialization. While attributes are given as name value pairs inside of opening tags in an XML document, they are given in round braces following a qualified name in Xcerpt^{XML}. Moreover, the beginning and end of an element are specified by opening and closing brackets (or braces). Namespace prefixes are declared outside of the data terms, which disallows redefinition of namespace prefixes. Nevertheless all XML documents conforming to the Namespace recommendation [25] can also be represented as an Xcerpt^{XML} data term. Finally, text nodes are enclosed within quotation marks in order to be differentiated from empty element nodes.

Xcerpt^{XML} Queries: Pattern-based Filtering of Search Results. Consider the task of finding people and their curriculum vitae who study or have studied at the university of Munich. Searching for the term "LinkedIn" and "Munich" with a decent search engine returns among other search results links to pages of personal profiles of persons living in that city. The following Xcerpt query can be used to filter out other links in the search result page of Google.[8]

```
  html{{
2   desc div((id="res"))[[
      h2((class="hd")){ "Search Results" },
4     desc h3((class="r")){{
        or(
6         a((href=var Link as /.*linkedin\.com\/in\//)){{ }},
          a((href=var Link as /.*linkedin\.com\/pub\//)){{ }}
8       )
      ]]
10  }}
    }}
```

The following features of Xcerpt must be explained to understand the above query: (in)completeness in breadth for elements and attributes, incompleteness in depth, logical variables, regular expressions and query term disjunction.

– Curly braces are used to specify subterm relationship between an element and another element or a text node. The query h2{ "Search Results"} finds h2 elements with an enclosed text node with text "Search results". Double curly braces signify that more subterms may be present than are specified. If more than one subterm is specified within double curly braces, they must be mapped in an injective manner, i.e. they may not match with the same subterm of the data. This injectivity requirement can be avoided by using triple curly braces {{{ }}}. Square parentheses may be used instead of curly braces, if the order of the subterms appearing in the query is relevant. In the presence of zero or one subterm only, using square brackets or curly braces has the same semantics. A query that uses double or triple braces or brackets is termed *incomplete in breadth*, a query with single braces or brackets only is termed *complete in breadth*.

---

[8] We make use of the fact that all LinkedIn profile pages start either with
http://www.linkedin.com/pub/ or http://www.linkedin.com/in/

- XML Attributes and values are given in round parentheses directly following element names. Attribute names are followed by an "=" sign and by an attribute value in quotation marks. Double parentheses may be used to state that there may be more attributes present in the data than specified in the query. Since XML attributes are always considered to be unordered, there is no way of expressing an ordered query on attributes in Xcerpt. In case of double parentheses, the attributes are said to be specified *incompletely in breadth*.
- The `desc` keyword has the same semantics as the XPath descendant axis: The subterm following the `desc` may either be a direct child of the surrounding term or nested at arbitrary depth within one of the children. A term using the `desc` keyword is termed *incomplete in depth*, the other terms are said to be completely specified in depth. As the example above shows, incompleteness significantly eases query authoring, since requires only a very basic knowledge about the structure underlying the queried data.
- Logical variables are used to extract information from an HTML or XML document. In Xcerpt^{XML} terms, variables may bind either to entire XML elements, in which case they are called *term variables*, to the labels of elements only (*label variables*), to entire attributes (*attribute variables*) or to the values of attributes only(*label variables*). Variables may additionally feature a variable restriction initiated with the `as` keyword. Variable restrictions serve to lay a restriction on the possible bindings of variables.
- Regular expressions are delimited by the sign `'/'` and can be used at the place of labels to restrict the set of XML names that are matched by an Xcerpt^{XML} query term. The query term `/ab*/`, for example, will match with the labels a, ab, abb, etc. only.
- Queries may be composed using the boolean connectives `and`, `or`, and `not` which have the same intuitive semantics as in logic.

Mining Semantic data from Microformats embedded in personal profiles. Let us now turn to the second task of our use case. Having identified relevant URIs from the results of a search engine query, we now exploit microformats as a semantic enrichment for HTML pages to gather additional knowledge from web pages.

LinkedIn uses the microformats hcalendar, hresume, hcard, hAtom, and XFN to semantically enrich the contents of their pages. Unfortunately, the use of microformats has not been standardized, but evolves over time. Moreover, there is no underlying formal data model for microformat data as in RDF or XML. Microformats primarily use the XML attribute names `class` and `rel` for semantic information. In contrast to RDF, microformats do not use namespaces or globally unique identifiers, which makes it hard or sometimes even impossible to find out the exact semantics of an HTML fragment enriched by microformats. For example, both the hresume and the hcalendar specifications make use of a tag called `summary` for specifying either the summary of one's experience gained during a professional career or the summary of an event description.[9] With this

---

[9] Consult the descriptions of these microformats available online
`http://microformats.org/wiki/hresume` and
`http://microformats.org/wiki/hcalendar` for details.

deficiency in mind, the importance of query languages that transform semantic information embedded in HTML pages into a more precise RDF dialect becomes even more obvious. The fragment of a personal profile in Listing 1.3 pictures the use of microformats on LinkedIn and serves as further example data in this section.[10] One can observe that finding the semantic information within the HTML markup requires knowledge about the microformat standards, and that using the class attribute both for identifying elements to be formatted by stylesheets and for microformat predicate names is against the principle of separation of concerns coined by Dijkstra in [59].

```
  <html xmlns="http://www.w3.org/1999/xhtml" xml:lang="en-US"
      lang="en-US">
2 <head><title>John Doe - LinkedIn</title></head>
  <body>
4 <div class="hresume">
    <div class="profile-header">
6     <div class="masthead vcard contact portrait">
        <h1 id="name">
8         <span class="fn n">
            <span class="given-name">John</span>
10          <span class="family-name">Doe</span>
          </span>
12        </h1>
      </div>
14  </div>
    <div id="experience">
16    <h2>John Doe Experience</h2>
      <ul class="vcalendar">
18      <li class="experience vevent vcard">
          <h3 class="title">Research assistant</h3>
20        <h4 class="summary">
            <a href="...">University of Munich</a>
22        </h4>
          <p class="organization-details">(Research industry)</p>
24        <p class="period">
            <abbr class="dtstart" title="2000-02-01">February
                2000</abbr> until
26          <abbr class="dtstamp"
                title="2008-11-24">Present</abbr>
            <abbr class="duration" title="P8Y10M">(8 years 10
                months)</abbr>
28        </p>
        </li>
30    </ul>
    </div>
32 </body>
  </html>
```

Listing 1.3. A simplified personal profile page with embedded semantic information

The following Xcerpt query extracts the first and last name of a Person, if she has some experience as a research assistant in some organization in Munich. Aside from that, the query extracts the duration of the working relationship between the person and the organization if present. Unlike other query subterms, the relevant subterm for the

---

[10] The majority of the HTML markup serving presentation purposes and also most of the irrelevant content has been stripped out to shorten the presentation.

duration is marked optional, which means that the whole query is still successfull, if the optional subquery fails to match. Optional matching of subterms is only suitable if the subterm contains variables, and has also been proposed for SPARQL and other query languages. In contrast to SPARQL, however, the order of optional subterms within a query does not have any effect on the query result – see [70] for a more detailed discussion of this issue.

Listing 1.3 makes use of abbreviations for displaying information about the start, end and duration of an event. The actual date or duration is hidden within an XML attribute value that is meant for computational processing. [11]

```
1 html{{
    body{{
3    desc{{
        desc /.*/((class="given-name")){ var FirstName },
5       desc /.*/((class="family-name")){ var LastName }
      }},
7     desc /.*/((class=/.*experience.*/)){{
        /.*/((class="title")){ "Research assistant" },
9       /.*/((class="summary")){ /.*Munich.*/ },
        optional /.*/((class="period")){{
11        /.*/((class="duration" title=var Duration)){{ }}
        }}
13    }
     }}
15 }}
```

Listing 1.4. Finding research assistants from some organization in Munich

Listing 1.4 highlights the pecularities of matching HTML document with embedded microformat information. While element names have almost no relevance, the values of the class attributes is of primary importance. When querying plain HTML data, or XML dialects such as XMLSchema or DocBook, however, the role of the attributes will be less important, but element names will occur more often in the query. Another issue in extracting microformat information from documents is that the values of class attributes are often space separated lists of microformat predicate names such as vcard contact portrait. Up until now, Xcerpt has no specialized means for accessing these atomic strings in the attribute values, which results in excessive use of regular expressions. Therefore, it may be beneficial to invent a domain specific language or at least a class of query patterns that are specifically suited for querying microformat information and which would allow a less verbose notation of the query in Listing 1.4. In the following Section, we introduce the class of Xcerpt[RDF] query terms, which are geared at native and concise RDF querying.

---

[11] This convention was proposed by Tantek Çelik on his blog (http://tantek.com/log/2005/01.html#d26t0100) since humans prefer dates in a natural language description over a formal and concise notation, and may also deduce some information from the context.

2.3.2 RDF Queries—Examples and Patterns

In this section, the RDF processing capabilities of Xcerpt – united under the term Xcerpt^RDF – such as data, query and construct terms particularly geared towards RDF are introduced by example. This section is structured in four parts. In Section 2.3.2 Xcerpt^RDF data terms as a convenient way for representing RDF data are introduced. In Sections 2.3.2 and 2.3.2, Xcerpt^RDF query and construct terms are introduced as syntactic extensions to data terms. Section 2.3.4 compares Xcerpt^RDF to SPARQL, which is by far the most prominent RDF query language today.

Representation of RDF Graphs as Xcerpt^RDF Data Terms. Many serializations for RDF Data have been proposed (RDF/XML, Notation3, Turtle, NTriples, etc.), with their inventors pursuing a set of partially competing goals: On the one hand, (i) RDF serializations are supposed to be as short as possible, on the other hand, (ii) an optimal serialization should have a canonical and unique representation for each RDF graph – put more formally, there should be an isomorphism between the set of RDF graphs and the set of RDF graph serializations. Moreover, RDF serializations should be (iii) interchangeable between software systems on the Web, and at the same time (iv) easy to author and read by humans.

RDF/XML was proposed by the W3C with the first and the third aim in mind. Due to the encoding of RDF in XML, RDF/XML is easily exchanged over the Web, and standard XML tools, such as XPath, XQuery, XSLT processors and XML Schema validators can be used to process this serialization. Furthermore, the RDF/XML syntax allows for a plethora of syntactic sugar notations that significantly reduce the verbosity of an XML encoding of data. Unfortunately, RDF/XML does not perform well in the second and fourth discipline, i.e. it is not canonical, and it is not easy to read and write by humans. Due the availability of the syntactic sugar notations, there are many different possibilities for encoding the same RDF graph, which makes parsing XML/RDF into a set of triples a major challenge, and also requires more background knowledge about the serialization format by the user than other serializations do.

Notation3 was also proposed by the W3C with the first and fourth reason in mind. Due to its non-XML serialization format and some short hand notations, it is easier to read and write for human users, and is also quite dense in comparison with other serialization formats. Notation3 does not perform well, however, taking only the second and third end into account.

Turtle being a subset of Notation3, and NTriples being a minimal subset of Turtle (and thus also of Notation3), NTriples does not provide any short hand notations and is thus significantly more verbose and redundant than Notation3. Still, it is quite readable for human users and can be easily read into or serialized from a relational database containing only one single relation for all triples in an RDF graph[12]. Due to its simplicity, NTriples comes pretty close to fulfilling the second aim: An RDF graph being a set of triples, its possible NTriples serializations only differ in the order of the triples and in the naming of the blank nodes.

With Xcerpt^RDF data terms, we introduce yet another format for serializing RDF graphs. Besides the common goals stated above, Xcerpt^RDF data terms were invented

---

[12] This is a common schema for RDF stores.

with three other goals in mind: (a) compatibility with Xcerpt[XML] data terms, (b) extensibility to query terms involving variables and incompleteness constructs[13], and (c) support for RDF specificities such as containers and collections[14].

Consider the RDF graph displayed as an XML/RDF document in Listing 1.1 and as an Xcerpt[XML] data term in Listing 1.2. Its representation as an Xcerpt[RDF] data term is as follows:

```
  declare namespace rdf "http://www.w3 ... rdf-syntax-ns#";
2 declare namespace rdfs "http://www.w3 ... rdf-schema#";
  declare namespace foaf "http://xmlns.com/foaf/0.1/"
4 declare namespace ex "http://www.example.org/"

6 ex:descriptions/Bill.foaf {
    rdf:type → foaf:PersonalProfileDocument,
8   foaf:maker → ex:#me,
    foaf:primaryTopic → ex:#me {
10    rdf:type → foaf:Person,
      foaf:givenname → "Bill",
12    foaf:mbox_sha1sum → "5e22c ... 35b9",
      foaf:depiction → base:images/bill.png,
14    foaf:knows {
        _:SomePerson {
16        rdf:type → foaf:Person,
          foaf:name → "Hillary",
18        foaf:mbox_sha1sum "1228 ... 2f5"
          rdfs:seeAlso → base:descriptions/Hillary.foaf
20 } } } }
```

Listing 1.5. A friend-of-a-friend-document written as an Xcerpt[RDF] data term

As another example consider Figure 6 from the RDF Primer [107]. Its representation as an Xcerpt[RDF] term is as follows:

```
  declare namespace exterms "http://www.example.org/terms/"
2 declare namespace exstaff "http://www.example.org/staffid/"

4 exstaff:85740 {
    exterms:address → _:A {
6     !http://www.example.org/terms/city → "Bedford",
      exterms:street → "1501 Grant Avenue",
8     exterms:state → "Massachusetts",
      exterms:postalCode → "01730"
10  }
  }
```

Listing 1.6. Example from the W3C RDF Primer in Xcerpt[RDF] notation

---

[13] Any Xcerpt[RDF] data term is per se also an Xcerpt[RDF] query.

[14] This last point has already been partially addressed by XML/RDF.

Similar to RDF/XML, Notation3, Turtle and SPARQL, Xcerpt^{RDF} data terms can be abbreviated using namespace prefixes in qualified names. Full URIs are distinguished from qualified names by prefixing an exclamation mark, blank nodes by the prefix _:, and literals by quotation marks.

In the RDF graph above, multiple statements have the blank node _:A as their common subject, which is factored out in the Xcerpt^{RDF} serialization. In many cases RDF statements do not only share the subject, but also the predicate, in which case also the predicate can be factored out:

```
declare namespace ex "http://www.example.org/"
2 declare namespace foaf "http://xmlns.com/foaf/0.1/"

4 ex:anna { foaf:knows → (ex:bob, ex:chuck) },
```

Xcerpt^{RDF} also supports the factorization of properties only, objects only, predicate and object, subject and object, and of all three elements – subject, predicate and object, in which case there will be one Xcerpt^{RDF} term for each RDF triple. Factoring out the predicate only could be used, for example, to represent a clique of friends, in which every member knows every other member and herself:

```
(ex:anna, ex:bob, ex:chuck) {
2   foaf:knows → (ex:anna, ex:bob, ex:chuck) },
```

The RDF graph in Listing 1.6 has only a single node without incoming edges, and therefore the choice of the root of the Xcerpt^{RDF} term is trivial. RDF graphs may, however, have multiple nodes without incoming edges or none at all, or may even be entirely disconnected. In the case of no nodes without incoming edges, one can arbitrarily pick a root node for the Xcerpt^{RDF} term representation, but in the case of multiple nodes without incoming edges, and in the case of a disconnected RDF graph, the graph cannot be serialized as a single Xcerpt^{RDF} term, but only as a conjunction of terms. Therefore, the keyword RDFGRAPH is introduced:

```
RDFGRAPH {
2   ex:anna { foaf:knows → ex:bob },
    ex:chuck { foaf:knows → ex:bob }
4 }
```

RDF Schema is a specification that "describes how to use RDF to describe RDF vocabularies" [113]. It therefore provides a set of URIs, with a semantics defined by RDFS entailment rules, and which are in popular use for defining new RDF ontologies. Xcerpt^{RDF} provides shorthand notations for the most common ones among them: rdf:type, rdfs:range, rdfs:domain, rdf:Property and rdfs:Resource.

```
ex:name { is [ex:Person → ex:Name] }
```

The Xcerpt^{RDF} term above is a shorthand for the following Xcerpt^{RDF} term:

```
1 ex:name {
    rdf:type → rdf:Property,
3   rdfs:domain → eg:Person,
    rdfs:range → eg:Name
5 }
```

If the domain and/or the range of a predicate shall be left unrestricted, then the restricting classes can be simply omitted as in the Xcerpt^RDF term eg:name{ is [eg:Person →] }. In Xcerpt^RDF this expands to the following term:[15]

```
  ex:name {
2   rdf:type → rdf:Property,
    rdfs:domain → eg:Person,
4 }
```

Besides the RDFS vocabulary, RDF distinguishes a set of URIs for expressing reification of RDF statements and containers and collections of Resources in RDF bags, sequences, alternatives or lists. Xcerpt^RDF provides syntactic sugar notations both for reifications on the one hand and RDF containers and collections on the other hand. Consider the following Xcerpt^RDF term:

```
  ex:bob { ex:believes →
2   _:Statement1 { < ex:anna{ foaf:knows → ex:bob } >
  }
```

The Xcerpt^RDF term enclosed in angle brackets is a reified statement, and thus the entire term is equivalent to the following, significantly more verbose one:

```
1 ex:bob { ex:believes →
   _:Statement1 {
3    rdf:type → rdf:Statement,
     rdf:subject → ex:anna,
5    rdf:predicate → foaf:knows,
     rdf:object → eg:tim
7  }
 }
```

Whereas bags, sequences and alternatives are termed as RDF containers, and are considered to be open (i.e. there may be other elements in the container, which are not specified in the present RDF graph), RDF collections (i.e. RDF lists) are considered to be completely specified. However, this intuitive semantics is in no way reflected within the RDF/S model theory. When using only Xcerpt^RDF shorthand notations for representing RDF graphs featuring RDF containers, collections or reification, one can be sure to respect this intuitive semantics. Xcerpt^RDF provides the reserved words bagOf, seqOf, altOf and listOf to reduce the verbosity serializing RDF containers and collections. To represent a research group, one might chose the following Xcerpt^RDF term, which would expand to four triples.

```
  _:Group1 { bagOf{ eg:anna, eg:bob, eg:chuck } }
```

Xcerpt^RDF Query Terms. Just as in Xcerpt^XML, Xcerpt^RDF data terms are augmented with constructs for specifying incompleteness to yield Xcerpt^RDF query terms. Such

---

[15] Note that under the RDFS entailment rules, also the triple ex:name rdfs:range-→ rdf:Resource would be implied. Xcerpt^RDF, however, does not enforce the RDFS semantics, since RDF/S entailment rules can be easily encoded in Xcerpt^RDF itself.

Table 1. Syntax of Xcerpt[RDF] data terms

term	= node | node '{' arc (',' arc)* '}' | reification
node	= blank | uri | literal | qname
arc	= uri '→' term | container | collection
blank	= attvalueW3C
literal	= '"' char* '"' | "'" char* "'"
uri	= '!' uriW3C
qname	= qnameW3C
collection	= bag | sequence | alternative
container	= 'listOf' '{ }' | 'listOf' '{' term (',' term)* '}'
bag	= 'bagOf' '{ }' | 'bagOf' '{' term (',' term)* '}'
sequence	= 'seqOf' '{ }' | 'seqOf' '{' term (',' term)* '}'
alternative	= 'altOf' '{ }' | 'altOf' '{' term (',' term)* '}'
reification	= '<' term '>'

constructs include the use of logical variables, subterm negation, subterm optionality, incompleteness in breadth and qualified descendant. While originally invented for XML processing, these constructs are also beneficial for querying RDF graphs as exemplified in Example 1.7. The query extracts variable bindings for all Persons and their nick names within an RDF graph, who know some Person with nick name 'Bill', who in case do *not* know any other Person named 'Hillary'. As in Xcerpt[XML], the optional keyword is used to bind the nick name to the variable var Nick whenever possible, but does not cause the query to fail if the nick name is not present. Also the semantics of double curly braces and the without keyword is analogous to Xcerpt[XML]. In Listing 1.7, the scope of the without and optional keyword is explicitly given by round parentheses. The scope of a without or optional does not have to be the entire subterm following the keyword, but may also be restricted to the edge only. Table 2 gives an intuition of the exact semantics of without with varying scopes by providing example data that does or does not simulate with the given query terms. The intuitive semantics for optional can be described by similar examples, but is left unspecified here for the sake of brevity. Note, however, that optional subterms are only useful if they contain variables for extracting data.

```
  var Person{{
2   optional (foaf:nick → var Nick),
    rdf:type → foaf:Person,
4   foaf:knows → _:X{{
      foaf:nick → 'Bill', rdf:type → Person,
6     without (
        foaf:knows → {{ _:Y{{ foaf:nick → 'Hillary' }} }}
8     )
    }}
10 }}
```

Listing 1.7. An Xcerpt[RDF] query term

Table 2. Query term simulation with different scopes for <u>without</u>

query term	simulating data terms	non-simulating data terms
a{{ <u>without</u> (b →) c }}	a{ d → c} a{ b → e, d →c }	a{ b → c} a{ b → d, b → c } a{ b → d }
a{{ <u>without</u> (b → c) }}	a{ } a{ b → d }	a{ b → c } a{ b → d, b → c }
a{{ b → <u>without</u> c }}	a{ b → d } a{ e → c, b → f }	a{ b → c } a{ }
a{{ b → <u>without</u> c {{ d → e }} }}	a{ b → c } a{ b → c{ d → f } }	a{ b → c{ d → e } } a{ }
a{{ b → (<u>without</u> c) {{ d → e }} }}	a{ b → f{ d → e } }	a{ b → f } a{ b → c }

Although the Xcerpt^{XML} constructs for specifying incomplete queries mentionend above retain their semantics in Xcerpt^{RDF}, there are some different requirements in XML and RDF processing that are also reflected in the way that Xcerpt^{RDF} variables are used in Xcerpt^{RDF} query terms.

An obvious difference between matching RDF graphs and matching XML documents is that while extracting entire subtrees from an XML document is a very common task, extracting entire RDF subgraphs from an RDF graph is less frequently used, since this may often result in the whole RDF graph being returned. Therefore, the default variable binding mechanism in Xcerpt^{RDF} is not *subgraph extraction* but *label extraction*. Therefore, the most common form of variables used in Xcerpt^{RDF} query terms are *node* and *predicate* variables. Node and predicate variables are written using the keyword <u>var</u>. A node (predicate) variable binds to a single node (arc) of the queried graph. *graph variables* are identified by the keyword graphVar and bind – similarly to Xcerpt^{XML} term variables – to entire subgraphs. Finally, *CBD-variables* (identified by the keyword cbdVar) bind to concise bounded descriptions[16].

Another difference is that once an RDF node in an RDF graph has been identified by a query and has been bound to a variable, the very same node can be easily recovered in a subsequent query, since both URI nodes and blank nodes are uniquely named in an RDF graph, whereas an XML Document may very well contain multiple nodes having the same tag name and even the same content. XQuery and Xcerpt 2.0 deal with this problem by introducing node identity for XML elements and attributes, thereby allowing the comparison of variable bindings not only by deep equality, but also by shallow equality [69]. This distinction is not necessary in RDF processing, since the *value* of a node is already a global (in the case of resources) or local (in the case of blank nodes) identifier.

---

[16] http://www.w3.org/Submission/CBD/

For the representation of complex values, however, the simplistic data model of RDF graphs as sets of triples is not well-suited. Here, blank nodes are used to group atomic attributes of a node together to form a complex attribute. Often, these complex attributes shall be selected together and collected in a single variable binding. This need has been addressed by the W3C consortium with the introduction of a concept known as *Concise Bounded Descriptions*. Xcerpt^{RDF} supports concise bounded descriptions by providing a special kind of variable which does not bind to the value of a node, nor to the subgraph rooted at the node, but to the concise bounded description associated with that node. Table 3 gives an example driven overview of the different types of variables in Xcerpt^{RDF} and their binding mechanisms.

Table 3. Query term simulation with variables for nodes, predicates, graphs and concise bounded descriptions

query term	data term	substitution set	
var X	a{ b → c }	{{ X ↦ a }}	(1)
a{{ b → var O }}	a{ b → c, b → _:X }	{{ O ↦ c }, { O ↦ _:X }}	(2)
a{{ var P → var O }}	a{ b → c, b → e }	{{ P ↦ b, O ↦ c },	
		{ P ↦ b, O ↦ e }}	(3)
graphVar G	a{ b → c }	{{ G ↦ a{ b → c }}}	(4)
graphVar G as g{{ }}	a{ b → c}	{{ }}	(5)
a{{ graphVar G }}	a{ b { a }, c }	{{ G ↦ b{ a { b, c }}}}	(6)
graphVar G as var L	a{ b → c}	{{ G ↦ a{ b → c }, L ↦ a }}	(7)
cbdVar G	_:X{ b → c{ d → e }}	{{ G ↦ _:X{ b → c }}}	(8)
cbdVar G	_:X{ b → _:Y{ d → e }}	{{ G ↦ _:X{ b → _:Y{ d→ e }}}} }}	(9)

Rows 1 and 2 show the simulation of a simple Xcerpt^{RDF} variable in subject and object position. Compare the binding of the graph variable G in row 4 with the one of the label variable X in row 1 under simulation with the same data term. Row 3 shows a variable in predicate position, row 5 a graph variable with a restriction, which has the same semantics as in Xcerpt^{XML} (since the label g of the restriction does not appear within the data, the substitution set is empty).

An interesting case is row 6. Since the queried graph d is not a tree, but a graph, the binding for variable G is not a subterm of d, but a subgraph.

Row 7 shows the contemporary use of a graph and label variable, and rows 8 and 9 illustrate the semantics of variables for concise bounded descriptions.

Table 4 shows the syntax of Xcerpt^{RDF} query terms as a context free grammar with terminal symbols in single quotes and the usual semantics of the meta-symbols * + ? and

Table 4. Syntax of Xcerpt^{RDF} query terms

term	::= '<u>desc</u>'? node | '<u>desc</u>'? node '{ {' arc (' , ' arc)* ' } }' | '<u>desc</u>'? reification
node	::= blank | uri | literal | qname | variable | graphVar | cbdVar
variable	::= '<u>var</u>' varname
varname	::= [A-Z][A-Za-z0-9*]
graphVar	::= 'graphVar' varname | 'graphVar' varname as term
cbdVar	::= 'cbdVar' varname | 'cbdVar' varname as term
arc	::= uri '→' term | rpe '→' term | container | collection
blank	::= attvalueW3C
literal	::= ' " ' char* ' " ' | " ' " char* " ' "
uri	::= '!' uriW3C
qname	::= qnameW3C
collection	::= bag | sequence | alternative
container	::= 'listOf' '{{ }}' | 'listOf' '{{' term (' , ' term)* ' }}'
bag	::= 'bagOf' '{{ }}' | 'bagOf' '{{' term (' , ' term)* ' }}'
sequence	::= 'seqOf' '{{ }}' | 'seqOf' '{{' term (' , ' term)* ' }}'
alternative	::= 'altOf' '{{ }}' | 'altOf' '{{' term (' , ' term)* ' }}'
reification	::= '<' term '>'

|. The nonterminal symbols `uriW3C`, `attvalueW3C` and `qnameW3C` correspond to the syntactic definition of URIs, attribute values and qualified names in the W3C recommendation for XML[27]. The non-terminal symbol `rpe` denotes an Xcerpt^{RDF} regular path expression, whose definition is omitted in this contribution for the sake of brevity.

Xcerpt^{RDF} Construct Terms and Rules. Consisting of a query part and a construct part, pure Xcerpt^{RDF} rules serve to transform RDF data. The query part is used to extract data from an RDF graph into sets of sets of variable bindings, also called substitution sets, and the construct part is used to reassemble these variable bindings within construct patterns, substituting bindings for variables.

Table 5 describes how substitution sets are applied to Xcerpt^{RDF} construct terms to yield Xcerpt^{RDF} data terms. Apart from the different kinds of variable bindings allowed in Xcerpt^{RDF} substitution sets, the algorithm differs from the application of Xcerpt^{XML} substitution sets to Xcerpt^{XML} terms in the following ways:

– In accordance with the most famous RDF query languages such as SPARQL [141] and RQL [92,29], URIs are treated as unique identifiers within an RDF graph and do not have any object identity besides the identity given by the URI itself. This convention has as an implication that a substitution set applied to different construct terms may result in semantically equivalent data terms. To see this consider rows 1 and 5 in Table 5. Although the Xcerpt^{RDF} construct terms are syntactically different, the data terms resulting from the application of the substitution set are equivalent RDF graphs. As a result, the use of <u>all</u> within construct terms made up of URIs only does not change the semantics of a rule.

Table 5. Application of substitution sets to Xcerpt^{RDF} construct terms

substitution set	construct term	Xcerpt^{RDF} result	
{ { O ↦ c }, { O ↦ d } }	a{ b → <u>var</u> O }	a{ b → c } a{ b → d }	(1)
{ { O ↦ c }, { O ↦ d } }	_:X{ b → <u>var</u> O }	_:X1{ b → c } _:X2{ b → d }	(2)
{ { S ↦ c }, { S ↦ d} }	<u>var</u> S{ b → a }	c{ b → a } d{ b → a }	(3)
{ { S ↦ c }, { S ↦ d} }	<u>var</u> S{ b → _:X }	c{ b → _:X1 } d{ b → _:X2 }	(4)
{ { O ↦ c }, { O ↦ d } }	a{ <u>all</u> b → <u>var</u> O }	a{ b → c, b → d }	(5)
{ { O ↦ c }, { O ↦ d } }	_:X{ <u>all</u> b → <u>var</u> O }	_:X{ b → c, b → d }	(6)
{ { O ↦ c }, { O ↦ d } }	a{ b → <u>var</u> O{ e → f } }	a{ b → c{ e → f } } a{ b → d{ e → f } }	(7)
{ { G ↦ a{ b → c } } }	graphVar G	a{ b → c }	(8)
{ { G ↦ a{ b → c } } }	d{ e → graphVar G }	d{ e → a{ b → c } }	(9)

- Just as RDFLog [33,34], but unlike SPARQL and other RDF query languages, Xcerpt^{RDF} supports arbitrary construction of blank node identifiers. While the majority of RDF query languages does not allow blank node construction at all or only blank nodes depending on all universally quantified variables of a rule (see [34] for details), Xcerpt^{RDF} and RDFLog support also construction of blank nodes that depend only on some or none of the universally quantified variables of a rule. RDFLog does this by explicit quantifier alternation, Xcerpt^{RDF} on the other hand achieves the same goal by using Xcerpt's <u>all</u> grouping construct. To see the difference consider rows 2 and 6 in Table 5. In row 2 the construct contains the free variable <u>var</u> O, whereas in row 6 the construct term does not contain any free variable. Thus in the first case, the substitution set is divided into two substitution sets according to the binding of variable <u>var</u> O, and each of the substitution sets is applied to the construct term. In the second case, however, the substitution set is not divided at all, but applied as a whole to the construct term.

Special care must be taken that the result of the application of a substitution set to an Xcerpt^{RDF} construct term is again an RDF graph. Guaranteeing that pure Xcerpt^{RDF} programs convert RDF graphs into valid RDF graphs allows easy composition of Xcerpt programs.

Providing the same input and output format for a language is a feature of many modern query languages and is usually referred to as *answer closedness*. Popular XML query languages in general are only weakly answer closed – which means that they allow for easy authoring of programs that again produce valid XML documents, but that it still *is* possible to generate non-XML data. A notable exception to this rule is Xcerpt^{XML}, which is strongly answer closed in the sense that every outcome of an Xcerpt^{XML} program is an XML fragment. On the other hand, the W3C languages XPath, XQuery and XSLT can also be used to output non-XML content such as PDF, Postscript, or comma separated values.

Definition 1 (Answer Closedness). *A web query language is called* answer closed, *if the following conditions are fulfilled:*

1. *data in the queried format can be used as queries*
2. *the result of queries is again in the same format as the data*

A web query language is called weakly answer closed, *if condition (2) is possible; it is called* strongly answer closed, *if condition (2) is always enforced.*

The assurance of answer closedness in Xcerpt^{RDF} must take the following two thoughts into account:

- *Abidance of RDF triple constraints.* The evaluation of query terms may bind node variables to literals or blank nodes. RDF graphs, however, do not allow literals in subject or predicate position or blank nodes in predicate position.
- *Abidance of RDF graph constraints.* Xcerpt^{RDF} supports four different kinds of variables: node variables, predicate variables, graph variables and concise bounded description variables. In general, it is only safe to substitute variables in construct terms by bindings of variables of the same type. Depending on the data, bindings for node, graph and concise bounded description variables may degenerate to plain URIs, and therefore it may be safe to substitute them for predicate or node variables.

With the above two restrictions in mind, there are three different possibilities for implementing answer closedness in Xcerpt^{RDF}.

- *Static Checking of Bindings:* Before an Xcerpt^{RDF} program is run, it is checked that predicate variables in the construct term are also used as predicate variables in the query term, and the same for graph variables, node variables and CBD variables. To be more precise, the semantics of graph and CBD variables only differ within the query term, and thus a CBD variable binding may be substituted for a graph variable in the construct term. Moreover, the binding of a predicate variable may be substituted for a label variable in the construct term, since predicate variables always bind to URIs. On the other hand, bindings of node variables, may *not* be substituted for predicate variables. While static checking of variable bindings ensures that all terms constructed by Xcerpt^{RDF} programs are valid RDF graphs, certain tasks, such as using URIs of nodes of a source graph in predicate position in the target graph, are impossible to achieve with this technique.
- *Dynamic Checking of Variable Bindings:* Dynamic checking of variable bindings is a sensible choice if there is reason to assume that the query author has some knowledge about the data to be queried. It is more flexible than static checking in the sense that a larger number of tasks can be realized, but is less reliable in the sense that runtime errors may occur.

Table 6. Application of substitution sets to Xcerpt^{RDF} construct terms with casting of variable bindings

substitution set	construct term	Xcerpt^{RDF} result
{ { V ↦ _:X } }	a{ var V → b }	
{ { V ↦ _:X } }	a{ var V → b, c → d }	a{ c → d }
{ { V ↦ 'literal1' } }	a{ var V → b, c → d }	a{ c → d }
{ { G ↦ a{ b → c } } }	graphVar G{ d → e }	a{ b → c, d → e }
{ { G ↦ a{ b → c } } }	d{ var G → e }	d{ a → e }
{ { G ↦ _:X{ b → c } } }	d{ var G → e }	_:X{ b → c }
{ { L ↦ 'literal1' } }	a{ b → { var L{ c → d } } }	a{ b → _:X { c → d }, b → 'literal1' }

– *Casting of Variable Bindings* unites the best of static checking of variable bindings (i.e. no runtime errors) and dynamic checking of variable bindings (i.e. a higher degree of flexibility). Consider the sources of runtime errors that may occur with dynamic checking of variable bindings – examples for each case are given in Table 6.

  • A literal or blank node bound to a node variable is substituted for a predicate variable in a construct term. Such triples are simply omitted from the resulting RDF graph.
  • A subgraph bound to a CBD or graph variable is substituted for a node variable in a construct term. In this case the subgraph rooted at the occurrence of the node variable in the construct term and the binding of the variable are merged.
  • A subgraph g rooted at a URI u and bound to a graph variable is substituted for a predicate variable in a construct term. The graph g is cast to u.
  • A subgraph g rooted at a blank node b and bound to a graph variable or CBD variable is substituted for a predicate variable. Since blank nodes may not appear in subject positions, the resulting triple is not included in the Xcerpt^{RDF} result.
  • A literal lit is substituted for a node variable L appearing in subject position in the construct term. In this case a fresh blank node B is substituted for the variable instead of the literal. If L additionally appears in object position, also the literal itself is substituted for L, but the triples containing lit in subject position are omitted.

Since the last alternative gives an operational semantics to programs which would be either considered invalid under the first approach or would throw runtime errors under the second, Xcerpt^{RDF} favors the casting of variable bindings. We acknowledge, however, that the first approach may make more sense for unexperienced users in that it is easier to understand, and that the second approach may uncover errors in the authoring of Xcerpt^{RDF} programs, which would pass unnoticed by the third approach.

2.3.3   Rules—Separation of Concern and Reasoning

Having introduced queries for both XML and RDF data, this section combines both features to realize the truly versatile use case already sketched in Section 2.3.1. Starting out from the result pages for the terms "LinkedIn Munich" of a popular Web search engine, links to relevant LinkedIn profile pages are extracted by the use of rich XML query patterns with logical variables. In a second step, the profile pages are retrieved and semantic microformat information is exploited to gather reliable information about the users. Finally, in a third step, this information is enriched by semantic information from FOAF profiles in RDF format using the RDF processing capabilities of Xcerpt.

In this use case Xcerpt's capability of handling XML query terms and RDF construct terms in the same rule (and the other way around) comes in particularly handy. As in pure XML querying and in pure RDF querying, the interface between querying and construction is a substitution set. Substitution sets generated by XML query terms differ in the allowed variable types from substitution sets generated by RDF query terms. As a result, there must either be a way to transform XML substitution sets to RDF substitution sets and reversely, or the application of XML substitution sets to RDF construct terms and the application of RDF substitution sets to XML construct terms must be defined. While both ways are feasible, we present here the first alternative, since it is less involved.

Xcerpt^{XML} Query Terms and Xcerpt^{RDF} Construct Terms and vice versa in the Same Rule. Note that it is Xcerpt's underlying principle of clear separation of querying and construction that allows for, e.g, an XML query term in a rule body and an RDF construct term in the head of the same rule. The applicability of this design principle remains untouched if further types of query and construct terms are introduced (e.g. for topic maps or queries aimed at specific microformats or at pages of a Semantic Wiki). The only requirement for these new types of queries and construct terms are the definition of the following four algorithms: (1) a simulation algorithm matching queries with data and returning a substitution set (a set of set of variable bindings),[17] (2) an application algorithm for substitution sets that fills in bindings for logical variables occuring in a construct term[18], (3) a mapping from variable bindings in the new format to variable bindings in the other formats (until now only XML and RDF) and finally (4) a mapping from XML and RDF variable bindings to variable bindings in the new format.

The following list defines informally the mapping of XML bindings to RDF and reversely.

– The Xcerpt^{RDF} URI `!http://www.example.org/#foo` is mapped to the Xcerpt^{XML} qualified name `eg:#foo` with the namespace prefix `eg` bound to the namespace `http://www.example.org/`. We adopt the convention that the Xcerpt^{RDF} URI is split into namespace and local name at the last '/', but other methods are also conceivable.
– The Xcerpt^{RDF} blank node `_:B` is mapped to the Xcerpt^{XML} element name `_:B`.

---

[17] See Tables 2 and 3 for an informal description of this algorithm for Xcerpt^{RDF}.
[18] Table 5 gives the relevant ideas for this algorithm in Xcerpt^{RDF}.

- The Xcerpt[RDF] literal "some literal" maps to the Xcerpt[XML] text node "some literal"[19].
- The Xcerpt[RDF] qualified name `eg:anna` is mapped to the Xcerpt[XML] qualified name `eg:anna`. An appropriate namespace binding is added to the Xcerpt[XML] term. Implementations may choose to expand the qualified name to a URI u, and map u instead.
- The Xcerpt[RDF] term `a{ b → c }` maps to the Xcerpt[XML] term `a{ b{ c } }` in correspondance to past work on querying XML serializations of RDF with Xcerpt [21]. Similarly, the Xcerpt[RDF] term `_:X{ a → b{ c → ''another literal'' } }` is mapped to the Xcerpt[XML] term `_:X{ a { b { c { ''another literal'' } } } }`.
- The Xcerpt[RDF] shorthand notation `ex:name{ is [ex:Person → ex:Name ] }` is expanded to its corresponding unabbreviated term as introduced in Section 2.3.2. Then this longer notation is mapped to an Xcerpt[XML] term as described above.
- The Xcerpt[RDF] reification term `a{ believes → _:S{ < b{ c → d } } }` is mapped to the Xcerpt[XML] term `a { believes _:S { xcrdf:reification { b { c { d } } } } }` with the namespace prefix `xcrdf` bound to `http://www.xcerpt.org/xcrdf`.
- The Xcerpt[RDF] term `_:X { bagOf { a, b, c } }` is mapped to the Xcerpt[XML] term `_:X { xcrdf:bag { a, b, c } }`. Expansion to the normalized RDF syntax and applying the standard mapping to Xcerpt[XML] terms could also be introduced. The choice of the conversion is, however, not of primary importance, as long as all information present in the Xcerpt[RDF] term is preserved. Additional transformation rules can be easily written to change the XML outcome and be provided as an Xcerpt[XML] module (See [11] for more about Xcerpt modules). Xcerpt[RDF] sequences, alternatives and lists are treated in the same manner.
- The Xcerpt[XML] qualified name `eg:a` is mapped to the Xcerpt[RDF] qualified name `eg:a` and the binding for the namespace prefix `eg` is preserved.
- The Xcerpt[XML] unqualified name `a` is mapped to the Xcerpt[RDF] qualified name `xcxml:a` with the namespace prefix `xcxml` bound to the namespace `http://www.xcerpt.org/xcxml`. Note that the RDF graph data model does not allow for local names other than blank nodes. The unqualified name is not mapped to a blank node to avoid naming conflicts with other resources that may be contained in the resulting RDF Graph.
- The Xcerpt[XML] term `eg:a[ eg:b, eg:c ]` is mapped to the Xcerpt[RDF] term `eg:a{ xcxml:child → eg:b, xcxml:child → eg:c }`, and the binding for the namespace prefix `eg` is preserved. Note that since RDF graphs are always considered to be unordered, Xcerpt[RDF] does not provide square brackets, and the information about the order is lost in this mapping. Encodings of XML terms as RDF graphs that preserve the order are conceivable.
- The Xcerpt[XML] term `eg:a(id="2"){ eg:b }` is converted to the Xcerpt[RDF] term `eg:a{ xcxml:child → eg:b }`, i.e. XML attributes are not mapped to Xcerpt[RDF] terms. Attribute names and values may, however, also be inserted into an

---

[19] We leave the details of treating typed RDF literals and literals with a language tag as future work.

RDF graph by binding them to label variables. Also in this case, a different kind of mapping may be chosen, but it turns out that for the applications considered in this report, this simple mapping suffices.

Transforming LinkedIn embedded Microformat information to DOAC and FOAF
Reconsider the Xcerpt^{XML} query term in Listing 1.4. It extracts bindings for the variables FirstName, LastName and Duration. It is easy to construct RDF data from those variable bindings with an Xcerpt rule featuring the construct term in Listing 2.3.3.

```
  declare namespace doac "http://ramonantonio.net/doac/0.1/"
2 declare namespace foaf "http://xmlns.com/foaf/0.1/"

4 _:Person {
    rdf:type → foaf:Person,
6   foaf:firstName → var FirstName,
    foaf:surname → var LastName,
8   all doac:experience → _:Exp {
      doac:title → "Research Assistant",
10     doac:duration → var Duration
    }
12 }
```

Note the semantics of the all construct in Listing 2.3.3. The all construct serves to collect a set of variable bindings within a data term to be constructed. The number of data terms generated for construct term *c* preceded by an all construct depends on the set of free variables inside of *c*, and the substitution set which is applied to the construct term. A variable *v* is free within a term *t*, if it does not occur within the scope of an all construct inside of *t*. Thus the variable Duration is free within the term doac:duration ..., but not inside of the entire construct term of Listing 2.3.3. The set of free variables in the term *c* :=doac:experience → _:Exp { ... } following the all keyword is the unary set {*Duration*}. The substitution set applied to the construct term is thus separated according to the bindings of the variable Duration only. Then each of the resulting substitution sets is applied to *c* independently and included as a subterm of the outermost foaf:Person label. Whenever a substitution set is applied to a term with a blank node, a new instantiation of this blank node is created, as showcased in Table 5. This is a major difference to application of substitution sets to terms starting with URIs.

Alternatively, one might want to create a single RDF bag enumerating the working relationships a person has had. This could be achieved by the following Xcerpt^{RDF} construct term:

```
  _:Person {
2  rdf:type → foaf:Person,
   foaf:firstName → var FirstName,
4  foaf:surname → var LastName,
   _:Experiences {
6    bagof {
       all _:Exp {
8        doac:title → var Title,
```

```
              doac:duration → var Duration
10        }
        }
12   }
    }
```

Once the microformat information from the LinkedIn page is transformed to the more precise RDF representation at the aid of this rule, it can be combined with RDF data located anywhere on the Web. These FOAF documents can be discovered in a very similar fashion as has been done for the LinkedIn profile pages in Section 2.3.1.

Since LinkedIn does not provide the hash sums of email-addresses or other globally unique identifiers for persons within their profile pages, combining the extracted RDF information will rely on simple joins over the names of people, which is not particularly reliable – see [99] for an overview of the problems that may occur.

With OpenID [131] becoming the de facto standard for distributed authentication and single-sign-on on the Web and with the largest corporations involved in online activities such as Google, Yahoo, Microsoft, etc already joining the bandwagon, it seems likely that also LinkedIn will provide an open identifier within its profiles. Also the extension of the FOAF vocabulary to provide for OpenIDs within FOAF profiles is already discussed. In the presence of this information, the combination of the collected microformat data and other RDF resources can easily and reliably achieved using Xcerpt^{RDF}.

2.3.4 State of the Art: The SPARQL Query Language and Its Extensions

With the publication of the SPARQL W3C recommendation on January 2008, SPARQL has become the first query language that has been standardized by a major standardization body. In contrast to most other languages that have been proposed for RDF querying, SPARQL is, due to its triple syntax, quite easy to understand and use for programmers familiar with relational query languages.

In this section, SPARQL is introduced by example, its semantics according to [124] is recapitulated, and several extensions to SPARQL are presented. Throughout the presentation, the commonalities and differences to Xcerpt^{RDF} are highlighted.

A SPARQL query consists of the three building blocks *pattern matching part, solution modifiers* and *output*. In addition there are four different kinds of query forms. Arguably the most popular one is the *select* query form, which is inspired by SQL and returns so-called solution sets, the counterpart of Xcerpt substitution sets in SPARQL. An example of a *select* query is given in Listing 1.8. In case of a select query, the output part of the query is a selection of distinguished variables, i.e. the specification of the variables of interest in the query. If no variable bindings are of interest, the *ask* query form is to be used. It simply gives a yes/no answer to the question if a given query pattern is entailed by the RDF graph being queried. A useful query form for RDF *graph transformations* is the *construct* query form, which does not return single values, but entire RDF graphs as a result. There are, however certain limitations to the blank node construction (in database theory termed *value invention*) in the SPARQL construct query form, see [39].A final query form is given by the *describe* key word which pays attribute to the fact that a blank node identifier returned as a variable binding in a SPARQL ask-query is somewhat useless, since it only asserts the fact that something exists, and cannot be reused in a follow-up query to extract further information about the resource in question. When using the *describe* query form, not only single identifiers are returned as variable bindings, but also *descriptions* of resources. The exact nature of a resource description is left unspecified in the SPARQL recommendation, but a sensible solution would be the one of *Concise Bounded Descriptions* [143].

The SPARQL query form which is most similar to Xcerpt^{RDF} rules is the *construct* query form. Xcerpt^{RDF} does not distinguish between query forms, but is strongly answer closed in the sense that every Xcerpt^{RDF} data term is also a Xcerpt^{RDF} query, and in that every result of an Xcerpt^{RDF} query is again an RDF graph. While SPARQL *construct* queries are answer closed, the remaining query forms are not. However, SPARQL *ask* and *select* queries can be simulated by *construct* queries. Similarly, boolean queries can be formulated in Xcerpt^{RDF} by interpreting the empty RDF graph as false and all other RDF graphs as true, and tuple-generating queries can be expressed in Xcerpt^{RDF} by wrapping the tuples within RDF containers or similar constructs. *Describe* queries are expressed in Xcerpt^{RDF} by using concise-bounded-description variables.

All four SPARQL query forms make use of the pattern matching part, which is described next.

SPARQL graph patterns. SPARQL is weakly answer closed in the sense that any RDF graph is also a valid SPARQL graph pattern. But only in the case of the construct query form, also the result of a SPARQL query is again an RDF graph. The syntax of SPARQL graph patterns resembles the one of Turtle, but is augmented with variables. Listing 1.8 (from [141]) shows a query to retrieve the name and email address of persons within an RDF graph using the FOAF vocabulary. With the term *graph pattern*, one refers to the set of triples within curly braces in lines 4 to 5. The select-clause serves to specify the *distinguished* variables of the query. Any variable appearing within the graph pattern, but not within the select-clause is called a *non-distinguished* variable. The terms *distinguished* and *non-distinguished* variables have thus the same meaning as in conjunctive queries in database theory.

```
1 PREFIX foaf: <http://xmlns.com/foaf/0.1/>
  SELECT ?name ?mbox
3 WHERE
    { ?x foaf:name ?name .
5     ?x foaf:mbox ?mbox }
```

Listing 1.8. A simple SPARQL query

SPARQL allows the selection of variables that do not appear within the graph pattern as shown in Listing 1.9. The empty query pattern matches with any RDF graph, and the variable ?x in the select clause does not appear within the query pattern. In database theory, such rules are said to violate the principle of range-restrictedness. In fact the intuitive semantics of non-range-restricted rules is unclear and varies from one language to another. While according to [141] Listing 1.9 is supposed to return a single solution with no binding for the variable ?x, unbound variables are forbidden within construct clauses of SPARQL queries. In Prolog, on the other hand, the non ground fact p(X) simply remains uninstantiated and can be unified with ground bodies of other queries such as p(a).

```
SELECT ?x
2 WHERE {}
```

Listing 1.9. A non-range-restricted SPARQL query matching with arbitrary RDF graphs

Since queries such as the one in Listing 1.9 can also be expressed with the SPARQL ask query form, and since SPARQL does not allow any kind of rule-chaining, non-range-restricted queries do not add to the expressive power of the SPARQL language, but cause the semantics of the language to be more complex than it needs to be.

The graph pattern in Listing 1.8 is termed a *basic graph pattern*. It consists of two *triple patterns*, which are ordinary RDF triples except that subject, predicate and object may be replaced by SPARQL variables. Basic graph patterns may contain *filter expressions* in addition to a set of triple patterns. Filter expressions use the boolean predicates '=', 'bound', '`isIRI`' and others to construct atomic filters. Additionally the logical connectives '&&' for logical conjunction, '| |' for logical disjunction and '!' for logical negation are used to construct compound filters from atomic ones. Atomic and compound filters are used to eliminate sets of variable bindings that do not fulfill the filter requirements.

Besides basic graph patterns, SPARQL provides group graph patterns that may either be *unions of graph patterns*, *optional graph patterns* or *named graph patterns*. *Unions of graph patterns* are similar to disjunctions in the bodies of rules in logic programming. For the query to succeed, only one of the graph patterns in the union must be successful, and the solution sets from all graph patterns in the union are collected to yield the solution set for the union. *Optional graph patterns* are patterns that may bind additional variables besides the ones present in the non-optional parts of a graph pattern, not causing the entire query to fail if the optional graph pattern fails. In contrast to unions of graph patterns, the non-optional part is obliged to match. *Named graph patterns* are introduced into the SPARQL language, because Semantic Web databases may hold multiple RDF graphs, each identified by a URI. To explain the concept of querying named graphs in SPARQL, the notion of a *dataset* must be introduced. A *dataset* is a pair (d, N) where d is the default graph to be queried, and N is a set of named graphs. Datasets are specified by the FROM and FROM NAMED clauses in SPARQL. Whereas the default graph is the merge of all RDF graphs specified in the FROM clause, the FROM NAMED clauses specify the set N of named graphs, and remain unmerged. The GRAPH key word must subsequently be used to refer to named graphs in a WHERE clause as Listing 1.10 (taken from [127]) illustrates.

```
SELECT ?N WHERE { ?G foaf:maker ?M .
2        GRAPH ?G { ?X foaf:name ?N } }
```

Listing 1.10. Querying named graphs in SPARQL

As [127] points out, the query in Listing 1.10 is somewhat unintuitive, since SPARQL engines compliant with the W3C specification will search for answers to the triple pattern ?X foaf:name ?N only in named graphs, but not in the default graph.

The notion of *named graphs* is discussed in more detail in [45], and can be compared to grouping XML data in XML documents.

Blank nodes in SPARQL graph patterns. Blank nodes in SPARQL graph patterns act in the same way as non-distinguished variables, and therefore cannot be used to reference specific blank node identifiers within an RDF graph. Hence, one could substitute an arbitrary blank node for the variable ?x in Listing 1.8 and still obtain the same result.[20]

Before proceeding, we will quickly discuss this treatment of blank nodes in SPARQL. When issuing a query with a blank node, newcomers to the SPARQL language may have five different expectations in mind:

- *Syntactic equality:* The blank node in the query is supposed to match only with the data that uses exactly the same blank node identifier, as it is the case for URIs in graph patterns. While this is a valid desire, it would fall into the domain of syntactic processing of RDF data. A query on two equivalent RDF graphs should obviously return equivalent answers. But what is a sensible notion of equivalence in this context? As with all data items in information processing, one may introduce several equivalence relationships for RDF graphs. One such equivalence relationship is bi-entailment, and it is arguably the most sensible one for RDF graphs. Another such equivalence relationship would be syntactic equality, and there is certainly the necessity to compare RDF graphs for syntactic equality, but then we could also simply consider them as plain text files and run a UNIX `diff` command to test them for equality. With the decision for *syntactic equality* for blank nodes in queries, one would obtain different results for equivalent RDF graphs (under bi-entailment), and for this fact the decision of SPARQL not to use *syntactic equality* is a sensible one.
- *Treatment as non-distinguished variables:* The blank node is supposed to act as a non-distinguished variable as explained above. One minor problem with this understanding is that there are two alternative ways of specifying the same query, which may be confusing for new-comers to the language. Another more important issue with this solution is that while SPARQL remains answer closed in the sense that any RDF graph can be used as a SPARQL query, the answer to such a query would not only be graphs that are equivalent or contain an equivalent graph, but also graphs that are more specific. The simple SPARQL graph pattern `_:X a b` will also return true on the RDF graph `a b c`.
- *Banning of blank nodes within queries*: As the inclusion of blank nodes within queries does not add expressive power to SPARQL graph patterns, an obvious approach is to ban blank nodes from graph patterns. This approach has the advantage that SPARQL users cannot be fooled to assume a different semantics of blank nodes in graph patterns other than non-distinguished variables. On the other hand, this approach has the obvious drawback that SPARQL is not answer closed in the sense that an RDF graph containing blank nodes cannot be viewed as a SPARQL query.
- *Treatment as ordinary variables:* Since blank nodes are viewed as existentially quantified variables in RDF graphs, one might view them as plain variables in

---

[20] Note that one could *not* use a blank node at the place of the other two variables in Listing 1.8, since they are distinguished.

queries as well, and specify in the select-clause if they are to be treated as distinguished variables or non-distinguished variables. This solution has the plain advantage that any RDF graph can be viewed as a query, but shares the same deficiencies with respect to answer closedness as treating them as non-distinguished variables. Clearly this approach would mean that there is no longer the necessity for SPARQL variables.

– *Matching blank nodes only:* A final intuition query authors may have in mind is that blank node identifiers in queries must be mapped to blank node identifiers in the data only. None of the above approaches can express this semantics. The graph pattern _:X b c would thus return true when evaluated on the graphs _:X b c and _:Y a b, but it would not match with a b c. Thus with answer closedness in mind, this approach ensures that an RDF graph q considered as a SPARQL query only matches with RDF graphs that are equivalent or have a subgraph equivalent to q. The major drawback of this solution is, however, that the same query may once return true for an RDF graph g_1 and false for an equivalent (under bi-entailment) RDF graph g_2. To see this, consider again the query _:X b c and the graphs $g_1 :=$ _:Y b c, a b c and a b c. Under the light of this deficiency and with the availability of the filter predicate isBlank in SPARQL that can be used for imitating this blank node semantics, it is a good choice not to adopt this treatment of blank nodes in SPARQL graph patterns.

Testing RDF Graphs for Equivalence in SPARQL. None of the above solutions are completely satisfactory in that they do not allow the specification of a query q that returns true on exactly the equivalence class $\Sigma_{\leftrightarrow}(g)$ induced by RDF bi-entailment for an arbitrary graph g containing a blank node.

Note that SPARQL query patterns cannot express the above query even in the absence of blank nodes. Consider the RDF graph g a b c consisting of a single triple. Evaluating g as a SPARQL query pattern will yield all RDF graphs that *contain* g, but there is no way of expressing a query that will find all *equivalent* graphs.

In other words, a SPARQL basic graph pattern q returns true on an RDF graph g iff g rdf-entails[21] $n(q)$ where the normalization operator n replaces variables in q by blank nodes (multiple occurrences of the same variable by the same blank node identifier, and distinct variables by distinct blank nodes, that do not occur anywhere else in q). Hence, with basic SPARQL graph patterns it is only possible to demand that something *be entailed* by the graph g to be queried, but not to restrict the entailments of g. The development of the language Xcerpt[RDF], on the other hand, is influenced by the assumption that query authors would like to both demand some entailments from a graph as well as demand that something is *not* entailed by it.

There is, however, the possibility to express such queries in SPARQL at the aid of optional graph patterns, SPARQL filter constructs, and the SPARQL bound predicate. The query in Listing 1.11 only returns true for the one-triple graph a b c. For all other graphs it returns false. The graph pattern first ensures that the triple a b c is in fact contained in the RDF graph. Secondly it uses an optional pattern to find other triples

---

[21] There are different variants of RDF entailment. In this section we mean simple RDF entailment when when speaking of RDF entailment only.

in the graph. The filter inside the optional pattern makes sure that the optional pattern does not match with a triple other than a b c. The second filter expression makes sure that the optional graph pattern was unsuccessful by testing for a binding of the variable ?x.

```
  PREFIX foaf: <http://xmlns.com/foaf/0.1/>
2 ASK
  WHERE { a b c .
4     OPTIONAL { ?x ?y ?z
        FILTER ( ?x != a || ?y != b || ?z != c )
6     }
      FILTER (!bound(?x))
8  }
```

Listing 1.11. A query that only matches with a graph consisting of a single triple (a b c)

Before proceeding to the study of the complexity and semantics of SPARQL, we will quickly discuss how to test for equivalence with RDF graphs containing blank nodes. Consider the graph $g = \_:X\ b\ c\ .\ a\ b\ d\ .$ consisting of two triples only with a single occurrence of a single blank node. When formulating a SPARQL query to return true on exactly the set of RDF graphs equivalent to g, one first needs to test for the presence of the two triples and then for the absence of triples that are different from the two ones given in the graph. While the query in Listing 1.12 is all but trivial to figure out, testing graphs for equivalence in SPARQL becomes even more complex in the presence of multiple occurrences of the same blank node identifier, since in this case it does not suffice to test for the absence of single triples only, but one has to test for the absence of multiple triples connected via blank nodes.

```
   PREFIX foaf: <http://xmlns.com/foaf/0.1/>
2  ASK
   WHERE {
4     a b c .
      ?blank b d
6     OPTIONAL { ?x ?y ?z
        FILTER ( ( ?x != a || ?y != b || ?z != c ) ) &&
8              ( !(isBlank(_?x1)) || ?y1 != b || ?z != d ) )
      }
10    FILTER (!bound(?x1))
   }
```

Listing 1.12. A query that only matches with a graph consisting of a single triple (a b c)

Obviously the queries in Listing 1.11 and 1.12 are much more complicated than they need to be. This is due to the absence of explicit negation in SPARQL, a design decision that makes implementation easier and circumvents the non-monotonicity of negation as failure.

Semantics and Complexity of SPARQL. [124] recursively defines the semantics of SPARQL query patterns in terms of relational algebra operators as follows:

- The semantics $[[t]]_G$ of a possibly non-ground triple t evaluated over an RDF graph G is the set of mappings μ such that the domain of μ is the set $Var(t)$ of variables in t and the application $\mu(t)$ of the mapping μ to t is in G. The application of a mapping μ to a triple pattern t is simply the triple pattern with the variables in t replaced by their bindings in μ.
- The semantics $[[(P_1 \text{ AND } P_2)]]_G$ of a conjunction of query patterns evaluated over the RDF graph G is defined as the set $\{[[P_1]] \bowtie [[P_2]]\} = \{\mu_1 \cup \mu_2 \mid \mu_1 \in [[P_1]], \mu_2 \in [[P_2]], \mu_1 \text{ and } \mu_2 \text{ are compatible}\}$ of unions of compatible pairs of mappings of P_1 and P_2. In this context two mappings are termed *compatible* if they coincide on the bindings of their common variables. The semantics of the conjunction can thus be thought of as the natural join over the relations defined by the conjuncts.[22]

  In [127] the notion of compatibility of pairs of mappings is refined to *brave compatibility*, *cautious compatibility* and *strict compatibility*. While in the absence of unbound variables within mappings, all three notions of compatibility coincides, in the presence of unbound variables, only the brave compatibility coincides with compatibility as understood by [124].

  - Two mappings σ_1 and σ_2 are *bravely compatible* if they coincide on the bindings of their common bound variables. Brave compatibility hence does not restrict the bindings of variables that are unbound in either σ_1, σ_2 or both.
  - σ_1 and σ_2 are *cautiously compatible* if for all common variables – no matter if bound or unbound – the bindings coincide.
  - σ_1 and σ_2 are *strictly compatible* if they are cautiously compatible and if additionally there is no common variable of σ_1 and σ_2 which is unbound in both.
- The semantics of a graph pattern $[[P_1 \text{ OPT } P_2]]_G$ including an optional construct over an RDF graph G is defined as the left outer join between $[[P_1]]$ and $[[P_2]]$.
- Finally the semantics $[[P_1 \text{ UNION } P_2]]$ of a union of two graph patterns is defined as the union of $[[P_1]]$ and $[[P_2]]$.

[124] extend the semantics to SPARQL queries including filter expressions and show some important properties of SPARQL queries:

- Generally the expressions

$$(P1 \ \underline{AND} \ (P2 \ OPTIONAL \ P3))$$

  and

$$(P1 \ \underline{AND} \ P2)OPTIONAL \ P3))$$

  are not semantically equivalent, but they are equivalent for the class of *well-defined* graph patterns introduced in the same work.
- In the presence of optional patterns, AND is only commutative for well-designed graph patterns.

Some results on the complexity of query evaluation in SPARQL from [124] are the following:

---

[22] Note that the terms *relation* and *sets of mappings* can be used interchangeably here.

- The combined complexity of SPARQL graph patterns involving only AND and FILTER expressions is in $O(|P| \cdot |D|)$ where $|D|$ is the size of the data and $|P|$ is the size of the query. This result is based on the assumption that the application of a mapping μ to a triple t is achieved in a constant amount of time, independently of the number of variables in μ.
- The combined complexity of SPARQL graph patterns involving AND, FILTER and UNION is NP-complete. The proof is by polynomial reduction of the satisfiability problem of propositional logic formulas in conjunctive normal form to SPARQL queries.
- The combined complexity of SPARQL graph patterns including AND UNION and OPTIONAL is PSPACE-complete, independently of the presence or absence of FILTER expressions.
- The data complexity of SPARQL graph patterns is in LOGSPACE.

Extensions of SPARQL. SPARQL being the most popular RDF query language and the only one which has been standardized by some standardization organization such as the W3C, it has received considerable attention from the research community. Its expressiveness and complexity has been formally studied, and as a result of its limited expressiveness, extensions of SPARQL in different directions have been proposed. With the absence of path expressions in SPARQL, nSPARQL[125] has been suggested to enhance the expressive power of SPARQL into this direction. The necessity of combined processing of XML and RDF has been acknowledged by XSPARQL[7], an extension of XQuery to RDF processing at the aid of SPARQL WHERE and CONSTRUCT clauses. Just as SQL allows the deletion and insertion of data and creation of new tables, SPARQL update [140] and SPARLQ+[23] extend SPARQL with facilities to manipulate and create RDF graphs. Finally [44], [127] and [136] eliminate the restriction of SPARQL to single rules by allowing possibly recursive multi-rule programs.

nSPARQL. nSPARQL[125] is an extension of SPARQL to support arbitrary-depth navigation in SPARQL queries. It arose from the need to answer queries for finding all nodes reachable from a given node via a given predicate name, a disjunction of predicate names or simply for finding all transitively connected nodes. The syntax of nSPARQL is heavily influenced by the syntax of XPATH, and nSPARQL borrows the notions of axes, node tests, reverse axes, step expressions, and path predicates from XPATH. While path expressions in XPATH evaluate to a set of nodes of an XML document, path expressions in nSPARQL evaluate to a set of *pairs* of nodes within of an RDF graph. This is due to the fact that XPATH expressions are always evaluated with respect to a context node, while this is not necessarily the case for nSPARQL expressions.

The following examples illustrate the syntax and semantics of nSPARQL path expressions evaluated over an RDF graph G:

- next::a allows the navigation from one node in an RDF graph to another node via an edge labelled a in a composed nSPARQL path expression. It evaluates to all pairs of nodes connected via a predicate labeled a: $\{(x,y) \mid (x,a,y) \in G\}$. The axis next^{-1} can be used to navigate in the reverse direction.

---

[23] http://arc.semsol.org/home

- `edge::a` allows the navigation from a node x to an edge y within an RDF graph, if the graph contains the triple (x, y, a). It evaluates to $\{(x, y) \mid (x, y, a) \in G\}$. The axis $edge^{-1}$ is used to navigate from predicates of triples to their subjects.
- `node:a` allows the navigation from an edge x to a node y if the corresponding triple has subject a. It evaluates to $\{(x, y) \mid (a, x, y) \in G\}$. $node^{-1}$ is used for navigating in the reverse direction.
- nSPARQL path expressions are combined just like XPATH step expressions by the / sign: The nSPARQL expression `next::a/next::b` finds pairs of nodes connected via two triples with predicate names a and b over an arbitrary intermediate node. The URI of the intermediate node can be checked by using the self axis: `next::a/self::c/next::b`.
- The evaluation of nested nSPARQL path expressions is more complex. The semantics of `edge::[exp]` is given by $\{(x, y) \mid \exists z, w.(x, y, z) \in G \land (z, w) \in [[exp]]_G\}$, where $[[exp]]_G$ is the semantics of exp over G. Nested path expressions including the axes `self`, `next` and `node` are similarly involved.

SPARQLeR. A different approach for extending SPARQL with regular path expressions is taken by the language SPARQLeR described in [104]. In contrast to nSPARQL, entire paths are bound to so-called path variables, which are distinguished from ordinary SPARQL variables in that they are prefixed by % instead of ?. The bindings of path variables are themselves represented as RDF sequences, which allows to put further restrictions on the bindings in SPARQL WHERE clauses, as the following example from [104] demonstrates:

```
  CONSTRUCT %path
2 WHERE  { r %path s .  %path rdfs:_1 p .  }
```

Listing 1.13. A simple SPARQLeR path query

The query in Listing 1.13 finds all directed paths between a resource r and a resource s that have p as the first predicate. Bindings for the path variable %path in the above query are of the form $p_1, n_1, p_2, n_2, \ldots, p_i, n_i, p_{i+1}$, such that the triples (r, p_1, n_1), (n_1, p_2, n_2), ..., (n_{i-1}, p_i, n_i) and (n_i, p_{i+1}, s) are in the queried graph. Since these bindings are represented as RDF sequences (as exemplified in Listing 1.14), triples in the same WHERE clause can be used to put restrictions on the bindings to path variables.

```
  _:Path1 rdfs:_1 p₁,
2 _:Path1 rdfs:_2 n₁,
  _:Path1 rdfs:_3 p₂,
4 ...
```

Listing 1.14. The RDF representation of bindings to SPARQLeR path variables

Since bindings to SPARQLeR path variables are represented as RDF sequences represented by blank nodes, the use of path variables within SELECT query forms hardly makes sense. Imagine Listing 1.13 with the SELECT keyword at the place of the CONSTRUCT keyword. The result of this query is a list of blank nodes generated by the SPARQLeR query generator, which means that the only information returned is the number of paths found within the queried graph. To deal with this inconvenience, SPARQLer introduces a `list` operator that extracts all resources from the paths. In the case of multiple bindings for a path variable, however, the application of the list operator merges the resources from all paths into a single list, thereby preventing the user from recognizing the actual paths.

SPARQLeR provides a second method for constraining paths at the aid of a ternary `regex` method to be used within FILTER clauses of SPARQLeR queries. The first argument to this method is the name of the path variable whose bindings are to be constrained, the second one is a regular path expression, and the third are options specifying whether the path must be directed, if it must be made up of schema classes, instances, or literals, and if `rdfs:subPropertyOf` inferencing is to be considered. SPARQLeR regular path expressions allow alternatives, concatenation, Kleene's star, wildcards, negations and reverse predicates. The SPARQLeR `length` method is used to find paths of a minimal, maximum or exact length.

While SPARQLeR seems to be a sensible suggestion for an extension of SPARQL, there are two obvious points of criticism:

– The fact that predicate names can be specified within path expressions, but subjects and objects cannot, seems to be an arbitrary design choice which is not motivated in [104].
– Representing bindings to variables as RDF sequences that are not part of the original RDF graph and allowing these RDF sequences to be queried within the SPARQLeR WHERE clause may be confusing for novices in that the WHERE clause is successfully evaluated on a graph which does not entail every single triple of the clause.

XSPARQL. [7] advocates the reuse of plain XML and HTML data of the Web as RDF data on the Semantic Web, and vice versa and introduces the notions of *lifting* – i.e. transforming "syntactic" XML data into "semantic" RDF data – and *lowering* – transforming RDF data into XML. Starting out from the insight that current tools and languages are not adequate for translating between syntactic and semantic web data, they propose an integration of SPARQL into XQuery, which they dub XSPARQL, together with use-cases and a formal semantics. Since it aims at being data-versatile in the same sense as Xcerpt does, we take a closer look at XSPARQL in this section.

```
  <relations>
2   <person name="Alice">
      <knows>Bob</knows>
4     <knows>Charles</knows>
    </person>
6     <knows>Charles</knows>
    </person>
8   <person name="Charles/>
  </relations>
```

Listing 1.15. XML example data

```
1 @prefix foaf: <...foaf/0.1/>.
  _:b1 a foaf:Person;
3    foaf:name _:b2;
     foaf:knows _:b3 .
5 _:b2 a foaf:Person;
     foaf:name "Bob";
7    foaf:knows _:b3 .
  _:b3 a foaf:Person;
9    foaf:name "Charles" .
```

Listing 1.16. RDF example data

Listing 1.17 shows how the lifting task is solved in XSPARQL for the example data given in Listings 1.15 and 1.16. In line 3 all names of persons of the XML input file are selected. Names are either given as the `name` attribute of a `person` element or as XML text nodes within `knows` elements. In order to make sure that the list `$persons` contains each name exactly once, duplicates are elminitated in the `where` clause by testing the absence of elements on the `following` axis that contain the same name. In this way duplicates are eliminated and only the last occurrence of a name is selected. In line 6, a numeric identifier is computed for each person which serves to construct unique blank nodes in the SPARQL construct pattern starting at line 8. The `construct` keyword is not part of the XQuery syntax, but newly introduced in XSPARQL to mark the beginning of a SPARQL construct pattern. Inside of SPARQL construct patterns, XQuery code is embedded within curly braces. In this way nested XSPARQL queries are constructed. While the outer XSPARQL query (lines 3 to 10) serves to represent the persons found in the XML source as RDF blank nodes with associated names and type, the inner SPARQL query translates the acquaintance relationships. Note that the triples constructed in line 18 are duplicates of the ones constructed in line 10, i.e. this line is superflous.

```
   declare namespace foaf="...foaf/0.1/";
   declare namespace rdf="...-syntax-ns#";
 3 let $persons := //*[@name or ../knows] return
   for $p in $persons
   let $n := if ( @p[@name] ) then $p/name else $p
 6 let $id := count($p/preceding::*) + count($p/ancestor::*)
   where not(exists($p/following::*[@name=$n or data(.)=$n]))
   construct
 9 _:b{$id} a foaf:Person;
             foaf:name { data($n) }.
     { for $k in $persons
12   let $kn := if ( $k[@name] ) then $k/@name else $k
     let $kid := count($k/preceding::*) + count($k/ancestor::*)
     where $kn = data(//*[@name=$n/knows) and
15         not(exists($kn/../following::*[@name=$kn or
                 data(.)=$kn]))
     construct
       _:b{$id} foaf:knows _:b{$kid} .
18     _:b{$kid} a foaf:Person .
   }
```

Listing 1.17. Lifting in XSPARQL

XSPARQL does not set out to be a query language that natively supports XML and RDF querying in an intuitive and coherent way. Instead it explores how SPARQL can be integrated into XQuery, how the semantics of this integration can be defined and proposes an implementation on top of existing XQuery and SPARQL engines. XSPARQL succeeds in its coherent treatment of schema heterogeneous RDF/XML files, and due to the large expressiveness of XQuery it allows the formulation of many queries not expressible in SPARQL alone. On the other hand it suffers from the following deficiencies:

- *Intertwined querying and construction.* As can be observed in Listing 1.17, there is no clear separation of querying and construction in XSPARQL queries, a deficiency which is inherited from XQuery. While it is clear that there are queries that cannot be expressed by a single rule with a single query and construction pattern, this is not the case for the query above.
- *Complicated blank node construction.* An RDF query language should support automatic construction of blank nodes without the need of computing blank node identifiers within a program. Since blank node construction is essentially the same as the introduction of skolem terms within logic programs, languages such as RD-FLog and Xcerpt achieve the same result in a much easier and straightforward way.
- *Absence of path patterns.* While XSPARQL inherits the complexity of XQuery, it suffers also from the limitations of SPARQL such as no support for containers, collections and reification, and limited support for negation. Above all, XSPARQL lacks rich path patterns to navigate RDF graphs at arbitrary depth, such as the ones proposed by nSPARQL and SPARQLeR.
- *Jumbling of query paradigms.* Due to the popularity of XQuery as an XML query language and SPARQL as an RDF query language, Listing 1.17 is easy to understand for most people familiar with (Semantic) Web querying. For people unfamiliar with one or both of these languages, it may be confusing that a functional language such as XQuery is intermingled with a rule based language such as SPARQL. With Xcerpt^{RDF} we introduce a purely rule based language based on the clear design principles of Xcerpt.

SPARQL update. Similar as for the XML query language XQuery, SPARQL has been conceived primarily as a data *selection* language, not as a data *manipulation* language. In fact, the SELECT, DESCRIBE and ASK query forms of SPARQL can only be used to *extract* parts of a graph, not to manipulate data or construct new data. The SPARQL CONSTRUCT query form allows limited transformations between one RDF dialect to another, but cannot be used to modify existing RDF stores. The W3C member submission *SPARQL update* sets out to elminate this restriction.

SPARQL update consists of two sets of directives – one for updating graphs and the other for graph management. The set of directives for updating existing RDF graphs with SPARQL update constists of the following seven commands:[24]

- The DELETE DATA FROM directive is used to delete a set of ground triples from a named or the default graph. In the latter case, the FROM keyword is omitted.
- The INSERT DATA INTO statement is used to insert a new set of ground triples into an existing graph identified by a URI. If the triples are to be inserted into the default graph, then the INTO keyword is omitted.
- The MODIFY operation consists of a delete and an insert statement (see below) issued on the same graph.
- The DELETE FROM ... **WHERE** operation is used to delete a set of triples from a graph. In contrast to the DELETE DATA operation discussed above, this command

---

[24] We only briefly sketch the commands for the sake of brevity.

may specify the triples to be deleted in a non-ground form, i.e. with SPARQL variables bound in the **WHERE** clause. If the WHERE clause consists of the empty graph pattern, this command is indeed equivalent to the DELETE DATA operation above. In case the FROM keyword is omitted, the default graph is manipulated.

- INSERT FROM ... **WHERE** is the non-ground version of the INSERT DATA command. Its relationship to INSERT DATA is analogous to the relationship from DELETE FROM ... **WHERE** to the DELETE DATA operation. Together with the DELETE FROM ... **WHERE** operation, this operation can be used to move data from one RDF graph to another.
- The LOAD primitive copies all RDF triples from one named graph to another named graph or the default graph.
- The CLEAR primitive removes all triples from the default graph, or a named graph. It can be simulated by a DELETE FROM ... **WHERE** operation selecting all triples of a graph.

Graph management in SPARQL update is achieved by the two operations CREATE GRAPH and DROP GRAPH which have the exact same semantics as the SQL operations CREATE TABLE and DROP TABLE. Only when a graph has been created by the CREATE GRAPH operation it is available for modification by one of the seven above mentioned manipulation directives.

To sum up, SPARQL update is a straight-forward extension of SPARQL to include mechanisms for creating new and changing existing RDF graphs, much inspired by SQL. The difference between the Web considered as a huge database and ordinary databases is, however, that the Web is open and generally readable and processable by any person or computer connected to the Internet. As a result RDF graphs will more likely be reasoned with and transformed than updated. Write access to RDF graphs is restricted to the content provider, but deriving new knowledge from existing one, which is the fundamental use case for Semantic Web use-cases, is possible for all Web users and will be achieved with rule languages, not update languages. Under these considerations, update primitives have been excluded from Xcerpt^{RDF}.

SPARQL and Rules. [127] defines translation rules for SPARQL rules to datalog rules and thus opens up the possibility to rule chaining, i.e. the translation of multiple SPARQL rules to Datalog and the combined evaluation of the resulting rule set by a logic programming engine, thus allowing intermediate results to be constructed and queried. This extension gives SPARQL an obvious boost in expressivity (recursion) and affects its termination properties. In the following, the translation procedure from SPARQL to Datalog given in [127] is quickly illustrated by an example, as it opens up the possibility for easy implementations also of single rule SPARQL queries on top of existing logic programming engines.

For this purpose reconsider the SPARQL query in Listing 1.18 and the RDF graph in Listing 1.19 available via the URL http://www.example.org/bob. The result of the translation is given in Listing 1.20.

```
  PREFIX foaf: <http://xmlns.com/foaf/0.1/>
2 SELECT ?name ?mbox
  FROM http://example.org/bob
4 WHERE
   { ?x foaf:name ?name .
6    ?x foaf:mbox ?mbox }
```

Listing 1.18. A simple SPARQL select-query

```
  _:B foaf:name bob .
2 _:B foaf:nick bobby .
  _:B foaf:mbox bob@example.org .
```

Listing 1.19. RDF Graph with some FOAF information

```
1 triple(S, P, O, default) :- rdf(http://example.org/bob, S, P,
    O) .
  answer_1( (Name, Mbox), default) :-
3 answer_2(vars(Name, X), default),
  answer_3(vars(Mbox, X), default) .
5 answer_2(vars(Name, X), default) :- triple(X, foaf:name,
    Name, default) .
  answer_3(vars(Mbox, X), default) :- triple(X, foaf:name,
    Mbox, default) .
```

Listing 1.20. Translation of the SPARQL query in Listing 1.18 to Datalog with external predicates

The translation makes use of the external predicate `rdf` that takes four arguments: the graph to be queried as input, and the subject, predicate and object of triples as output. The external predicate `rdf` can thus be used to enumerate all triples within an RDF graph given by the input URI. The first rule in Listing 1.20 defines the 4-ary relation `triple`. In the case of multiple FROM or FROM NAMED clauses in the original SPARQL query, the relation `triple` will obviously be defined by the corresponding number of clauses. Since Listing 1.18 only contains conjunctions of triple patterns, but no UNION, OPTIONAL or FILTER expression, the translation remains of manageable size, and we focus the discussion of the tranlsation procedure on conjunctive triple patterns.

As can be observed in Listing 1.20, each triple pattern in the SPARQL query translates to a single Datalog rule, and each conjunction of triple patterns translates to a rule with body atoms referencing the rules obtained by the translation of its conjuncts. As expected, disjunctions (UNION) of triple patterns are translated to sets of rules. For details on the tranlsation procedure, involving more complex SPARQL queries with FILTER and OPTIONAL, the interested reader is referred to [127].

While reusing existing rule languages together with the enormous body of knowledge about their semantics, evaluation methods and complexity is certainly a sensible way for designing a rule language for the Semantic Web, the approach taken in [127] is not completely satisfactory for the following reasons:

- Blank node construction in rule heads has been largely ignored, especially the different modes of blank node construction as pointed out by [34].
- This approach inherits the weakness of SPARQL concerning negation: implicit negation as failure is provided by the combination of the OPTIONAL directive and the unbound predicate. For newcomers to the language this feature is hard to discover, and should be better declared as what it is.
- The expressivity of SPARQL graph patterns is limited when compared to languages that allow possibly recursive path expressions such as Versa on RDF graphs or Conditional XPath[112,111] and Xcerpt on XML documents. This limitation is obviously inherited by all rule extensions to SPARQL.
- Rule extensions of SPARQL remain pure RDF query languages and therefore cannot deal with the versatility requirements for modern Web query languages.

2.4   Versatile Semantics

Having given an informal, example-driven introduction to the language Xcerpt, its evaluation principles and intuitive semantics in the preceding sections, this section introduces the precise semantics for Xcerpt *query terms* through a formal definition of *query term simulation* (Section 2.4.1), and *programs* through an iterative fixpoint procedure (Section 2.4.2). Previous publications on the semantics of Xcerpt have considered the class of *stratified Xcerpt programs* only. Section 2.4.2 extends the semantics of Xcerpt programs to the class of *locally stratified programs*, which is a true superset of the set of stratifiable Xcerpt programs, and which is inspired by the notion of local stratification in logic programming [52]. In Section 2.4.3 the well-founded semantics for general logic programs is adapted to Xcerpt, thereby also giving a semantics to programs that are not locally stratified. Although not formally proven, we argue that locally stratifiable Xcerpt programs have a two-valued well-founded model which coincides with the model computed by the iterative fixpoint procedure over its local stratification.

While this section transfers the notion of local stratification and well-founded semantics to Xcerpt only, the proposed method can be applied to any other rule-based language with non-monotonic term negation and disjunction-free heads, that provides the same interface to terms as Xcerpt does (defined in Section 2.3).

2.4.1   Simulation as Foundation for a Semantics of Versatile Queries

Simulation between Xcerpt terms is inspired by rooted graph simulation [116,86], but is by far more involved since Xcerpt terms feature constructs for specifying incompleteness in depth, breadth, and order, allow variables, regular expressions and negated subterms. This section formally defines a subset of Xcerpt^{XML}[25] variables, descendant constructs, subterm negation, incompleteness in breadth and with respect to order, multiple variables, multiple occurrences of the same variable, and variable restrictions. In

---

[25] Chapter 2.3 introduces both Xcerpt^{RDF} and Xcerpt^{XML} query, construct and data terms. In this section we concentrate on Xcerpt^{XML} terms, but most of the results and design principles also apply to Xcerpt^{RDF} terms. We write "Xcerpt term" to denote the abstract concept of terms in both Xcerpt^{RDF} and Xcerpt^{XML}, and "Xcerpt^{XML} term" to refer to Xcerpt^{XML} terms only.

comparison to full Xcerpt^{XML} query terms as described in [69,134] and for the sake of brevity, this definition does not include term identifiers and references, non-injective subterm specifications, optional subterms, qualified descendants, label variables, and the new syntax for XML attributes. Based on this definition of Xcerpt^{XML} query, construct and data terms, ground and non-ground query term simulation is defined as the formal semantics for the evaluation of Xcerpt^{XML} query terms on semi-structured data.

Definition 2 (Xcerpt^{XML} query term). *Query terms over a set of labels N, a set of variables V, and a set of regular expressions R are inductively defined as follows:*

- for each label $l \in N$, $l\{\{\ \}\}$ and $l\{\ \}$ are atomic query terms. l is a short hand notation for $l\{\{\ \}\}$. The formal treatment of square brackets in query terms is omitted in this contribution for the sake of brevity.
- for each variable $X \in V$, var X is a query term
- for each regular expression $r \in R$, $/r/\{\{\ \}\}$ and $/r/\{\ \}$ are query terms. $/r/$ is a shorthand notation for $/r/\{\{\ \}\}$. With $\mathcal{L}(r)$ we denote the set of labels matched by r, i.e. the language defined by the regular expression.
- for each variable $X \in V$ and query term t, var X as t is a query term. t is called a *variable restriction* for X.
- for each query term t, desc t is a query term and called *depth-incomplete* or *incomplete in depth*.
- for each query term t, without t is a query term and called a *negated subterm*.
- for each query term t optional t is an *optional query term*.
- for each label or regular expression l and query terms $t_1, \ldots, t_n$ with $n \geq 1$,

$$q_1 = l\{\{\ t_1,\ \ldots,\ t_n\ \}\}$$
$$q_2 = l\{\ t_1,\ \ldots,\ t_n\ \}$$

are query terms. q_1 is said to be *incompletely specified in breadth*, or simply *breadth-incomplete*, whereas q_2 is *completely specified in breadth*, or simply *breadth-complete*.

A variable X is said to *appear positively* in an Xcerpt^{XML} query term q, if it is included in q not in the scope of a without construct. It *appears negatively* within q if it is included within the scope of a without construct. Note that the same variable may appear both positively and negatively within q – e.g. X within a$\{\{$ var X, without var X $\}\}$.

Definition 3 (Xcerpt^{XML} data terms). *An Xcerpt^{XML} data term is a ground Xcerpt^{XML} query term that does not contain the constructs* without, optional, desc, *regular expression and double braces.*

Definition 4 (Xcerpt^{XML} construct terms). Xcerpt^{XML} construct terms *over a set of variables V and a set of labels $\mathcal{L}$ are defined as follows:*

- *an Xcerpt^{XML} data term d over $\mathcal{L}$ is a construct term*
- *for each variable $X \in V$,* var X *is a construct term*
- *for a construct term c,* all c *is a construct term*
- *for a construct term c,* optional c *is a construct term*

- *for a construct term c, and a sequence of variables $X_1, \ldots, X_k \in \mathcal{V}$* all c group
  by $\{X_1, \ldots, X_k\}$ *is a construct term*
- *for a label $l \in \mathcal{L}$ and set of construct terms $c_1, \ldots, c_n$, $l\{c_1, \ldots, c_n\}$ is a construct*
  term.

In the following, we let $\mathcal{D}$ and Q denote the set of all Xcerpt[XML] data and query terms, respectively.

A query term and a data term are in the simulation relation, if the query term "matches" the data. Matching Xcerpt[XML] query terms with data terms is very similar to matching XPath queries with XML documents – apart from the variables and the injectivity requirement in query terms. The formal definition of simulation of a query term with semi-structured data is somewhat involved. To shorten the presentation, we first introduce some notation:

Definition 5 (Injective and bijective mappings) [26]
Let $I := \{t_1^1, \ldots, t_k^1\}$, $J := \{t_1^2, \ldots, t_n^2\}$ *be sets of query terms and $\pi : I \Rightarrow J$ be a mapping.*

- *π is injective, if all $t_i^1, t_j^1 \in I$ satisfy $t_i^1 \neq t_j^1 \Rightarrow \pi(t_i^1) \neq \pi(t_j^1)$.*
- *π is bijective, if it is injective and for all $t_j^2 \in J$ there is some $t_i^1 \in I$ such that $\pi(t_i^1) = t_j^2$.*

We use the following abbreviations to reference parts of a query term q:

$l(q)$: the string or regular expression used to build the query term. For a variable v, $l(v)$ is undefined.
$ChildT(q)$: the set of direct subterms of q
$ChildT^+(q)$: the set of positive direct subterms (i.e. those direct subterms which are not of the form *without*...),
$ChildT^-(q)$: the set of negated direct subterms (i.e. the direct subterms of the form *without*...),
$Desc(q)$: the set of direct descendant subterms of q (i.e. those of the from *desc*...),
$SubT(q)$: the direct or indirect subterms of q, i.e. all direct subterms as well as their subterms.
$ss(q)$: the subterm specification of q. It can either be *complete* (single curly braces) or *incomplete* (double curly braces).
$vars(q)$: the set of variables occurring somewhere in q.
$pos(q)$: q', if q is of the form without q', q otherwise.

Definition 6 (Label subsumption). *A term label l_1 subsumes another term label l_2 iff l_1 and l_2 are strings and $l_1 = l_2$, or l_1 is a regular expression and l_2 is a string such that l_1 matches with l_2, or l_1 and l_2 are both regular expressions and l_1 matches with any label that l_2 matches with.*

---

[26] This definition of injectivity and bijectivity concerns the subterms – or nodes – of a query term only. Therefore it is also referred to as *node injectivity*. In previous publications about Xcerpt, we have used *position injectivity* instead, which concerns the edges between parent and child terms. In the absence of references (as in Definition 4), however, node and position injectivity are semantically equivalent. Therefore, and for the sake of simplicity, we use node injectivity in this contribution.

Definition 7 (Ground query term simulation). *Let q be a ground query term[27] and d a data term. A relation $S \subseteq (SubT(q) \cup \{q\}) \times (SubT(d) \cup \{d\})$ is a simulation of q into d if the following holds:*

- $q\, S\, d$
- *if $q := l_1\{\{q_1,\ldots,q_n\}\}\ S\ l_2\{d_1,\ldots,d_m\} =: d$ then l_1 must subsume l_2, and there must be an injective mapping $\pi : ChildT^+(q) \to ChildT^+(d)$ such that $q_i\, S\, \pi(q_i)$ for all $i \in ChildT^+(q)$. Moreover, there must not be a $q_j \in ChildT^-(q)$ and $d_l \in ChildT^+(d) \setminus range(\pi)$ such that $pos(q_j) \leq d_l$ (note the recursive reference to '$\leq$' here).*
- *if $q := l1\{q_1,\ldots,q_n\}\ S\ l2\{d_1,\ldots,d_m\} =: d$ then l_1 must subsume l_2, and there must be a bijective mapping $\pi : ChildT^+(q) \to ChildT^+(d)$ such that $q_i\, S\, \pi(q_i)$ for all $i \in ChildT^+(q)$. We impose no further requirements on the set $ChildT^-(q)$ of negated direct subterms of q. The totality of π already ensures that there is no extension of π to some element $q_j \in ChildT^-(q)$ such that $pos(q_j) \leq d_l$ for some $d_l \in ChildT^+(d) \setminus range(\pi)$. Therefore the semantics of query terms is independent from the presence of negated direct subterms within breadth-complete query terms.*
- *if $q = desc\ q'\ S\ d$ then $q'\ S\ d$ or $q'\ S\ d'$ for some subterm d' of d.*

We say that q simulates into d (short: $q \leq d$) if and only if there is a relation S that satisfies the above conditions. To state the contrary we write $q \not\leq d$.

Since every Xcerpt^{XML} data term is also a query term, the above definition of simulation between a query term and a data term can be extended to a relation between pairs of query terms. For the sake of brevity this full definition of *extended ground query term simulation* is given in the appendix of [35].

The existence of a ground query term simulation states that a given data term satisfies the conditions encapsulated in the query term. Many times, however, query authors are not only interested in checking the structure and content of a document, but also in extracting data from the document, and therefore query terms may contain logical variables. To formally specify the data that is extracted by matching a query term with a data term, the notion of non-ground query term simulation is introduced (Definition 8). Substitutions are defined as usual, and the application of a substitution to a query term is the consistent replacement of the variables by their images in the substitution.

Definition 8 (Non-ground query term simulation). *A query term q with variables simulates into a data term d iff there is a substitution $\sigma : Vars(q) \to \mathcal{D}$ such that $q\sigma$ simulates into d.*

In some cases query terms are not expressible enough or inconvenient for specifying a query in the body of a rule. Conjunctions of query terms are needed if more than one resource is queried and the results are to be joined. Disjunctions of query terms are convenient to extract data from different resources and wrap them into a common XML fragment or RDF graph. Finally the absence of data simulating with a given query term is tested by query negation. The notion of a query combines conjunctions, disjunctions and negations of query terms:

---

[27] For the sake of brevity we assume that q does not contain any optional subterms.

Definition 9 (Xcerpt query)

- *an Xcerpt query term is an Xcerpt query*
- *for a set of Xcerpt queries* $q_1, \ldots, q_n$, *the conjunction* $C :=$ `and` $(q_1, \ldots, q_n)$, *the disjunction* $\mathcal{D} :=$ `or` $(q_1, \ldots, q_n)$ *and the negation* $\mathcal{N} :=$ `not` (q_1) *are Xcerpt queries. If a variable X appears positively within a* q_i $(1 \leq i \leq n)$ *then it also appears positively within C and* $\mathcal{D}$, *but negatively within* $\mathcal{N}$. *If X appears nevatively within* q_i, *it also appears negatively within C,* $\mathcal{D}$ *and* $\mathcal{N}$.

Definition 10 (Xcerpt rule, goal, fact, program). *Let q be a query over a set of labels* $\mathcal{L}$, *a set of variables* $\mathcal{V}$ *and a set of regular expressions* $\mathcal{R}$ *and c a construct term over* $\mathcal{L}$ *and* $\mathcal{V}$. *Then* **CONSTRUCT** *c* FROM *q* END *is an* Xcerpt rule, GOAL *c* FROM *q* END *is an* Xcerpt goal, *and* **CONSTRUCT** *c* END *is an* Xcerpt fact. *An Xcerpt program is a sequence of range-restricted Xcerpt rules, goals and facts.*[28]

The construct term *c* is called the *head* of an Xcerpt rule or goal, the query *q* is called its *body*. An Xcerpt fact can also be written as an Xcerpt rule with an empty body. An Xcerpt rule, goal or fact is called *range restricted*, if all variables that appear in its head also appear positively in its body. In a forward chaining evaluation of a program, the distinction between goals and facts is unnecessary. In a backward chaining evaluation, however, the goals are the starting point of the resolution algorithm. In contrast to Logic programming, goals are not a single term only, but an entire rule to ensure answer closedness of Xcerpt programs. Especially for the task of information integration on the Web, answer closedness is indispensable.

2.4.2 Rules with Negation and Versatile Queries: Local Stratification

While Section 2.4.1 defines the semantics of single query terms and queries, this section defines the semantics of Xcerpt rules and programs. Special attention is laid on the interplay between simulation unification and non-monotonic negation in rule bodies.

The problem of evaluating rule based languages with non-monotonic negation has received wide-spread attention throughout the logic programming community (See [10] and [31] for surveys). A multitude of semantics have been proposed for such languages (program completion semantics, stable-model semantics [75], well-founded semantics [148], inflationary semantics [105]). Especially the well-founded and stable-model semantics have been found to comply with the intuition of program authors and are therefore implemented by logic programming engines such as XSB [133] and DLV [61]. Several classes of logic programs have been defined for which some of the above mentioned semantics coincide. Among these classes are definite programs, stratifiable programs, locally stratifiable programs [130] and modularly stratifiable programs [132]. The well-founded semantics and the stable model semantics coincide on the class of locally stratifiable programs.

In the following we introduce stratifiable and locally stratifiable Xcerpt programs. In adapting these concepts to Xcerpt, one has to pay close attention to the differences introduced by the richer kind of unification employed.

---

[28] Since facts and goals are a kind of rules, we refer to Xcerpt programs as a sequence of rules in the following.

Definition 11 (Stratification). *A stratification of an Xcerpt program P consisting of the rules $r_1, \ldots r_n$ is a partitioning of $r_1, \ldots r_n$ into strata $S_1, \ldots, S_k$, such that the following conditions hold:*

- *All facts are in S_1.*
- *If a rule r_1 contains a positive query term q that simulates with the contstruct term c of another rule r_2, then r_1 positively depends on r_2, and r_1 is in the same or a higher stratum than r_2.*
- *If a rule r_1 contains a negated query term* not *q such that q simulates with the construct term c of another rule r_2, then r_1 negatively depends on r_2 and is in a strictly higher stratum than r_2.*

Given the stratification of a program P, its semantics can be defined by the iterative fix-point procedure suggested for general logic programs. For finite programs, stratification is decidable. However, there are Xcerpt programs, such as the one in Listing 1.21, that are not stratifiable, but which may be evaluated bottom up.

Listing 1.21 is a formulation of the single source shortest path problem over a directed social graph, which is given by the facts (lines 1 to 5) in Listing 1.21 and which is depicted in Figure 3. The program computes for each node n in a directed graph the shortest distance to some source node s, in this case anna.

This program uses a slight extension of Xcerpt's term syntax. The term

Acquaintance[anna, $\leq i$]

simulates with the data terms Acquaintance[anna, j] if and only if i and j are natural numbers and $j \leq i$. Furthermore, the terms Acquaintance[anna, $\leq i$] and Acquaintance[anna, i] simulate with Acquaintance[anna, $> j$] if and only if $i > j$. The symbol '>' can be interpreted as a hint by the programmer to the evaluation engine, that a rule can only be used to derive atoms with integer values greater than a certain natural number. The example in Listing 1.21 serves to illustrate the problems and challenges for defining the semantics and evaluation of possibly recursive rule programs with non-monotonic negation and rich unification. These challenges are encountered independent of the specific kind of rich unification, be it SPARQL query evaluation, Xcerpt query term simulation, or XPath query evaluation.

To see that Program P in Listing 1.21 is not stratifiable, consider the negated query term not q, with q = Acquaintance [var P, $\leq$ var D] in the body of the only rule of P. q simulates with the head h = Acquaintance [var P, D + 1 > 0] of the same rule. Thus the rule should be in a strictly higher stratum than itself, which is a contradiction.

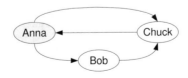

Fig. 3. Social graph corresponding to the facts in Listing 1.21

```
  CONSTRUCT knows[ anna, bob ] END
2 CONSTRUCT knows[ bob, chuck] END
  CONSTRUCT knows[ anna, chuck ] END
4 CONSTRUCT knows[ chuck, anna ] END

6 CONSTRUCT Acquaintance[ anna, 0 ] END

8 CONSTRUCT
    Acquaintance[ var P, var D + 1 ]
10 FROM
    and (
12   Acquaintance[ var P', var D ],
      knows[ var P, var P'],
14     not ( Acquaintance[ var P, ≤ D ] )
  END
```

Listing 1.21. Single source shortest path problem for the source node 'anna'

To see that P can nevertheless be evaluated in a bottom up manner, consider a ground instance g of the recursive rule in Listing 1.21. The term constructed by the head of g contains an integer value i which is exactly by one larger than the integer values of terms that may simulate with (negated or positive) query terms in the body of g. Thus, in a bottom up evaluation of the program, we may first compute the fixpoint of the program considering only terms containing the integer value 0, followed by the fixpoint computation for terms with the value 1, and so on. Since a valid rule application will only construct terms containing the value $n + 1$ using terms with values n, it may never be the case that the body of a rule once found true is invalidated by the derivation of a fact at a later point in time. Figure 4 visualizes the resulting stratification.

With the concept of *local stratification* we distinguish the class of *locally stratifiable Xcerpt programs*, which is a true superset of the class of stratifiable Xcerpt programs, and thereby introduce a more general characterization of Xcerpt programs that guarantees that these programs can be evaluated by an iterative fixpoint procedure in a bottom

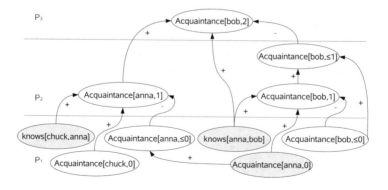

Fig. 4. Local stratification for Listing 1.21

up manner. A local stratification partitions the *Herbrand universe* of an Xcerpt program rather than the *rules* of the program into strata.

Definition 12 (Xcerpt Herbrand universe, Xcerpt Herbrand base). *The* Herbrand universe *of an Xcerpt program P are all Xcerpt data terms that can be constructed over the vocabulary of P.*[29] *Since Xcerpt programs consist only of terms without predicate symbols, the* Herbrand base *of P is defined to be the same as the Xcerpt Herbrand universe.*

Note that the above definition deviates from the Herbrand universe for logic programs as follows: While Prolog function symbols have always an associated arity, Xcerpt labels may be used to construct terms with arbitrary many children. Thus a program over the vocabulary $V = \{a\}$ has the Herbrand universe $\{$ a$\{$ $\}$, a$\{$ a $\}$, a$\{$ a$\{$ a $\}$ $\}$, a$\{$ a, a $\}$ ...$\}$. In the following discussion of the well-founded semantics we will, however, not consider the entire Herbrand universe for computing unfounded sets, but restrict them to the terms that occur in ground instances of the rules.

Definition 13 (Local stratification). *A local stratification of an Xcerpt program P is a partitioning of the Herband universe of P into strata such that the following conditions hold:*

- *All facts in P are in stratum 1.*
- *If a term q appears positively within the body of a rule R in the Herbrand instantiation of P, and c appears in its head, then q must be in the same or in a higher stratum than c.*
- *If a term q appears negatively within the body of R and c in its head, then q is in a strictly higher stratum than c.*
- *If a term q simulates into a term c, then q is in the same or in a higher stratum than q.*

The definition of local stratification of Xcerpt programs coincides with the definition of local stratification for general logic programs in the first three points. The fourth condition is necessitated by the richer unification relation induced by simulation unification in Xcerpt. While in logic programming two ground terms unify if and only if they are syntactically identical, this is not true for Xcerpt terms (consider e.g. the terms a$\{\{$ $\}\}$, a[[ ]] and a$\{$ b $\}$).

Example 1.22 underligns the necessity of the fourth condition in Definition 13: By Definition 13, Program *P* in Listing 1.22 is not locally stratifiable, but it would be, if the last condition were not part of the definition. In fact, the semantics for *P* is unclear, and it cannot be evaluated by an iterative fixpoint procedure. Figure 5 shows the dependency graph for Listing 1.22, which contains a cycle including a negative edge. The dependency graph for a ground Xcerpt program simply includes all rule heads and body literals as nodes, and all simulation relations between query and construct terms and negative and positive dependencies of rule heads on their body literals. The dependency graph for a non-ground Xcerpt program is the dependency graph of its Herbrand

---

[29] The vocabulary of *P* is the set of labels appearing in *P*.

Instantiation. An Xcerpt program P is locally stratifiable, if its dependency graph does not contain any negative cycles (i.e. cycles including at least one negative edge).

```
CONSTRUCT a{ b } FROM not(c{{ desc b{{ }} }}) END
2 CONSTRUCT c[ b ] FROM a{{ }} END
```

Listing 1.22. An Xcerpt program that is not locally stratifiable

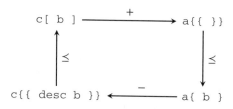

Fig. 5. Dependency graph for Listing 1.22

Since Listing 1.22 is not locally stratifiable, its semantics cannot be defined by a fixpoint procedure over its stratification. Similar programs – except for the simulation relation – have been studied in logic programming. For example, the logic program $\{(a \leftarrow \neg c),(c \leftarrow a)\}$ is not locally stratifiable, still the well-founded semantics of the program is given by the empty interpretation {}. To give Xcerpt programs a semantics, no matter if they are locally stratified or not, we adapt the well-founded semantics to Xcerpt programs in the Section 2.4.3.

2.4.3 Rules with Negation and Versatile Queries: Well-Founded Semantics

For the sake of simplicity this section only considers Xcerpt programs without the grouping constructs `all`. Moreover queries are assumed to be either simple query terms, negations of query terms or conjunctions of positive or negated query terms. In the absence of grouping constructs or aggregate functions, a rule involving a disjunction in the rule body can be rewritten into an equivalent set of rules that are disjunction free. Also negations of conjunctions can be rewritten to conjunctions with only positive or negative query terms as conjuncts.[30]

Definition 14 (Xcerpt literal). *An Xcerpt literal is either an Xcerpt data term or the negation not d of some Xcerpt data term d. For a set S of Xcerpt literals, pos(S) denotes the positive literals in S, neg(S) the negative ones.*

Definition 15 (Consistent sets of Xcerpt literals). *For a set of Xcerpt literals S we denote with $\neg \cdot S$ the set of terms obtained by negating each element in S. Let p and n =not d be a positive and negative literal, respectively, and let S be a set of literals. p and S are consistent, iff not p is not in S. n and S are consistent iff d is not in S. S is consistent, if it is consistent with each of its elements.*

---

[30] This normalization of Xcerpt rules is similar to finding the disjunctive normal form of logical formulae.

Definition 16 (Partial interpretation of an Xcerpt program (adapted from [147]))
Let P be an Xcerpt program, and HB(P) its Herbrand base. A partial interpretation I is a consistent subset of $HB(P) \cup \neg \cdot HB(P)$.

Definition 17 (Satisfaction of Xcerpt terms). *Let I be a partial interpretation for a program P. The model relationship between I and an Xcerpt term is defined as follows.*

- *Let q be a positive query term.*
    - *I satisfies q ($I \models q$) iff there is some data term $d \in pos(I)$ with $q \leq d$.*
    - *I falsifies q ($I \not\models q$) iff for all data terms $d \in HB_P$ holds $q \leq d \Rightarrow d \in neg(I)$.*
    - *Otherwise, q is undefined in I.*
- *Let q = not q' be a negative query term.*
    - *I satisfies q ($I \models q$) iff for all data terms d holds $q' \leq d \Rightarrow d \in neg(I)$.*
    - *I falsifies q ($I \not\models q$) iff there is some data term $d \in pos(I)$ with $q' \leq d$.*
    - *Otherwise, q is undefined in I.*

Definition 18 (Satisfaction of Xcerpt queries). *Let I be a partial interpretation and q a conjunction of Xcerpt terms. I satisfies q if I satisfies each conjunct in q.*[31]

Definition 19 (Xcerpt Unfounded Sets (adapted from [147])). *Let P be an Xcerpt program, HB_P its Herbrand base, and I a partial interpretation. We say $A \subseteq HB_P$ is an unfounded set of P with respect to I if each atom $p \in A$ satisfies the following condition. For each instantiated rule R of P with head p and body Q at least one of the following holds:*

1. *For some positive literal $q \in Q$ holds that for all $d \in HB_P$ holds $q \leq d \Rightarrow d \in A \vee d \in neg(I)$.*
2. *Some negative literal $q \in Q$ is satisfied in I.*

The greatest unfounded set of P with respect to an interpretation I is the union of all unfounded sets of P with respect to I.

Definition 20 (Well-founded semantics of an Xcerpt program). *The well-founded semantics of an Xcerpt program P is defined as the least fixpoint of the operator $W_P(I) := T_P(I) \cup \neg \cdot U_P(I)$ where U_P and I_P are defined as follows:*

- *a postive Xcerpt literal l is in $T_P(I)$ iff there is some ground instance R_g of some rule R in P with construct term l and query Q such that $I \models Q$.*
- *$U_P(I)$ is the greatest unfounded set of P with respect to I.*

Consider the program P in Listing 1.23. Its Herbrand base is $HB(P) = \{a\{ \}\}$. Starting with the empty interpretation I_0, $T_P(I_0) = \emptyset$, $U_P(I_0) = \emptyset$, and $I_1 := W_P(I_0) = \emptyset = I_0$. Thus the well-founded semantics of P is $\emptyset$.

---

[31] Xcerpt rules are assumed to be in disjunctive normal form. Therefore disjunctions need not be considered here. Satisfaction of negations is treated in Definition 17 above.

```
CONSTRUCT a{ } FROM not( a{{ }} ) END
```

Listing 1.23. Simple Negation through recursion and simulation (A)

```
1 CONSTRUCT a{ } FROM not( a{{ }} ), not( b{ } ) END
  CONSTRUCT a{ b } END
```

Listing 1.24. Simple Negation through recursion and simulation (B)

As a second example, consider program Q in Listing 1.24 with Herbrand base $HB(Q) = \{\,a\{\ b\ \},a\{\ \},b\{\ \}\}$. We obtain the following fix point calculation:

- $I_0 = \emptyset$
- $\mathbf{T}_Q(I_0) = \{\,a\{\ b\ \}\,\}$
- $\mathbf{U}_Q(I_0) = \{\,a\{\ \},b\{\ \}\,\}$
- $I_1 = \mathbf{W}_Q(I_0) = \{\,a\{\ b\ \},\text{not } a\{\ \},\text{not } b\{\ \}\,\}$
- $\mathbf{T}_Q(I_1) = \{\,a\{\ b\ \}\,\}$
- $\mathbf{U}_Q(I_1) = \{\,a\{\ \},b\{\ \}\,\}$
- $I_2 = \mathbf{W}_Q(I_1) = \{\,a\{\ b\ \},\text{not}(\ a\{\ \}\ ),\text{not}(\ b\{\ \}\ )\,\}\} = I_1$

As a final example, consider the stratified and locally stratified program R in Listing 1.25 with Herbrand universe $HB(R) = \{\,b\{\ \},a\{\ b\ \},a\{\ \},c\{\ c\ \}\,\}$.

```
  CONSTRUCT b{ } FROM not( a{{ }} ) END
2 CONSTRUCT a{ b } FROM not( c{{ }} ) END
  CONSTRUCT a{ } FROM not( c{{ }} ) END
4 CONSTRUCT c{ c } END
```

Listing 1.25. Simple Negation through recursion and simulation (C)

We obtain the following fixpoint calculation:

- $I_0 = \emptyset$
- $\mathbf{T}_R(I_0) = \{\,c\{\ c\ \}\,\}$
- $\mathbf{U}_R(I_0) = \emptyset$
- $I_1 = \mathbf{W}_R(I_0) = \{\,c\{\ c\ \}\,\}$
- $\mathbf{T}_R(I_1) = \{\,c\{\ c\ \}\,\}$
- $\mathbf{U}_R(I_1) = \{\,a\{\ \},a\{\ b\ \}\,\}$
- $I_2 = \mathbf{W}_R(I_1) = \{\,c\{\ c\ \},\text{not}(\ a\{\ \}\ ),\text{not}(\ a\{\ b\ \}\ )\,\}$
- $\mathbf{T}_R(I_2) = \{\,c\{\ c\ \},b\{\ \}\,\}$
- $\mathbf{U}_R(I_2) = \{\,a\{\ \},a\{\ b\ \}\,\}$
- $I_3 = \mathbf{W}_R(I_2) = \{\,c\{\ c\ \},b\{\ \},\text{not}(\ a\{\ \}\ ),\text{not}(\ a\{\ b\ \}\ )\,\}$
- $\mathbf{T}_R(I_3) = \mathbf{T}_R(I_2)$
- $\mathbf{U}_R(I_3) = \mathbf{U}_R(I_2)$
- $\mathbf{W}_R(I_3) = \mathbf{W}_R(I_2)$

It is immediate that the well-founded semantics of R coincides with the fixpoint calculated over the stratification of R – a fact that is true for every locally stratified Xcerpt program.

Theorem 1. *For a locally stratified Xcerpt program P, the well-founded semantics of P is total and coincides with the fixpoint calculated over the local stratification of P.*

In [129] the class of weakly stratified logic programs is introduced, which is a true superset of the class of locally stratified programs and has a well-defined, two-valued intended semantics. Put briefly, to decide whether a logic program is locally stratifiable one considers the dependency graph constructed from the *entire* Herbrand instantiation of the logic program. In contrast, the decision for weak stratification is based on the absence of negative cycles within the dependency graph constructed *from a subset* of the Herbrand interpretation. This subset excludes instantiated rules containing literals of extensional predicate symbols that are not given in the program. The standard example for a program that is weakly stratified but not locally stratified is the following:

$$win(X) : -move(X, Y) \land \neg win(Y)$$

A position X is a winning position of a game, if there is a move from X to position Y and Y is a losing position. As mentioned above, weak stratification depends on the extension of extensional predicate symbols (*move* in the above example), and the program above is only weakly stratifiable in the case that *move* has an acyclic extension. Obviously this program can be formulated also as an Xcerpt program, and the class of locally stratified Xcerpt programs could be extended to the class of weakly stratified Xcerpt programs in a straight-forward manner. We leave the formal definition of weak stratification for Xcerpt and the question on how the richer kind of unification employed in Xcerpt affects the applicability of weak stratification for future work.

2.4.4 A Relational Semantics for Versatile Queries

Versatile queries form the central innovation of XML and RDF query languages, as illustrated in the previous sections: They allow the query author to introduce controlled forms of incompleteness or "don't cares" such as "here don't care about the order" or "here don't care about the path between two nodes as long as there is one". They are controlled in that they have to be explicitly requested by the query user and in that they have a precise logical semantics (rather than being based on approximation or ranking as in Web search engines).

The *logical semantics of versatile queries* is the focus of the following section. Rather than directly assigning meaning to versatile Web queries using simulation (Section 2.4.1) and investigating the affects on the semantics of rule languages build upon such queries (Section 2.4.2 and Section 2.4.3), we show how to reduce versatile queries to standard first-order logic, more precisely to Datalog with negation value invention. This is an interesting and well understood fragment of first-order logic: though computationally as expressive as full first-order logic it provides more controlled means for the creation of new terms (or "complex values") and can be easily mapped to SQL which provides similarly constrained means for value creation.

Contributions. Casting the semantics of versatile queries in general and Xcerpt in particular in terms of Datalog allows us to compare and contrast them with previous database languages. In particular, we use this logical semantics of Xcerpt

1. to study the complexity and expressiveness of Xcerpt and several sub-languages of Xcerpt. In particular, we show that
   (a) Xcerpt expresses *all computable queries* modulo copy removal (Section 2.4.4);
   (b) the same applies already to *stratified Xcerpt* (Section 2.4.4);
   (c) weakly-recursive Xcerpt has the combined NEXPTIME-complete. Intuitively, a weakly-recursive Xcerpt program is an Xcerpt program that limits recursion to rules that do not increase either the nesting depth or the breadth. Thus we can rearrange the program to postpone value invention to the end of query evaluation and do not suffer complexity penalties for value invention (Section 2.4.4);
   (d) non-recursive Xcerpt on tree data has data complexity in NC_1 and program complexity PSPACE-complete (Section 2.4.4).
2. to implement versatile queries on top of *relational databases* by translating them into SQL (Section 2.4.4). For such a translation to be efficient, we also need a relational representation of versatile graph-shaped data that is both space efficient and provides efficient access to graph properties such as edge traversal or reachability. Such a representation (by means of a novel labeling scheme) with linear space and time complexity for evaluating acyclic Web queries on many graphs is provided by ClQcAG, see Section 2.5.1.
3. to provide a *common logical foundation* for versatile queries. This allows us, as shown in Section 2.4.4, to integrate different Web query languages such as XQuery, SPARQL, and Xcerpt and to evaluate them with the same query engine. This differs notably from other approaches for the integration of Web query languages where the evaluation of the integrated languages remains separate and enables cross-language optimization and planning. Yet, thanks to the novel graph representation with ClQcAG, we can evaluate each language as efficient as the best known approaches limited to that language.

Preliminaries

XML and RDF Data as Relational Structures. Following [16], we consider an **XML tree** as a relational structure: An XML tree is considered a relational structure T over the schema $((\text{Lab}^\lambda)_{\lambda \in \Sigma}, R_{\text{child}}, R_{\text{next-sibling}}, \text{Root})$. The nodes of this tree are labeled using the symbols from Σ which are queried using Lab^λ (note, that λ is a single label not a label set). The parent-child relations are represented by R_{child}. The order between siblings is represented by $R_{\text{next-sibling}}$. The root node of the tree is identified by Root. There are some additional derived relations, viz. $R_{\text{descendant}}$, the transitive, $R_{\text{descendant-or-self}}$ the transitive reflexive closure of R_{child}, $R_{\text{following-sibling}}$, the transitive closure of $R_{\text{next-sibling}}$, R_{self} relating each node to itself, and $R_{\text{following}}$ the composition of $R_{\text{descendant-or-self}}^{-1} \circ R_{\text{following-sibling}} \circ R_{\text{descendant-or-self}}$. Each node n is also related by R_{arity} to $|\{n' : R_{\text{child}}(n, n')\}|$. Finally, we can compare nodes based on their label using $\cong$ which contains all pairs of nodes with same label, based on their node identity using $=$ which relates each node only to itself, and based on their structure deep equality $=_{\text{deep}}$ which holds for two nodes if there exists an isomorphism between their respective sub-trees. The above ignores some XML specifics such as attributes, comments, or processing instructions but these can be added easily. For also allow an all-distinct$(n_1, \ldots, n_k)$ constraint as generalisation of $=$ from two nodes to k nodes.

For example, the XML document (using subscripts to indicate node identities)

```
<a>₁ <b/>₂ <c>₃<c/>₄</c> </a>
```

is represented as $T = (\mathsf{Lab}^a = \{1\}, \mathsf{Lab}^b = \{2\}, \mathsf{Lab}^c = \{3,4\}, R_{\mathsf{child}} = \{(1,2),(1,3),(3,4)\}$, $R_{\mathsf{next\text{-}sibling}} = \{(2,3)\}, \mathsf{Root} = \{1\})$ over the label alphabet $\{a,b,c\}$. All other relations can be derived from this definition.

In some contexts, a *graph view of XML* data is preferable as chosen in the description of Xcerpt in Section 2.3.1. This view does not affect the signature of the relational structure[32], but adds additional pairs of nodes to the extensions of R_{child} and $R_{\mathsf{next\text{-}sibling}}$ and all relations derived from them. Say we want to treat ID/IDREF links like child relations resulting from element nesting in the XML document. This adds additional pairs of referencing and referenced node to R_{child}.

In the following, we choose this graph view of XML unless explicitly stated otherwise. We also allow unions of such structures, i.e., graphs consisting in multiple connected components each with its own root node (graph view of "XML forests").

An **RDF graph** can be represented similarly as a relational structure. The main differences are the lack of order, the addition of edge labels, and the presence of node types such as literal, blank node, and resource: An RDF graph is considered a relational structure T over the schema $((\mathsf{Lab}^\lambda)_{\lambda \in \Sigma}, \circ\!\!\rightarrow, \rightarrow\!\!\circ, \mathsf{Edge}, \mathsf{Literal}, \mathsf{Blank}, \mathsf{Named})$. As in the case of XML, Lab^λ provides labels from $\Sigma = \mathsf{U} \cup \mathsf{L}$, but labels both nodes and edges. A label is either an URI or a literal. Nodes are typed by the three characteristic relations Edge, $\mathsf{Literal}$, Blank, and Named into edges, literals, blank nodes, and named resources. The four sets are pairwise disjoint. Following [126], we represent labeled edges as first class elements of the domain and provide separate relation for navigating from the *source* node of an edge to that edge $(\circ\!\!\rightarrow)$ and from that edge to its *sink* $(\rightarrow\!\!\circ)$ node. There are some additional derived relations, viz. $R_{\mathsf{child}} = \circ\!\!\rightarrow \circ \rightarrow\!\!\circ$, $R^\lambda_{\mathsf{child}} = \circ\!\!\rightarrow \circ \mathsf{Lab}^\lambda \circ \rightarrow\!\!\circ$ and $R^{(\lambda)}_{\mathsf{descendant}}$ the transitive closure of $R^{(\lambda)}_{\mathsf{child}}$. Each node n is also related by R_{arity} to $|\{e' : n\circ\!\!\rightarrow e')\}|$ and each edge e to $|\{n' : e\rightarrow\!\!\circ n')\}|$. Finally, we can compare nodes and edges with the same equality relations as in the XML case.

For example, the following RDF graph (using subscripts to indicate node or edge identities)

```
₁  @prefix ex: <http://example.org/libraries/#> .
   @prefix bib: <http://www.edutella.org/bibtex#> .
₃  ex:smith2005₁ ex:isPartOf₂ [₃ a₄ bib:Journal₅ ;
              bib:number₆ "11"₇; bib:name₈ "Computer Journal"₉ ] ;
```

is represented as $T = (\mathsf{Lab}^{ex:smith2005} = \{1\}, \mathsf{Lab}^{ex:isPartOf} = \{2\}, \mathsf{Lab}^{rdf:type} = \{4\}$, $\mathsf{Lab}^{bib:Journal} = \{5\}, \mathsf{Lab}^{bib:number} = \{6\}, \mathsf{Lab}^{11} = \{7\}, \ldots, \circ\!\!\rightarrow = \{(1,2),(3,4),(1,6),(1,8)\}$, $\rightarrow\!\!\circ = \{(2,3),(4,5),(6,7),(8,9)\}, \mathsf{Edge} = \{2,4,6,8\}, \mathsf{Literal} = \{7,9\}, \mathsf{Blank} = \{3\}, \mathsf{Named} = \{1,5\})$. All other relations can be derived from this definition.

Datalog with Value Invention. For investigating the formal properties of languages with versatile queries and for implementing them in a relational database, we use

---

[32] Though we might obviously also choose to provide both views of the XML data simultaneously by additional relations instead of modified extensions of the existing ones.

Datalog with negation and value invention (short $\text{Datalog}^{\neg}_{new}$) as a convenient, well-studied fragment of first-order logic [2]. $\text{Datalog}^{\neg}_{new}$ extends Datalog with negation and a means for creating new values.

Rule bodies are as in standard $\text{Datalog}^{\neg}$, though we also allow disjunction in rule bodies. Rule heads are extended with conjunction and a means for *value invention*. We use a value invention term $\text{new}(x_0, x_1, \ldots, x_n)$, i.e., a function that maps each binding tuple for the invention variables $x_0, \ldots, x_n$ to a unique new value. We will usually use some unique constant c domain for x_0 to distinguish different value invention terms. In this case, we write also $\text{new}_c(x_1, \ldots, x_n)$. It is easy to see that we can transform such value invention terms to the notation from [2]. In addition to the simple value invention, we also add a *deep copy* or clone facility. The deep clone term $\text{deep-copy}(x_0, x_1, \ldots, x_n)$ is also a function on the binding tuples of $x_0, \ldots, x_n$ that returns a unique new value t, but also adds t to all unary relations that contain x_n and a pair (t, t') to each binary relation containing a pair (x_n, x') where $t' = \text{deep-copy}(x_0, x_1, \ldots, x_{n-1}, x')$.[33]

For convenience, we allow *conditionals* in the head: some part of the head may depend on whether some variables are bound or not. A conditional rule has the form $h \wedge \texttt{if } X \texttt{ then } hc_1 \texttt{ else } hc_2 \longleftarrow b$ and can be rewritten to rules without conditional constructions as follows:

$$h \wedge hc_1 \longleftarrow b \wedge \texttt{bound(X).}$$
$$h \wedge hc_2 \longleftarrow b \wedge \texttt{not(bound(X)).}$$

Answer variables are variables that occur in the head outside of the condition of a conditional expression.

The usual safety restrictions for $\text{Datalog}^{\neg}$ apply to ensure that all rules are *range-restricted*, see [2]: For each negation, all answer variables must occur also in a positive expression in the rule body. For each disjunction, all nested expressions have the same answer variables. Finally, each answer variable must also occur in the body.

Adapting the notation of [89], we call an *invention atom* an atom containing new terms. The relation name of such an atom is called an *invention relation name*. A rule is a *non-invention rule*, if it contains no invention atom in the head, otherwise it is an *invention rule*.

Logical Semantics for Xcerpt. In Section 2.4.1, we give a semantics for versatile Xcerpt queries by using the notion of simulation. Though simulation provides us with an intuitive, concise notion for the semantics of Xcerpt queries, it is a non-standard notion specifically designed for Xcerpt. Here, we choose a different approach: a semantics based on $\text{Datalog}^{\neg}_{new}$, a well-understood and extensively investigated fragment of standard first-order logic.

To keep the presentation focus on the salient points of the translation, we will only consider a slightly simplified version of Xcerpt queries (a logical semantics for full Xcerpt is given in [71,69]). Specifically, we omit optional as well as sub-term negation (without) from query terms as they can be rewritten the queries with top-level negation (not), though potentially at exponential cost. We also omit construction of ordered terms, position, and label variables. Regular expressions to limited to ⋆ as label

---

[33] For cyclic graphs, each node is cloned only once using standard memoization.

wildcard (matching nodes with any label). For simplicity, we assume in the following that term identifiers and variables are disjoint.

Query Terms. To gently introduce the translation for Xcerpt, we start again with a few examples. In this section, we consider only query terms. Recall that Xcerpt query terms serve to select data from the input graph and to provide bindings for any contained variables. Intuitively, a query term can be seen like a pattern or example for the data to be selected. For details on Xcerpt query terms see Section 2.3.1. The translation of basic query terms is fairly straight-forward:

```
conference{{ desc paper{{ author{{}} }} }}
```

is translated to

$$\mathsf{Root}(v_1) \wedge \mathsf{Lab}^{\mathsf{conference}}(v_1) \wedge R_{\mathsf{descendant}}(v_1, v_2) \wedge \mathsf{Lab}^{\mathsf{paper}}(v_2) \wedge R_{\mathsf{child}}(v_2, v_3) \wedge \mathsf{Lab}^{\mathsf{author}}(v_3).$$

We ask for root nodes (bound to v_1) with label **conference** and their descendants (bound to v_2) with label **paper**. For these descendants we are also interested in **authors**.

The previous example contains only partial query terms with a single sub-term each. *Total* query terms are translated very similarly but with an additional constraint on the *arity* of the respective node. For instance, `paper{ author{{ }} }` (where paper is total rather than partial as above) is translated to

$$\mathsf{Lab}^{\mathsf{paper}}(v_2) \wedge R_{\mathsf{arity}}(v_2, 1) \wedge R_{\mathsf{child}}(v_2, v_3) \wedge \mathsf{Lab}^{\mathsf{author}}(v_3).$$

If we consider terms with more than one sub-term, we have to distinguish *ordered* and *unordered* terms. In an unordered term such as `paper{{ author{{ }}, title{{ }} }}` multiple sub-terms lead to node inequality constraints:

$$\mathsf{Lab}^{\mathsf{paper}}(v_2) \wedge R_{\mathsf{child}}(v_2, v_3) \wedge \mathsf{Lab}^{\mathsf{author}}(v_3) \wedge R_{\mathsf{child}}(v_2, v_4) \wedge \mathsf{Lab}^{\mathsf{title}}(v_4) \wedge v_3 \neq v_4.$$

In an ordered term such as `paper[[ author{{ }}, title{{ }} ]]` multiple sub-terms lead to order constraints:

$$\mathsf{Lab}^{\mathsf{paper}}(v_2) \wedge R_{\mathsf{child}}(v_2, v_3) \wedge \mathsf{Lab}^{\mathsf{author}}(v_3) \wedge R_{\mathsf{following\text{-}sibling}}(v_3, v_4) \wedge \mathsf{Lab}^{\mathsf{title}}(v_4).$$

Finally, Xcerpt allows multiple occurrences of query variables and requires that all occurrences are structurally (or deep) equal. For instance, the following Xcerpt term

```
conference{{ desc paper{{ var X → author }}, var X }}
```

is translated to

$$\mathsf{Root}(v_0) \wedge \mathsf{Lab}^{\mathsf{conference}}(v_0) \wedge R_{\mathsf{child}}(v_0, v_1) \wedge R_{\mathsf{descendant\text{-}or\text{-}self}}(v_1, v_2) \wedge$$
$$\mathsf{Lab}^{\mathsf{paper}}(v_2) \wedge R_{\mathsf{child}}(v_2, v_3) \wedge \mathsf{Lab}^{\mathsf{author}}(v_3) \wedge R_{\mathsf{child}}(v_1, v_4) \wedge v_1 \neq v_4$$
$$\wedge v_3 =_{\mathsf{deep}} v_4.$$

Notice also, how we split the $R_{\mathsf{descendant}}$ relation used above into R_{child} and $R_{\mathsf{descendant\text{-}or\text{-}self}}$ relations to allow for the inequality constraint amidst the children of conference. Though in this case, we can observe that bindings of v_1 and v_4 can never be the same, as bindings for v_1 must be labeled paper and bindings of v_4 (since it is deep equal to v_3) author. Therefore, we can simplify to

$$\text{Root}\,(v_1)\wedge\text{Lab}^{\text{conference}}\,(v_1)\wedge R_{\text{descendant}}\,(v_1,\ v_2)\wedge\text{Lab}^{\text{paper}}\,(v_2)\wedge R_{\text{child}}\,(v_2,$$
$$v_3)\wedge\text{Lab}^{\text{author}}\,(v_3)\wedge R_{\text{child}}\,(v_1,\ v_4)\wedge v_3 =_{\text{deep}} v_4 .$$

To provide an easier to grasp manner in which denote more complex $Datalog_{new}^{\neg}$ expressions we introduce a graphical notation for queries (that also needed later to define structural properties queries). The two last $Datalog_{new}^{\neg}$ expressions are shown in Figure 6.

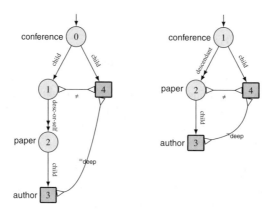

Fig. 6. Translation of Xcerpt variables with $=_{\text{deep}}$

This representation of queries as graphs is used throughout this section and Section 2.5.1: Query variables are represented as nodes with labels. Root constraints are denoted by an incoming arrow. Two nodes are connected if there is an atom involving the two variables. The edge is labeled with the respective relation name. Answer variables are marked by darker rectangles whereas normal variables are indicated by lighter circles.

Formally, we define the translation of Xcerpt query terms to $Datalog_{new}^{\neg}$ expressions by means of the tq_{term} function shown in Table 7. Xcerpt contains two context-sensitive features: multiple occurrences of Xcerpt variables as well as references (defined using @ and referenced using ^). Occurrences of Xcerpt variables and references are managed in a *environment* $\mathcal{E}$ that contains always the Datalog variable associated with the last occurrence of an Xcerpt variable or reference (if there is any). With $\mathcal{E}[X \leftarrow v]$ we denote the assignment (possibly overwriting existing values) of X to v in $\mathcal{E}$. Otherwise, the translation function is defined by structural recursion over the Xcerpt query term grammar. It returns for each Xcerpt term the $Datalog_{new}^{\neg}$ expression corresponding to the given term as well as the modified environment and the top-level $Datalog_{new}^{\neg}$ variable. The top-level variable is needed to express the different semantics of ordered versus unordered term lists.

Table 7. Translating Xcerpt query terms

query term	$\text{Datalog}^{\neg}_{new}$ expression
$\text{tq}_{term}(\mathcal{E})\langle\lambda\{\{t_1,\ldots,t_n\}\}\rangle$	$= (\mathcal{E}_n, v', \text{Lab}^\lambda(v') \wedge F_1 \wedge \ldots \wedge F_n \wedge R_{\text{child}}(v',v_1) \wedge \ldots \wedge$ $R_{\text{child}}(v',v_n) \wedge \text{all-distinct}(v_1,\ldots,v_n))$ **where** v' new variable $(\mathcal{E}, v, F) = \text{tq}_{term}(\mathcal{E}, v')\langle t_1\rangle$ $\vdots$ $(\mathcal{E}_n, v_n, F_n) = \text{tq}_{term}(\mathcal{E}_{n-1}, v')\langle t_n\rangle$
$\text{tq}_{term}(\mathcal{E})\langle\lambda\{t_1,\ldots,t_n\}\rangle$	$= (\mathcal{E}', v', F' \wedge R_{\text{arity}}(v',n))$ **where** $(env', v', F') = \text{tq}_{term}(\mathcal{E}, v)\langle\lambda\{\{t_1,\ldots,t_n\}\}\rangle$
$\text{tq}_{term}(\mathcal{E})\langle\lambda[[t_1,\ldots,t_n]]\rangle$	$= (\mathcal{E}_n, v', \text{Lab}^\lambda(v') \wedge F_1 \wedge \ldots \wedge F_n \wedge R_{\text{child}}(v',v_1) \wedge$ $R_{\text{following-sibling}}(v_1,v_2) \wedge \ldots \wedge R_{\text{following-sibling}}(v_{n-1},v_n))$ **where** v' new variable $(\mathcal{E}_1, v_1, F_1) = \text{tq}_{term}(\mathcal{E}, v')\langle t_1\rangle$ $\vdots$ $(\mathcal{E}_n, v_n, F_n) = \text{tq}_{term}(\mathcal{E}_{n-1}, v')\langle t_n\rangle$
$\text{tq}_{term}(\mathcal{E})\langle\lambda[t_1,\ldots,t_n]\rangle$	$= (\mathcal{E}', v', F' \wedge R_{\text{arity}}(v',n))$ **where** $(env', v', F') = \text{tq}_{term}(\mathcal{E}, v)\langle\lambda[[t_1,\ldots,t_n]]\rangle$
$\text{tq}_{term}(\mathcal{E})\langle\text{var } X \to t\rangle$	$= (\mathcal{E}'[X \leftarrow v'], v', Q \wedge \begin{cases} (v' =_{\text{deep}} v'') & \text{if } (X,v'') \in \mathcal{E} \\ \top & \text{else} \end{cases})$ **where** $(\mathcal{E}', v', Q) = \text{tq}_{term}(\mathcal{E}, v)\langle t\rangle$
$\text{tq}_{term}(\mathcal{E})\langle\text{var } X\rangle$	$= (\mathcal{E}'[X \leftarrow v'], v', \begin{cases} (v' =_{\text{deep}} v'') & \text{if } (X,v'') \in \mathcal{E} \\ \top & \text{else} \end{cases})$ **where** v' is a new variable
$\text{tq}_{term}(\mathcal{E})\langle tid \,@\, t\rangle$	$= (\mathcal{E}'[tid \leftarrow v'], v', Q \wedge \begin{cases} (v' = v'') & \text{if } (tid,v'') \in \mathcal{E} \\ \top & \text{else} \end{cases})$ **where** $(\mathcal{E}', v', Q) = \text{tq}_{term}(\mathcal{E}, v)\langle t\rangle$
$\text{tq}_{term}(\mathcal{E})\langle \hat{\ } tid\rangle$	$= (\mathcal{E}'[tid \leftarrow v'], v', \begin{cases} (v' = v'') & \text{if } (tid,v'') \in \mathcal{E} \\ \top & \text{else} \end{cases})$ **where** v' is a new variable
$\text{tq}_{term}(\mathcal{E})\langle\text{desc } t\rangle$	$= (\mathcal{E}', v_1, R_{\text{desc-or-self}}(v_1,v_2) \wedge Q)$ **where** v_1 is a new variable $(\mathcal{E}', v_2, Q) = \text{tq}_{term}(\mathcal{E}, v_1)\langle t\rangle$
$\text{tq}_{term}(\mathcal{E})\langle\text{"}string\text{"}\rangle$	$= (\mathcal{E}, v', \text{Lab}^{string}(v') \wedge R_{\text{arity}}(v',0))$ **where** v' is a new variable

The translation function tq_{term} is given in three parts, the first showing the translation of terms with sub-term specification, the second the translation of variables and references, and the third the remaining base cases. A term with sub-term specification is translated by assigning a new Datalog variable v', adding atoms for any label restriction, and translating all its sub-terms. The top-level variables returned by the translation of its sub-terms are collected and associated with v': If it is an ordered term, the top-level variable of the first child is connected to v' with R_{child}, the remaining chained with successive $R_{\text{following-sibling}}$ relations (which imply that they are also children of v'). If it is an unordered term, all top-level variables are connected to v' using R_{child} and an all-distinct constraint between all top-level variables is added.

Variables and references are translated roughly in the same way: If the environment already contains a Datalog variable for the Xcerpt variable or reference, an equality constraint between the two variables is added. The environment is updated in any case (thus only linear many equality constraints are created). Variables and references differ in the choice of the equality: Variables result in a structural or deep equality constraints ($=_{deep}$), references in node equality constraints ($=$). If we also add label variables (that are omitted here for conciseness), also label equality constraints ($\cong$) are generated, see [69].

To keep the translation concise, the resulting $Datalog^\neg_{new}$ expressions are not always minimal. For instance, we add an atom R_{child} followed by a $R_{desc-or-self}$ atom even when there is only a single sub-term (prefixed with desc). However, it is easy to remove these redundancies, in particular to remove all occurrences of $\top$, Lab^*, and to compact relations where possible.

Construct Terms. Construct terms serve in Xcerpt to reassemble new graphs given variable bindings obtained in related query terms. As above, we start with a few examples illustrating the translation of Xcerpt construct terms. The following assumes that we have obtained an environment $\mathcal{E}$ from the associated query term containing the mappings (X, v_1) and (Y, v_2), i.e., the representative $Datalog^\neg_{new}$ variable for the Xcerpt variable X (Y) is v_1 (v_2). We also abbreviate $new_i(v_1, \ldots, v_n)$ with $i(v_1, \ldots, v_n)$.

Again translating basic construct terms is fairly straight-forward:

```
authors{ author{ var X }, paper{ var Y, "best" } }
```

is translated to the following (conjunctive) $Datalog^\neg_{new}$ rule head:

$Root(1()) \wedge Lab^{authors}(1()) \wedge R_{child}(1(), 2()) \wedge Lab^{author}(1()) \wedge R_{child}(2(),$ deep-copy$_3(v_1)) \wedge R_{child}(1(), 4()) \wedge Lab^{paper}(4()) \wedge R_{child}(4(),$ deep-copy $_5(v_2)) \wedge R_{child}(4(), 6()) \wedge Lab^{best}(6())$

Graphically we denote heads of $Datalog^\neg_{new}$ rules similarly as their bodies (but in blue hues rather than red ones). Use of query variables for copying and grouping is indicated by dotted resp. dashed arrows. Figure 7 shows the graphical representation of the above rule head. The rule head specifies that there is a new value to be added to the Root relation. That same value $(1())$ is labeled authors and stands in R_{child} relation to two other new values. One of those is labeled author and contains a single child, the deep copy of the query variable v_1. The other is labeled paper and contains two children, one the deep copy of the query variable v_2, the other a new value labeled best.

In the translation, we only give the immediate binary relations R_{child} (and $R_{following-sibling}$ if considering ordered construct terms). Derived relations can be either automatically added to each rule head or be derived by additional rules, see [69].

Beyond what is shown in the first example, the main additional feature of construct terms is the possible presence of grouping expressed using all. The following is a simple example of such a construct term, where all bindings of X are listed (rather than only one as above), each wrapped in an author element which are all inside the same authors element:

```
authors{ all author{ var X } group-by(var X) }
```

This is translated very similarly as above, but now value invention (and deep copy) terms depend on query variables. More specifically, each value invention term depends on the grouping variables in whose scope the corresponding construct term occurs:

1   $\mathsf{Root}\,(1()) \wedge \mathsf{Lab}^{\mathsf{authors}}\,(1()) \wedge R_{\mathsf{child}}\,(1(),\ 2(v_1)) \wedge \mathsf{Lab}^{\mathsf{author}}\,(1(v_1)) \wedge R_{\mathsf{child}}$
     $(2(v_1),\ \mathsf{deep\text{-}copy}_3\,(v_1,\ v_1)\,)$

Obviously, with nested groupings this becomes more involved as in the following, final example: Here we create one pair of author and paper elements for each unique binding of X. Within paper we group all bindings of Y for the current binding of X:

```
1   authors{ all( author{ var X }, paper{
                          all var Y group-by (Y), "best" }
3             ) group-by (X) }
```

The translation makes the dependence of the terms inside the second grouping on Y *and* X explicit:

1   $\mathsf{Root}\,(1()) \wedge \mathsf{Lab}^{\mathsf{authors}}\,(1()) \wedge R_{\mathsf{child}}\,(1(),\ 2(v_1)) \wedge \mathsf{Lab}^{\mathsf{author}}\,(1(v_1)) \wedge R_{\mathsf{child}}$
     $(2(v_1),\ \mathsf{deep\text{-}copy}_3\,(v_1,\ v_1)\,) \wedge R_{\mathsf{child}}\,(1(),\ 4(v_1)) \wedge \mathsf{Lab}^{\mathsf{paper}}\,(4(v_1)) \wedge$
     $R_{\mathsf{child}}\,(4(v_1),\ \mathsf{deep\text{-}copy}_5\,(v_1,\ v_2,\ v_2)\,) \wedge R_{\mathsf{child}}\,(4(v_1),\ 6(v_1)) \wedge \mathsf{Lab}^{\mathsf{best}}$
     $(6(v_1))$

Formally, we define the translation from Xcerpt construct terms to $\mathsf{Datalog}^{\neg}_{\mathsf{new}}$ by means of the function $\mathsf{tc}_{\mathit{term}}$ shown in Table 8. As for query terms, we use an environment $\mathcal{E}$ to store associations between Xcerpt query variables or references and Datalog variables. Additionally, $\mathcal{E}$ also holds the current sequence of grouping variables, which is initially empty. $\mathsf{tc}_{\mathit{term}}$ returns, similar to $\mathsf{tq}_{\mathit{term}}$, the updated environment, the top-level construct variable, and the $\mathsf{Datalog}^{\neg}_{\mathsf{new}}$ (conjunctive) head formula. Again, the definition is divided in three part. The first case describes the semantics of unordered terms (ordered terms are omitted here) and empty terms. The second part that of variables and references and the final third part that of grouping terms. Grouping terms are responsible for modifying the initially empty sequence of iteration variables $\mathcal{E}.\mathsf{iter}$: For its

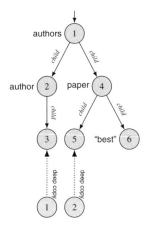

Fig. 7. Translation of Xcerpt construct term

Table 8. Translating Xcerpt construct terms

construct term	Datalog$^-_{new}$ expression
$\mathsf{tc}_{term}(\mathcal{E})\langle tid\,@\,\lambda\{t_1,\ldots,t_n\}\rangle$	$= (\mathcal{E}_n, v, \mathsf{Lab}^\lambda(v) \wedge R_{\mathsf{child}}(v,v_1) \wedge C_1 \wedge \ldots \wedge R_{\mathsf{child}}(v,v_n) \wedge C_n)$
	$\mathbf{where}\ v = \begin{cases} v' & \text{if } (tid,v') \in \mathcal{E} \\ \mathsf{id}(\mathcal{E}.\mathsf{iter}) & \text{with id new identifier} \end{cases}$
	$(\mathcal{E}_i, C_i, n_i) = \mathsf{tc}_{term}(\mathcal{E}_{i-1})\langle t_i\rangle$ with $\mathcal{E}_0 = \mathcal{E}[tid \leftarrow v]$
$\mathsf{tc}_{term}(\mathcal{E})\langle\text{"}\,string\,\text{"}\rangle$	$= (\mathcal{E}, v, \mathsf{Lab}^{string}(v))$
	$\mathbf{where}\ v = \mathsf{id}(\mathcal{E}.\mathsf{iter})$ with id new identifier
$\mathsf{tc}_{term}(\mathcal{E})\langle\texttt{var}\ X\rangle$	$= (\mathcal{E}, \mathsf{deep\text{-}copy}(\mathcal{E}.\mathsf{iter}, \mathcal{E}(X)), \top)$
$\mathsf{tc}_{term}(\mathcal{E})\langle\widehat{}\ tid\rangle$	$= (\mathcal{E}[tid \leftarrow v], \begin{cases} v' & \text{if } (tid,v') \in \mathcal{E} \\ \mathsf{id}(\mathcal{E}.\mathsf{iter}) & \text{with id new identifier} \end{cases}, \top)$
$\mathsf{tc}_{term}(\mathcal{E})\langle\texttt{all}\ t$ $\qquad\texttt{group-by}\,(X_1,\ldots,X_n)\rangle$	$= \mathsf{tc}_{term}(\mathcal{E}')\langle t\rangle$ $\mathbf{where}\ \mathcal{E}' = \mathcal{E}$ with $\mathcal{E}'.\mathsf{iter} = \mathcal{E}.\mathsf{iter} \circ [\mathcal{E}(X_1),\ldots,\mathcal{E}(X_n)]$

Table 9. Translating Xcerpt rules

Xcerpt	Datalog$^-_{new}$ expression
$\mathsf{tr}_{\mathsf{Xcerpt}}\langle\texttt{CONSTRUCT}\ head$ $\qquad\qquad\texttt{FROM}\ body\ \texttt{END}\rangle$	$= C \longleftarrow Q\ \mathbf{where}\ (\mathcal{E}, Q) = \mathsf{tq}(\emptyset)\langle query\rangle$ $\qquad\qquad C = \mathsf{tc}(\mathcal{E})\langle body\rangle$
$\mathsf{tc}(\mathcal{E})\langle cterm\rangle$	$= \mathsf{root}(v) \wedge C\ \mathbf{where}\ (\mathcal{E}', v, C) = \mathsf{tc}_{term}(\mathcal{E})\langle cterm\rangle$
$\mathsf{tq}(\mathcal{E})\langle\texttt{and}\,(t_1,t_2)\rangle$	$= (\mathcal{E}_2, (Q_1 \wedge Q_2))\ \mathbf{where}\ (\mathcal{E}_1, Q) = \mathsf{tq}(\mathcal{E})\langle t_1\rangle,\ (\mathcal{E}_2, Q) = \mathsf{tq}(\mathcal{E}_1)\langle t_2\rangle$
$\mathsf{tq}(\mathcal{E})\langle\texttt{or}\,(t_1,t_2)\rangle$	$= (\mathcal{E}', ((Q_1 \wedge v_X = v_1) \vee (Q_2 \wedge v_X = v_2)))$
	$\mathbf{where}\ (\mathcal{E}_1, Q) = \mathsf{tq}(\mathcal{E})\langle t_1\rangle,\ (\mathcal{E}_2, Q) = \mathsf{tq}(\mathcal{E})\langle t_2\rangle$
	$\qquad \mathcal{E}' = \mathcal{E}_2[X \leftarrow v_X]$ for all X with $(X, v_1) \in \mathcal{E}_1, (X, v_2) \in \mathcal{E}_2$
$\mathsf{tq}(\mathcal{E})\langle\texttt{not}\,(t)\rangle$	$= (\mathcal{E}', \neg(Q))\ \mathbf{where}\ (\mathcal{E}', Q) = \mathsf{tq}(\mathcal{E})\langle t\rangle$
$\mathsf{tq}(\mathcal{E})\langle qterm\rangle$	$= (\mathcal{E}', \mathsf{root}(r) \wedge Q)$
	$\mathbf{where}\ (\mathcal{E}', r, Q) = \mathsf{tq}_{term}(\mathcal{E})\langle qterm\rangle$

sub-terms the input $\mathcal{E}.\mathsf{iter}$ is extended by its grouping variables $X_1,\ldots,X_n$. Thus value invention (and deep copy) terms inside that grouping term depend also on $X_1,\ldots,X_n$.

Grouping in Xcerpt is always modulo structural or deep equivalence, i.e., all node invention and deep copy terms produce a new value only for each equivalence class modulo deep equal over the binding tuples. In other words, if there are two binding tuples where the bindings for all grouping variables are deep equal, we only produce a single new value.

Queries and Rules. Based on the logical semantics for construct and query terms established in the previous sections, we can finally conclude the semantics by considering full Xcerpt rules. Rules are translated using $\mathsf{tr}_{\mathsf{Xcerpt}}$ as shown in Table 9. It delegates the translation of rule bodies and heads to different functions which each create root atoms where necessary.

An Xcerpt rule body is translated using tq which takes care of all top-level disjunction, conjunction, or negations. Note, that for conjunctions we propagate the environment returned by the translation of the first operand to the translation of the second operand, thus ensuring that matches are deep equal (as within the same query term). For disjuncts, however, we do not propagate variable mappings, but rename answer variables (i.e., variables that occur in both disjuncts) to gather bindings from both disjuncts into one variable (v_X for answer variables X). This assumes that, as usual, that non-answer variables are standardized apart for each disjunct.

The following proposition is an immediate consequence of the above construction: Variables can occur negatively only in parts of the query resulting from a negated query term where the same safety restrictions apply as for $\text{Datalog}^{\neg}_{new}$.[34]

Proposition 1. *Let R be a range-restricted Xcerpt rule. Then $tr_{Xcerpt}(R)$ is a safe $Datalog^{\neg}_{new}$ rule.*

It is easy to verify that in each step of the above translations the resulting $\text{Datalog}^{\neg}_{new}$ expression is linear in the input query term. Furthermore, each case treats one or more input constructs. Therefore we can surmise:

Theorem 2. *The size of the $Datalog^{\neg}_{new}$ expression Q returned by tr_{Xcerpt} for a given Xcerpt rule P is linear in the size of P.*

To complete the semantics we also need to consider Xcerpt goals. Goals are treated the same as normal rules, but root nodes of goals are constructed in the relation answer-root rather than in Root. This also prevents the result of goals to partake in the rule chaining (observe that rule bodies only match data starting with a node in Root).

Definition 21 (Logical Semantics of Xcerpt). *Let $P = \{R_1, \ldots, R_n\}$ be an Xcerpt program. Then $P_d = tr_{Xcerpt}(R_1) \cup \ldots \cup tr_{Xcerpt}(R_n)$ is a safe $Datalog^{\neg}_{new}$ program. The logical semantics of P is the relational structure obtained by removing the Root relation and all references to nodes not reachable from a node in answer-root from the semantics of P_d (as defined in [2]).*

Example of the Full Translation. To conclude the discussion of the logical semantics for Xcerpt, we give a final example of the semantics. The following Xcerpt goal selects papers containing "Cicero" as author and "puts them in a shelf".

```
1 GOAL
    shelf{ all var X group-by(var X) }
3 FROM
    conference{{
5     var X → paper{{
        desc author{{ "Cicero" }} }} }}
7 END
```

Applying tr_{Xcerpt} to that rule yields the following $\text{Datalog}^{\neg}_{new}$ program, also depicted in Figure 8:

---

[34] We use inequalities outside of the translation of negated query terms, but only in a safe manner, see Table 7.

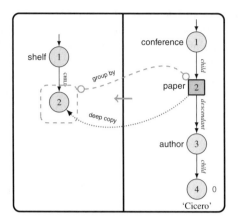

Fig. 8. Example of rule translation

Root $(1())$ $\wedge$ Lab^{shelf} $(1())$ $\wedge$ R_{child} $(1()$, deep-copy $(v_2, v_2))$
$\longleftarrow$ Root (v_1) $\wedge$ Lab^{conference} (v_1) $\wedge$ R_{child} (v_1, v_2) $\wedge$ Lab^{paper} (v_2) $\wedge$ $R_{\text{descendant}}$
(v_2, v_3) $\wedge$ Lab^{author} (v_3) $\wedge$ R_{child} (v_3, v_4), Lab^{Cicero} (v_4) $\wedge$ R_{arity} $(v_4, 0)$.

The query variable v_2 is used in the head to specify which part of the data to copy and how often. Recall that deep-copy(v_2, v_2) indicates that, for each unique binding of v_2, a new node should be created that is a deep copy of v_2 itself.

Outlook: Xcerpt^{RDF}. The above treatment of Xcerpt is focused on Xcerpt^{XML}. Though extending the translation to Xcerpt^{RDF} is not difficult, it requires a number of adjustments that are briefly summarized in the following.

- Most importantly, the above translation considers only XML data. If we also want to query *RDF data* we have to extend the translation rules to the specifics of that data model. Section 2.4.4 outlines how to represent RDF data in a relational structure that can be queried using Datalog$^{\neg}_{\text{new}}$.
- RDF and Xcerpt^{RDF} distinguish different node types such as blank nodes, named resources, and literals and contain named edges. All these features require slight adaptations to the translation. To give a flavor of these adaptations consider the following Xcerpt^{RDF} query term:

```
var X {{ foaf:knows → _:1{{ foaf:name → "Julius
    Caesar"}} }}
```

It queries for persons that know someone who is named "Julius Caesar". Its translation uses $\circ\!\!\rightarrow$ and $\rightarrow\!\!\circ$ for named edge traversal (rather than R_{child} as for unnamed edge traversal in XML) and requires each Datalog$^{\neg}_{\text{new}}$ to be a specific kind of RDF node or edge.

Named (n_1) $\wedge$ Lab$^{\text{ex:anna}}$ (n_1) $\wedge$ $\circ\!\!\rightarrow$ (n_1, e_1) $\wedge$ Edge (e_1) $\wedge$ Lab$^{\text{foaf:knows}}$ (e_1) $\wedge$
$\rightarrow\!\!\circ$ (e_1, n_2) $\wedge$ Blank (n_2) $\wedge$ $\circ\!\!\rightarrow$ (n_2, e_2) $\wedge$ Edge (e_2) $\wedge$ Lab$^{\text{foaf:name}}$
(e_2) $\wedge$ $\rightarrow\!\!\circ$ (e_2, n_3) $\wedge$ Literal (n_3) $\wedge$ Lab$^{\text{Julius Caesar}}$ (n_3)

– Xcerpt^{RDF} contains not just term (there called "graph") variables as discussed in the previous sections, but a number of additional variable kinds. Node and predicate (label) variables can easily be added to the above semantics. Essentially they are treated the same as label variables in [69]. More challenging are CBD-variables that select *concise bounded descriptions* of a matching node. A concise bounded description is similar to a term variable in that it binds to a structure (rather than just to a single label as label variables). But that structure may be only an excerpt of the actual sub-graph rooted at a matching node: It includes only all paths up to and including the first named resource on that path. CBD-variables can be added to the translation *without any changes to the target language Datalog$_{new}^{\neg}$* as they can be expressed through recursive Datalog$_{new}^{\neg}$ rules. However, such a realisation is likely to be inefficient. Therefore adding a specific operator for these variables is preferable.

In the following, we will continue considering only Xcerpt^{XML}, except for Section 2.4.4 where we briefly revisit the integration of Xcerpt^{RDF} and Xcerpt^{XML}. However, it is easy to check that all the results below transfer also to Xcerpt^{RDF} as all of the added or changed features can be expressed in Datalog$_{new}^{\neg}$ over relational structures representing RDF graphs. Also the features can be translated to Datalog$_{new}^{\neg}$ expressions in linear time and space, as in the case of Xcerpt^{XML}.

This concludes the definition of the logical semantics of Xcerpt by translation to Datalog$_{new}^{\neg}$. The following sections exploit this semantics to prove complexity and expressiveness properties of Xcerpt and several sub-languages of Xcerpt (Section 2.4.4) and to implement Xcerpt on top of relational database (Section 2.4.4).

Expressiveness and Complexity of Xcerpt. From the previous section, we obtain a linear translation of Xcerpt programs to Datalog$_{new}^{\neg}$ programs. Here we show how to use that translation to adapt or extend a number of existing results on expressiveness and complexity of Datalog$_{new}^{\neg}$ to Xcerpt and some interesting sub-languages of Xcerpt.

Xcerpt: Query Complete. First, let us consider full Xcerpt. The above translation establishes that we can find a Datalog$_{new}^{\neg}$ program to compute the semantics of any Xcerpt program and that this translation is linear. What about the other direction? It turns out, that we can encode each Datalog$_{new}^{\neg}$ program in an equivalent Xcerpt program of at most quadratic size:

Theorem 3. *Xcerpt has the same expressiveness, complexity, and completeness properties as Datalog$_{new}^{\neg}$ (and thus ILOG [89]).*

Proof. By the translation above, we can give a Datalog$_{new}^{\neg}$ program for each Xcerpt program.

On the other hand, each Datalog$_{new}^{\neg}$ program P can be encoded as an Xcerpt program P' preserving the semantics of P in the following way:

Each atom in the body is represented as an ordered, total Xcerpt query term with the predicate symbol as term label, replacing Datalog$_{new}^{\neg}$ variables by Xcerpt variables and Datalog$_{new}^{\neg}$ constants c by "c". The head atom is represented as an ordered, total Xcerpt construct term, replacing Datalog$_{new}^{\neg}$ variables by Xcerpt variables and

Datalog$^\neg_{new}$ constants c by $"c"$. An invention symbol in the head is replaced by the Xcerpt term new$[t_1,\ldots,t_n]$ where $t_1,\ldots t_n$ are the non-invention variables or constants in that head and new is a unique symbol not otherwise used in the program or data. Thus a resulting term simulates only with other instances of the same head by virtue of the unique term label new. Essentially we generate a new term for each unique binding tuple of $t_1,\ldots,t_n$ (modulo deep equal).

It is easy to see that if $P \models p(t_1,\ldots,t_n)$ then there is an isomorphism κ from Xcerpt terms with new labels to invention constants such that $p[t'_1,\ldots,t'_n]$ can be derived with the rules in P', $t'_i = t_i$ if t_i is a normal constant, and $t'_i = \kappa(t_i)$ otherwise.

This translation is at worst quadratic in the size of the Datalog$^\neg_{new}$ program: Invention symbols may lead to duplication of variable occurrences, but since only *non-invention* variables and constants are ever included this duplication does not lead to exponential size.

For the following corollary we exploit several results on ILOG [89], a syntactic variant of Datalog$^\neg_{new}$. First, we call two answers equivalent up to "copy removal" if they differ only in invented values and those invented values are deep or structurally equivalent. Second we recall the class of (list) constructive queries from [42] which are designed to capture precisely the queries expressible in languages such as Datalog$^\neg_{new}$, ILOG, or Xcerpt. It coincides with the class of queries where the new domain elements in the output can be viewed as hereditary finite lists constructed over the domain elements of the input. Hereditary finite lists are lists constructed over a given set U of "ur-elements" from the input domain such that each element of the list is either from U or a hereditary finite list over U. With this definitions and respective results on ILOG from [89] and [43] we obtain that

Corollary 1. *1. Xcerpt is Turing complete. 2. Xcerpt is query complete modulo copy removal, i.e., it expresses all computable queries modulo copy removal. 3. Xcerpt is (list) constructive complete.*

The reason for the limitation to "modulo copy removal" is that Xcerpt uses deep or structural equivalence as equivalence relation for grouping and can not distinguish between two terms that are deep equal.[35]

This result shows that while Xcerpt is indeed Turing complete that expressive power is justified as it can express all computable queries modulo deep equal. Since the whole language is Turing complete, it is worth investigating sub-languages with better computational properties. Before we turn to that question, let us briefly consider the effect of stratification on expressiveness and complexity:

Stratified Xcerpt: Still Query Complete. In the translation above as well as most parts of Section 2.3 we only consider programs with a limited form of negation, viz. stratified negation.

---

[35] This issue is closely related to the issue of lean vs. non-lean RDF graphs as answers in languages such as SPARQL or RDFLog [38]: That Xcerpt is complete "modulo copy removal" means that it can not create answers (or groupings) with several instances of the same, structurally equivalent graph, i.e., it can only produce the term equivalent of lean answers.

Recall the definition of stratified Xcerpt from Section 2.4.2, here recast using the dependency graph on Xcerpt rules, as we make use of that notation also for defining some sub-languages of Xcerpt below. Given an Xcerpt rule R we define its top-level query terms as usual: A query terms in R is called *top-level* if it occurs inside the body of R nested only inside (arbitrary combinations) of `and`, `or`, and `not`.

Definition 22 (Dependency graph for Xcerpt programs). *Let $P = R_1, \ldots, R_n$ be an Xcerpt program. Then the dependency graph $D(P) = (N, E)$ for P is defined as follows:*

- *The nodes of $D(P)$ are the rules of P.*
- *There is an edge from R_i to R_j in $D(P)$ iff one of the top-level query terms q of R_i simulates with the construct term of R_j. The edge is negative if q occurs inside a `not` in R_i, otherwise it is positive.*

Using the notion of dependency graph, we can define stratified Xcerpt programs as follows:

Definition 23 (Stratified Xcerpt). *Let $P = R_1, \ldots, R_n$ be an Xcerpt program. Then P is called* stratified, *if there is a partitioning of P into strata $S_1, \ldots, S_k$ such that there is no negative edge from an $R \in S_i$ to an $R' \in S_j$ with $i < j$.*

Proposition 2. *Let P be a stratified Datalog$_{new}^{\neg}$ program. Then the Xcerpt encoding of P by the proof of Theorem 3 is a stratified (and therefore locally stratified) Xcerpt program.*

Proof. A stratification of P immediately gives us a stratification of its Xcerpt encoding P' as any negated query term t in P' yields from a negated atom in P and the corresponding rule can all construct terms in lower strata have top-level labels that are different from the top-level label of t and thus do not unify. Otherwise the predicate that construct term is the encoding of depends on the negated atom already in P, yet is in a lower stratum in contrast to the assumption that P is stratified.

From this result and [43] which shows that stratified Datalog$_{new}^{\neg}$ has the same expressive power as full Datalog$_{new}^{\neg}$ and thus can express all computable queries modulo copy removal we can deduce the same observation for Xcerpt:

Corollary 2. *Stratified Xcerpt already expresses all computable queries modulo copy removal.*

In other words, the class of queries expressible in Xcerpt does not shrink if we constrain ourselves to stratified programs. This contrast to the case of Datalog$^{\neg}$ without value invention where stratification is indeed a limitation on the kind of queries expressible in the language.

Weakly-Recursive Xcerpt: Finite Models. A first decidable sub-language of Xcerpt is inspired by the notion of weakly acyclic Datalog$_{new}^{\neg}$ [89] that is also used extensively, e.g., in the data exchange setting. Essentially combining recursion and value invention is dangerous as we can no longer give a bound on the number of ground terms entailed by a program (in other words the active domain is no longer finite). The notion of weak acyclicity is a sufficient condition to guarantee finite active domains: We allow recursion but only if no new values can be created on by the recursive rules. In terms of the dependency graph of the program: we allow cycles in the dependency graph as long as they are not through invention atoms.

Directly applying this notion to Xcerpt is unsatisfying as every Xcerpt rule generates new nodes. However, since we generate new nodes modulo deep equality, we only need to ensure that the number of different terms a program can generate is finite.

Given two terms t and t'. We define the *nesting depth* of t in t' as usual: If $t = t'$ then the nesting depth is 0. Otherwise, if t' is nested inside a term t'' in t with nesting depth d then t' has nesting depth $d + 1$. If t' occurs several times in t then its nesting depth is the minimum of the nesting depths of its occurrences.

Definition 24 (Weakly-recursive Xcerpt). *Let $P = R_1, \ldots, R_n$ be an Xcerpt program. Then P is called* weakly-recursive, *if for each edge (R_i, R_j) in $D(P)$ the following holds:*

- *The construct term c of R_j does not contain any grouping terms (no* all*).*
- *For each variable in c the nesting depth of its occurrence in c is less or equal to the nesting depth in any top-level query term q in R_i that simulates with c.*

Weakly-recursive Xcerpt is the fragment of Xcerpt containing all such programs.

Roughly speaking the absence of grouping terms prevents terms with unbounded breadth and the second condition places a bound on the breadth of terms.

Theorem 4. *Weakly-recursive Xcerpt is decidable and the combined complexity of its evaluation is* NEXPTIME*-complete.*

Proof. Weakly-recursive Xcerpt is NEXPTIME-hard as we can reduce weakly-recursive Datalog$_{new}^{\neg}$ which is known to be NEXPTIME-complete [43] to weakly-recursive Xcerpt by the construction in the proof of Theorem 3. The resulting Xcerpt programs are indeed weakly-recursive, as the construction never creates grouping terms and the nesting depth only increases when translating invention atoms.

On the other hand, weakly-recursive Xcerpt is also obviously in NEXPTIME: For a given input program we can compute the maximum depth and breadth of a term as well as the number of distinct labels for a given input term of depth d, breadth b, and number of distinct labels l.

We generate each of the $O(b^d \cdot l)$ different terms that can be generated with these bounds. Then we compute the Xcerpt program by a standard fixpoint operator, but instead of generating new terms we only mark those terms we have already derived. If there are no more rules that mark additional terms or all terms are marked, the evaluation terminates. A single derivation step is obviously in NP. Since each step marks at least one term, there are at most $O(b^d \cdot l)$ steps

Non-Recursive Xcerpt: Parallelizable. Though weakly-recursive Xcerpt is decidable it is still fairly expensive to evaluate. An obvious further restriction is to allow no recursion in Xcerpt at all:

Definition 25 (Non-recursive Xcerpt). *Let* $P = R_1, \ldots, R_n$ *be an Xcerpt program. Then P is called* non-recursive, *if its dependency graph* $D(P)$ *is acyclic. Non-recursive Xcerpt is the fragment of Xcerpt containing all such programs.*

Though this restriction limits the construction of new values, it turns out that even the application of a single Xcerpt rule can be potentially expensive due to the use of **deep-equal**. For arbitrary Xcerpt terms this is as hard as graph isomorphism for which no polynomial time algorithms are known. Therefore, we also limit ourselves to trees-shaped data as input and disallow references in rules.

It turns out that with these two restriction, we obtain a sub-language that is efficiently parallelizable (wrt. data complexity):

Proposition 3. *Non-recursive Xcerpt on trees has data complexity in* NC$_1$ $\subseteq$ L *and program complexity* PSPACE-*complete.*

Proof. We can obtain a non-recursive Datalog$^\neg$ program with **deep-equal** by 1. computing the (now acyclic) dependency graph, 2. indexing all relations in the head of each rule with a unique identifier, 3. replacing references to the relation in the body of each rule with a disjunction referencing the indexed relations of all rules they may depend on. The resulting program is a non-recursive Datalog$^\neg$ program with **deep-equal** and is at most exponential in the size of the input Xcerpt program. A Datalog$^\neg$ program with **deep-equal** can be evaluated with data complexity in NC$_1$ and program complexity in PSPACE(which is thus not affected by the exponential translation size) since **deep-equal** on trees is NC$_1$-complete [91].

From Xcerpt to SQL: A Foundation for a Relational Implementation. With the translation to Datalog$^\neg_{new}$ for Xcerpt programs, we not only achieve a purely logical semantics, but also the foundation for a relational implementation: First notice, that each stratified Datalog$^\neg$ program can be translated into a, possibly recursive, SQL expression. SQL recursion (introduced in SQL:1999 and refined in SQL:2003) is expressed using `with` and is limited to monoton recursion: A relation P may be defined by means (including negation) of a relation Q only if adding tuples to Q cannot cause any triple of P to be deleted. Fortunately, stratified Datalog$^\neg$ programs are designed to be allow only monoton recursion.

With the addition of ranking operators in SQL:1999 controlled value generation has been standardized as well. When translating a Datalog$^\neg_{new}$ program to SQL, we employ the **ROW_NUMBER** or **DENSE_RANK** function to generate new node IDs based on the invention variables. For details see [82] where these are used in the context of XQuery iteration.

The chief disadvantage of translating Xcerpt (in this or other ways) to SQL for implementation is that the nave relational representation discussed in Section 2.4.4 does not perform well in practice. This has been observed frequently and, for tree data, labeling schemes such as the pre-/post-encoding [81], ORDPATH [122], or BIRD [153] have been suggested to provide better XML storage. They provide linear time and space processing of XML tree queries on tree data. However, when querying graph data these approaches do not immediately apply. Therefore, we have developed a labeling scheme for graph data that not only provides linear time and space evaluation for tree data but also for many graphs, see Section 2.5.1.

Versatile Semantics: Adding XPath, XQuery, and SPARQL. The above translation has been focused so far on Xcerpt with a brief outlook to Xcerpt^{RDF}. However, Datalog$^\neg_{new}$ together with the relation representations for RDF and XML data from Section 2.4.4 can form a common basis for analysing and evaluating a far larger set of query languages.

In fact in [71,69], we show how to map not only Xcerpt but also XPath, XQuery, and SPARQL to Datalog$^\neg_{new}$. Combined with the labeling scheme and evaluation for tree and graph data discussed in Section 2.5.1 this allows us the use of the same, efficient evaluation engine for all this languages. In particular, we can integrate queries written in these very different languages. Though such integration has been suggested previously (e.g., in [128]), none of the previous approaches achieves language integration also on the level of the evaluation engine. By translating both languages to Datalog$^\neg_{new}$ we open up opportunities for cross language optimization. Furthermore, the labeling scheme propose in Section 2.5 allows for such integration without sacrificing efficiency for the more restricted languages (e.g., for XPath on tree data).

Example: Language Integration. As illustration let us consider an example of such language integration where we allow XPath and SPARQL queries to occur in the body of an Xcerpt rule. XPath queries are always only filters, i.e., they do not provide variable bindings. SPARQL queries may provide variable bindings, though such variables are always *label variables* only. The same variables may be used in body parts of different languages and are understood as multiple variable occurrences in Xcerpt. However, if Xcerpt term variables are used in SPARQL or XPath only their top-level label is considered. For simplicity we assume that XPath and Xcerpt query the same XML data, but SPARQL queries a separate RDF graph. Of course, we could also access different data sets in each language.

The following example selects the names of authors of conference papers in a variable X if they contain "Cicero" in an Xcerpt query. An XPath filter constraints these bindings by requiring that there is also a member of the organizers from Plato's "Akademia" with the same name. Finally, we also select all resources in the RDF data whose dc:creator has the same full-name.

```
1 GOAL
    shelf{ all author { var X, all var A group-by A } group-by X
        }
3 FROM
    and(
5   conference{{
      paper{{
7       desc author{{ var X → /.*/ }} }} }},

9     //organizers/member[affiliation[text() = 'Akademia']
          name[text()=$X]],

11    SELECT ?A
      WHERE { ?A dc:creator ?P AND ?P vcard:FN ?X }
13  )
    END
```

The translation to Datalog$^{\neg}_{new}$ is very much along what we have seen in Section 2.4.4:

Root $(1())$ $\wedge$ Lab^{shelf} $(1())$ $\wedge$ R_{child} $(1(),\ 2(v_2))$ $\wedge$ Lab^{author} $(2(v_2))$ $\wedge$ R_{child} $(2(v_2),$
     deep-copy $(v_2,\ v_2))$ $\wedge$ R_{child} $(2(v_2),\ $ deep-copy $(v_2,\ s_1,\ s_1))$

2   $\longleftarrow$ Root (v_1) $\wedge$ Lab^{conference} (v_1) $\wedge$ R_{child} $(v_1,\ v_2)$ $\wedge$ Lab^{paper} (v_2) $\wedge$ $R_{\text{descendant}}$
     $(v_2,\ v_3)$ $\wedge$ Lab^{author} (v_3) $\wedge$ R_{child} $(v_3,\ v_4)$, R_{arity} $(v_4,\ 0)$ $\wedge$

4   Root (x_1) $\wedge$ $R_{\text{descendant-or-self}}$ $(x_1,\ x_2)$ $\wedge$ Lab^{organizers} (x_2) $\wedge$ R_{child} $(x_2,\ x_3)$ $\wedge$
     Lab^{member} (x_3) $\wedge$ R_{child} $(x_3,\ x_4)$ $\wedge$ Lab^{affiliation} (x_4) $\wedge$ R_{child} $(x_4,\ x_5)$ $\wedge$
     Lab^{Akademia} (x_5) $\wedge$ R_{child} $(x_3,\ x_6)$ $\wedge$ Lab^{name} (x_6) $\wedge$ R_{child} $(x_6,\ x_7)$ $\wedge$
     $x_7 \cong v_4$ $\wedge$

6     $\circ\!\!\rightarrow (s_1,\ e_1)$ $\wedge$ Edge (e_1) $\wedge$ Lab$^{\text{dc:creator}}$ (e_1) $\wedge$ $\rightarrow\!\!\circ (e_1,\ s_2)$ $\wedge$ $\circ\!\!\rightarrow (s_2,\ e_2)$ $\wedge$
     Edge (e_2) $\wedge$ Lab$^{\text{vcard:FN}}$ (e_2) $\wedge$ $\rightarrow\!\!\circ (e_2,\ s_3)$ $\wedge$ $s_3 \cong v_4$.

Obviously, in this case the use of XPath affords little gain compared to Xcerpt only queries, but the same technique can be applied to integrate XPath into SPARQL or SPARQL into XQuery. Full translations for SPARQL, XPath, and XQuery can be found in [69].

Outlook. Xcerpt and versatile Web queries in general are a powerful and convenient tool for accessing Web data. In this section, we show that, both their semantics and evaluation, can nevertheless be cast in terms of existing logical foundations and technology. In particular, we show how Xcerpt can be translated to Datalog$^{\neg}_{new}$ and use that translation to proof several formal properties of Xcerpt and interesting sub-languages thereof. Perhaps even more important is that the suggested translation can also be achieved for such diverse Web query languages as SPARQL, XPath, or XQuery. Not only does that provide us with a playground for comparing and investigating these languages, it also allows us, as discussed in the last two Sections, to integrate and implement these languages in a common engine. We have only outlined first ideas towards this integration here. There remain a plethora of open issues such as the right mapping of variable bindings. One of the most crucial of these issues is the question whether the use of such a common engine does not sacrifice performance for the more restricted languages such as XPath. The following Section 2.5 essentially answers this question negative: We can provide a common engine for these languages based on a uniform evaluation of tree and graph data, that nevertheless provides a linear time and space (and thus optimal) evaluation for XPath (tree queries on tree data). It even extends this complexity to many graphs in contrast to all previous approaches.

2.5 Versatile Evaluation

In the previous sections, we have shown how versatile query languages advance the state-of-the-art for querying the Web, where often the same application needs access to data published in different formats.

    Employing a versatile query language may be convenient, but what about the cost? If the price is that we have to forgo efficient evaluation methods that exploit the specific

Table 10. Complexity of graph labeling schemes for reachability test on arbitrary graphs (labeling size is per node). $n \geq e_g$: number of non-tree edges.

approach	reachability time	labeling time	labeling size
2-Hop [53]	$O(n)$	$O(n^4)$	$O(n)$
HOPI [137]	$O(n)$	$O(n^3)$	$O(n)$
Graph labeling [6]	$O(n)$	$O(n^3)$	$O(n)$
SSPI [48]	$O(n)$	$O(n^2)$	$O(n)$
Dual labeling [151]	$O(1)$	$O(n^3)$	$O(n)$
GRIPP [145]	$O(n)$	$O(n^2)$	$O(n)$

limitations of the involved data formats, versatile query languages may often not be practical.

Fortunately, we show in this section that in two crucial aspects this concern is not justified: First, we present a uniform evaluation algorithm that is capable of dealing with arbitrary graphs (as they occur in RDF data), but (seamlessly) processes trees and even many non-trees

2.5.1 Evaluating Queries: Structure Scaling with CIQCAG

What makes Web queries different from those used in centralized, relational databases is the emphasize on versatile, flexible structure conditions. Web queries are often written against data, where neither the exact shape of the children of a node nor of the paths connecting two nodes is known. This observation leads to emphasize on a flexible representation of Web data, be it tree- or graph-shaped, where we can not assume a fixed, recursion- and repetition-free schema.

Labeling schemes have become a popular means for providing efficient queries to Web data, in particular if that Web data is represented relationally. Labeling schemes assign each node a unique (constant[36]) label such that we can decide whether two nodes stand in a certain structural relations given only their labels. For tree data, several labeling schemes with constant time membership test have been proposed [81,122,153].

On arbitrary graph data testing adjacency (or reachability) in both constant time *and* constant per-node space is not possible.[37] Labeling schemes have therefore focused on heuristics for finding compact representations of reachability and adjacency. These heuristics come in roughly two kinds (here and in the following n, m are the number of nodes and edges in a given graph):

- For reachability in *arbitrary graphs*, the 2-hop cover [53] is a set of shortest paths such that for any two nodes there is a concatenation of two such paths that is a

---

[36] As most other works on labeling schemes, we consider label size to be bounded in practice and thus as constant. More precisely, label size is in $O(\log n)$ where n is the number of nodes in the document.

[37] As there are 2^{n^2} different graphs, yet constant per-node space allows only for 2^n different representations.

shortest path for those two nodes. Such a 2-hop cover can be exploited to assign labels for reachability testing: Each node v is labeled with two sets $(L_{out}(v), L_{out}(v))$ such that $L_{out}(v) \cap L_{in}(v')$ for each node v' reachable from v. However, finding an optimal 2-hop labeling NP-hard and there are graphs whose optimal 2-hop labeling is at least $\Omega(n \cdot m^{1/2})$ in size. Further work on 2-hop labeling has focused mostly on efficient approximation algorithms [137].

– Often sparse graphs are almost tree-shaped with only few non-tree edges. This is exploited by a several approaches [6,48,151,145] for extending tree labelings, mostly pre-/post-labelings [81], to graphs. However, for all of these approaches the largest interesting class of graphs where they can still guarantee constant time and per-node space reachability tests are trees. On arbitrary graphs they either degenerate in space or time complexity.

Roughly speaking all three four approaches extend pre-/post-tree labeling to arbitrary graphs by first labeling a spanning tree. They differ in how they deal with non-tree edges: In [6] nodes get additional pre-/post-intervals for descendants not reachable by tree edges at the cost of up to linear space per node. In [48] non-tree edges are iterated at query time at the cost of up to linear time for testing reachability. In [151] the transitive closure of non tree edges is computed and stored at the cost of up to linear space per node space. Finally, [145] presents a refined combination of [6] and [48] that performs on sparse graphs often significantly better, but does not improve the worst-case space or time complexity.

Table 10 summarizes these time and per-node space complexity.

Contributions. In this chapter, we present a *novel characterization* of a class of graphs that is a proper, non-trivial superclass of trees that still exhibits a *labeling scheme with constant time, constant per-node space* adjacency and reachability tests. Furthermore, we give a quadratic algorithm that computes, for an arbitrary graph, such a labeling if one exists.

Constant time membership test almost immediately yields linear time evaluation for existential acyclic conjunctive queries on tree data. However, nave approaches for n-ary universal queries take at least quadratic time in the graph size. We show how the above labeling scheme can be exploited to give an *algorithm for evaluating acyclic conjunctive queries* that is $O(n \cdot q)$ wrt. time and space complexity, i.e., linear in both data and program complexity. Furthermore, our algorithm guarantees iteration in the size of the related nodes rather than in all nodes.

Labeling Beyond Trees: Continuous-Image Graphs. Tree data, as argued above, allows us to represent relations on that data more compactly, e.g., using various interval-based labeling schemes. Here, we introduce a new class of graphs, called *continuous-image graphs* (or CIGS for short), that generalize features of tree data in such a way that we can evaluate (tree) queries on CIGS with the same time and space complexity as techniques such as twig joins [30] which are limited to tree data only.

Continuous-image graphs are a proper superset of (ordered) trees. On trees we require that each node has at most one parent. For continuous-image graphs, however, we only ask that we can find a single order on all nodes of the graph such that the

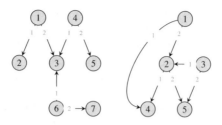

Fig. 9. Sharing: On the Limits of Continuous-image Graphs

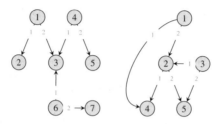

Fig. 10. Sharing: On the Limits of Continuous-image Graphs

children of each parent form a continuous interval in that order. Formally, we define a continuous-image graph by means of the image interval property (a generalization of corresponding properties of tree-shaped relations or closure relations of tree-shaped base relations. Recall that we denote with $R(n)$ for a node $n \in N$ and a binary relation R over the domain N the set $\{n' \in N : (n, n') \in R\}$.

Definition 26 (Continuous-image Graph). *Let R be a binary relation over a domain (of nodes) N. Then R is a continuous-image graph, short CIG, if it carries the image interval property: there is a total order $<_i$ on N with the induced sequence S over N such that for all nodes $n \in N$, $R(n) = \emptyset$ or $R(n) = \{S[s], \ldots, S[e] : s \leq e \in \mathbb{N}\}$.*

The definition of continuous image graphs allows graphs where some or all children of two parents are "shared" (in contrast to trees where this is never allowed). However, it limits the degree of sharing: Figure 10 shows two minimal graphs that are *not* CIGs. Incidentally, both graphs are acyclic and, if we take away any one edge in either graph, the resulting graph becomes a CIG. The second graph is actually the smallest (w.r.t. the number of nodes and edges) graph that is not a CIG. The first is only edge minimal but illustrates an easy to grasp sufficient but not necessary condition for violating the image interval property: if a node has at least three parents and each of the parents has at least one (other) child not shared by the others then the graph can not be a CIG.

On continuous-image graphs we can exploit similar techniques for representing structural relations as on trees, most notably we can label each node with a single, continuous interval for its children and/or descendants. Together with a simple index to represent that nodes position in the underlying order, we obtain constant space labels (three integers), yet can test adjacency and/or reachability in constant time (with two integer comparisons).

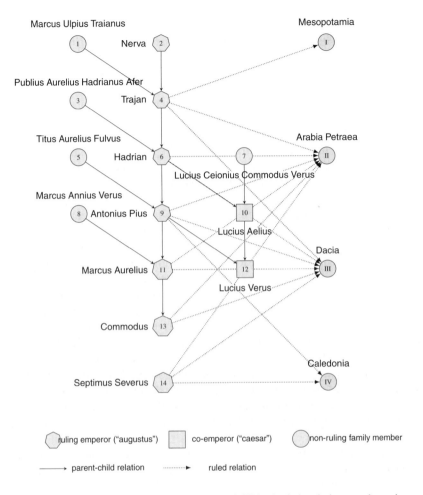

Fig. 11. "The Five Good Emperors" (after Edward Gibbon), their relations, and provinces

Testing for CIGs: consecutive ones property. Moreover, whether a given graph is a CIG (and in what order its node must be sorted to arrive at continuous intervals for each parent's children) is just another way of saying that its adjacency matrix carries the *consecutive ones* property [68]. For the consecutive-ones problem [24] gives the first linear time (in the size of the matrix) algorithm based on so called PQ-trees, a compact representation for permutations of rows in a matrix. More recent refinements in [84] and [88] show that simpler algorithms, e.g., based on the PC-tree [87], can be achieved. We *adapt* these algorithms to obtain a linear time (in the size of the adjacency matrix) algorithm for deciding whether a graph is a CIG and computing a CIG-order.

From a practical perspective, CIGs are actually quite common, in particular, where time-related or hierarchical data is involved: If relations, e.g., between Germany and kings, are time-related, it is quite likely that there will be some overlapping, e.g., for periods where two persons were king of Germany at the same time. Similarly, hierarchical data often has some limited anomalies that make a modelling as strict tree data impossible. Figure 11 shows actual data[38] on relations between the family (red nodes, non-ruling member ①, co-emperor or heir designate $\boxed{10}$, emperors ②) of the Roman emperors in the time of the "Five Good Emperors" (Edward Gibbon) in the 2nd century. It also shows, for actual emperors, which of the four new provinces (①) added to the roman empire in this period each emperor ruled (the other provinces remained mostly unchanged and are therefore omitted). Arrows between family members indicate, natural or adoptive, fathership[39]. Arrows between emperors and provinces show rulership, different colors are used to distinguish different emperors. Despite the rather complicated shape of the relations (they are obviously not tree-shaped and there is considerable overlapping, in particular w.r.t. province rulership).

The previous example also highlights the intuition behind continuous-image graphs: we allow some overlapping between among the children of different nodes, but only in such a way that the images can still be represented (over some order on the nodes) as continuous intervals. Figure 12 illustrates the intervals on the Roman provinces for representing the ruled provinces of each emperor: With the given order on the provinces, each image is a single interval (e.g., Trajan I–III and Septimus Severus II–IV) even though the data is clearly not tree-shaped (or a closure relation of a tree-shaped relation).

How continuous-image graphs differ from tree-shaped data (or closure relations over tree-shaped data) is further detailed in Figure 13: Tree data carries the image disjointness property as, under the order on the nodes induced by a breadth-first traversal, the nodes in the image of any parent node in the tree form a continuous, non-overlapping interval. Closure relations over tree data (i.e., relations such as XPath's descendant) carry the image containment property as, e.g., under the order on the nodes induced by a depth-first traversal, again the nodes in the image of any parent node form a continuous interval and overlapping is limited: either two such intervals do not overlap at all or one is contained within the other.

Continuous-image graphs (as shown in the right of Figure 13) carry, as stated above, the image interval property, i.e., there is some order on the nodes such that the nodes in the image of each parent form a continuous interval. Here, the intervals may overlap arbitrarily as illustrated in Figure 13. However, in contrast to the tree or closure relation over tree case the required order on the nodes is no longer known a-priori but must be determined for each graph using, e.g., the above described algorithms.

Intermediary Answers as Interval Labels. When we evaluate acyclic conjunctive or tree queries, we can observe that for determining matches for a given query node only the match for its parent and child in the query tree are relevant. Intuitively, this "locality" property holds as in a tree there is at most one path between two nodes. To illustrate, consider, e.g., the XPath query //a//b//c selecting c descendants of b descendants of a's. Say there are n a's in the data nested into each other with m b's nested inside the a's and finally inside the b's (again nested in each other) l c's. Then a naive evaluation of

---

[38] The name and status of the province between the wall of Hadrian and the wall of Antonius Pius in northern Britain is controversial. For simplicity, we refer to it as "Caledonia", though that actually denotes all land north of Hadrian's wall.

[39] Note that all emperors of the Nervan-Antonian dynasty except Nerva and Commodus were adopted by their predecessor and are therefore often referred to as "Adoptive Emperors".

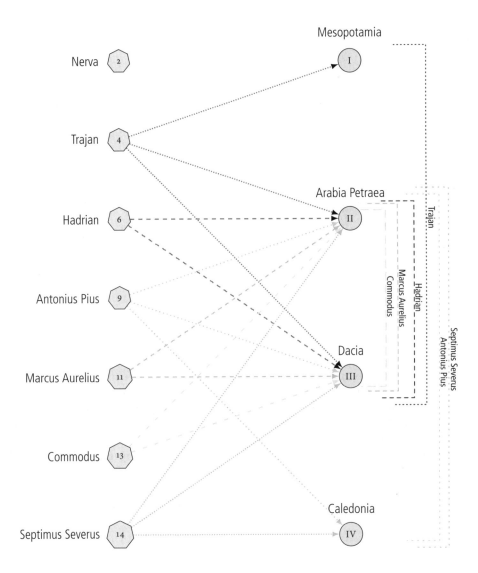

Fig. 12. Overlapping of province children in the "The Five Good Emperors" example, Figure 9

the above query considers all triples (a, b, c) in the data, i.e., $n \times m \times l$ triples. However, whether a c is a descendant of a b is independent of whether a b is a descendant of an a. If a b is a descendant of several a's makes no difference for determining its c descendants. It suffices to determine in at most $n \times m$ time and space all b's that are descendants of a, followed by a separate determination of all c's that are descendants of such b's in at most $m \times l$ time and space.

Indeed, if we consider the answer relation for a tree query, i.e., the relation with the complete bindings as rows and the query's nodes as columns, this relation always

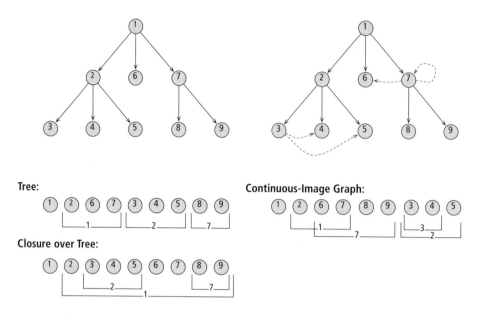

Fig. 13. Overlapping of images in trees, closure relations over trees, and continuous-image graphs

exhibits multivalue dependencies [65]: We can normalize or decompose such a relation for a query with n nodes into $n - 1$ separate relations that together faithfully represent the original relation (and from which the original relation can be reconstructed using $n - 1$ joins). This allows us to compact an otherwise potentially exponential answer (in the data size) into a polynomial representation.

This is the first principle of the algorithm: decompose the query into separate binding sequences for each query node with "links" or pointers relating bindings of different nodes. We thus obtain an exponentially more succinct data structure for (intermediary) answers of tree queries than if using standard (flat) relational algebra. In this sense, a sequence map can be considered a fully decomposed *column store* for the answer relation.

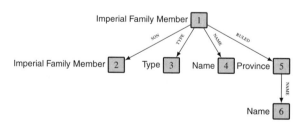

Fig. 14. Selecting sons, type, name, and ruled provinces for all members of the imperial family in the data of Figure 11

Imp-II	Type	Name	Son-II	Ruled-II	Ruled-Name
1	non-ruling	Marcus Ulpius Traianus	4	–	–
2	augustus	Nerva	4	–	–
3	non-ruling	P. Aurelius Hadrianus Afer	6	–	–
4	augustus	Trajan	6	I	Mesopotamia
4	augustus	Trajan	6	II	Arabia Petraea
4	augustus	Trajan	6	III	Dacia
5	non-ruling	Titus Aurelius Fulvus	9	–	–
6	augustus	Hadrian	9	II	Arabia Petraea
6	augustus	Hadrian	10	II	Arabia Petraea
6	augustus	Hadrian	9	III	Dacia
6	augustus	Hadrian	10	III	Dacia
7	non-ruling	L. Ceionius Commodus Verus	10	–	–
8	non-ruling	M. Annius Verus	11	–	–
9	augustus	Antonius Pius	11	II	Arabia Petraea
9	augustus	Antonius Pius	12	II	Arabia Petraea
9	augustus	Antonius Pius	11	IIi	Dacia
9	augustus	Antonius Pius	12	III	Dacia
9	augustus	Antonius Pius	11	IV	Caledonia
9	augustus	Antonius Pius	12	IV	Caledonia
10	caesar	Lucius Aelius	12	–	–
11	augustus	Marcus Aurelius	13	II	Arabia Petraea
11	augustus	Marcus Aurelius	13	III	Dacia
12	caesar	Lucius Verus	–	–	–
13	augustus	Commodus	–	II	Arabia Petraea
13	augustus	Commodus	–	III	Dacia
14	augustus	Septimus Severus	–	II	Arabia
14	augustus	Septimus Severus	–	III	Arabia
14	augustus	Septimus Severus	–	IV	Caledonia

Fig. 15. Answers for query from Figure 14, single, flat relation

To illustrate this, consider the query in Figure 14 on the data of Figure 11. The query selects sons and ruled provinces of members of the imperial family. We also record type and name of the family member and name of the province to easier talk about the retrieved data. The answers for such a query, if expressed, e.g., in relational algebra or any language using standard, flat relations to represent n-ary answers, against the data from Figure 11 yields the flat relation represented in Figure 15. As argued above, we can detect multivalue dependencies and thus redundancies, e.g., from emperor to province, from province to province name, from emperor (Imp-ID) to type and name.

To avoid these redundancies, we first decompose or normalize this relation along the multivalue dependencies as in Figure 16. For the sequence map, we use always a full decomposition, i.e., we would also partition type and name into separate tables as in a column store.

Storing Intermediary Results as Intervals. Once we have partitioned the answer relation into what subsumes to only link tables as in column stores, we can observe even more regularities (and thus possibilities for compaction) if the underlying data is a tree or continuous-image graph. Look again at the data in Figure 11 and the resulting answer representation in Figure 16: Most emperors have not only ruled one of the new provinces Mesopotamia, Arabia Petraea, Dacia, and Caledonia but several. However,

Imp-II	Type	Name
1	non-ruling	M. Ulpius Traianus
2	augustus	Nerva
3	non-ruling	P. A. Hadrianus Afer
4	augustus	Trajan
5	non-ruling	Titus Aurelius Fulvus
6	augustus	Hadrian
7	non-ruling	L. C. Commodus Verus
8	non-ruling	M. Annius Verus
9	augustus	Antonius Pius
10	caesar	Lucius Aelius
11	augustus	Marcus Aurelius
12	caesar	Lucius Verus
13	augustus	Commodus
14	augustus	Septimus Severus

Imp-II	Son-ID
1	4
2	4
3	6
4	6
5	9
6	9
6	10
7	10
8	11
9	11
9	12
10	12
11	13

Imp-II	Prov-ID
4	I
4	II
4	III
6	II
6	III
9	II
9	III
9	IV
13	II
13	III
14	II
14	III
14	IV

Prov-II	Name
I	Mesopotamia
II	Arabia Petraea
III	Dacia
IV	Caledonia

Fig. 16. Answers for query from Figure 14, no multivalue dependencies

Imp-II	Son Range
1	4
2	4
3	6
4	6
5	9
6	9–10
7	10
8	11
9	11–12
10	12
11	13

Imp-II	Prov Range
4	I–III
6	II–III
9	II–IV
13	II–III
14	II–IV

Fig. 17. Answers for query from Figure 14, multiple relations, interval pointers. The first table from Figure 16 remains unchanged.

since the data is a continuous-image graph there is an order (indeed, the order of the province IDs if interpreted as roman numerals) on the provinces such that the provinces ruled by each emperor form a continuous interval w.r.t. that order. Thus we can actually represent the same information much more compactly using interval pointers or links as in Figure 17 where we do the same also for the father-son relation (although there is far less gain since most emperors already have only a single son).

Instead of a single relation spanning 28 rows and 6 columns (168 cells), we have thus reduced the information to $5 \cdot 2 + 11 \cdot 2 + 14 \cdot 3 = 74$ cells. This compaction increases exponentially if there are longer paths in a tree query (e.g., if the provinces would be connected to further information not related to the emperors). It increases quadratically with the increasing size of the tables, e.g., if we added the remaining n provinces of the Roman empire ruled by all emperors in our data we would end up with $7 \cdot n$ additional rows of 6 columns in the first case (each of the 7 emperors in our data ruled all these provinces), but only $n \cdot 2$ additional cells when using multiple relations and interval pointers.

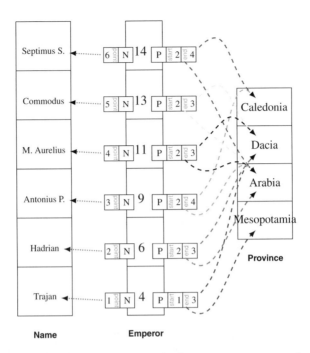

Fig. 18. Sequence Map: Example. For a query selecting roman emperors together with their name and ruled provinces on the data of Figure 11.

Formally, we represent (intermediary) answers to an acyclic query as a mapping from the set of query variables V to sequences of matches for that query node. A *match* for query node v in itself is the actual data node or edge v is matched with and a set of pairs of child nodes of v to start and end positions. Intuitively, it connects the match for v to matches of its child nodes in the tree query. We obtain in this way a data structure as shown in Figure 18 for a query selecting roman emperors with their names and ruled provinces on the data of Figure 11.

Note, that we allow for each child node of v *multiple* intervals. If the data is a CIG, it is guaranteed that only a single interval is needed and thus the overall space complexity of a sequence map is linear in the data size. However, we can also employ a sequence map for non-CIG graphs. In this case, we often still benefit from the interval pointers, but in worst-case we might need $|N \cup E|$ many interval pointers to relate to all bindings of a child variable. Overall, a sequence map for non-CIG graphs thus may use up to quadratic space in the data size.

Representing intermediary results: A Comparison. As stated above, the sequence map is heavily influenced by previous data structures for representing intermediary answers of tree queries. Figure 19 shows the most relevant influences. Complexity and supported data shapes are compared below after discussing the actual evaluation of tree queries using the interval labeling for data and intermediary results. Here, we illustrate that the above discussed choices when designing a data structure for intermediary

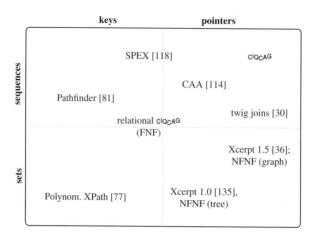

Fig. 19. Data structures for intermediary results (of a tree query)

answers of tree queries are actually present in many related systems: We can find systems such as Xcerpt 1.0 [135], many early XPath processors (according to [77]), and tree algebras such as TAX [90] that use exponential size for storing all combinations of matches for each query node explicitly. [77] shows that XPath queries can in fact be evaluated in polynomial time and space, which is independently verified in SPEX [118], the first streaming processor for navigational XPath with all structural axes. Like SPEX and our approach, complete answer aggregates [114] use interval compaction for relating matches between different nodes in a tree query. CAAs are also most closely related to our approach w.r.t. the decomposition of the answer relation: fully decomposed without multivalue dependencies. In contrast, Pathfinder [22] uses standard relational algebra but for the evaluation of structural joins a novel staircase join [83] is employed that exploits the same interval principles used in CAAs and our approach.

Streaming or cursor-based approaches such as twig join approaches [30,48] consider the data in a certain order rather than all at once. In such a model, it is possible and desirable to skip irrelevant portions of the input stream (or relations) and to prune partial answers as soon as it is clear that we can not complete such answers. Recent versions of SPEX [37,118] contain as most twig join approaches, mechanisms to skip over parts of the stream (at least for some query nodes) if there can not be a match (e.g., because there is no match for a parent node and we know that matches for parent nodes must come before matches for child nodes). Both twig joins and SPEX also prune results as soon as possible. However, twig joins are limited to vertical relations (child and **descendant**) whereas SPEX and c!QcAG can evaluate all XPath axes, though only on tree data.

To summarize, though our approach to representing intermediary answers is similar in its principles to several of the related approaches in Figure 19, it combines *efficient intermediary answer storage* as in CAAs with *fully algebraic* processing as in Pathfinder and *efficient skipping and pruning* as in twig joins.

Furthermore, where most of the related approaches are limited to tree data (with the notable exception of Xcerpt), our approach allows processing of *many graphs*, viz. cIGs, as efficient as previous approaches allow for trees.

Evaluating Tree Queries on Interval Labels. For evaluating acyclic conjunctive queries (or tree queries) the essential operation is a join that allows us to gather results of the evaluation of the constituent query atoms in a consistent manner: Only where variables are bound the same in answers to different query atoms those answers are "joined" together to form a larger answer.

There are some other operations needed for implementing full acyclic conjunctive queries, most importantly selection, but they are omitted here for clarity of presentation. For full details see [69].

The join operation for interval representations of relations is defined as follows:

Definition 27 (Sequence map join (disjoint edge covers)). *Let D be a relational structure, Q a tree query, and S_1, S_2 two interval representations for D over Q such that there is no edge in Q that is covered by both S_1 and S_2. Then $\bowtie_{[}](S_1, S_2)$ returns an interval representation S_3 such that*

1. *the induced relation of S_3 is the natural join of the induced relations of S_1 and S_2.*
2. *$S_3|_{domS_1 \cup domS_2} = S_3$ (S_3 contains bindings only for variables mapped either in S_1 or in S_2).*

Note that this definition yields an interval representation that leaves bindings for *non-shared* variables unchanged from the input representations. These variables occur only in one of the query parts covered by the input, but not in the other. For *shared* variables, only those bindings are retained that occur in both representations. This also applies to the (interval pointer) references from bindings of a parent variable v to a child variable v' of v: They are contained only in one of the sequence maps (due to the edge cover restriction), for the other sequence map the induced relation records *any* combination of bindings by definition.

The restriction on the edge covers on S_1 and S_2 is imposed to ensure that for any pair of variables v, v' only one of the interval representations may contain interval pointers from v to v', though both may contain *bindings* for v and v'. In other words, each *edge* of the query is enforced by at most one of the two sequence maps.

For a given tree query expression, the edge cover of each sub-expression can be determined statically, without knowledge about the data the expression is to be evaluated against. Thus, we can also statically determine whether a join expression is valid or violates the edge cover restriction defined above. For the evaluation of tree queries we never need joins with overlapping edge covers.

Algorithm 2.5.1 computes an interval representation that represents the join of the induced relations as demanded in the definition of $\bowtie_{[}]()$, but may be inconsistent: It "bombs" bindings not contained in both sequence maps rather than dropping them entirely. This has the effect that interval pointers can remain unchanged (but now point to an interval containing possibly bombed entries). Note, that interval pointers to bindings of a variable occur only in one of the two input interval representations as the incoming edge of each variable is unique (since the query is tree-shaped) and the edge covers are disjoint. This allows line 16 where we simply throw together intervals from both sequence maps. Finally, observe that by the definition of the initialization of a sequence map, bindings for the same query variable occur in the *same* order in all sequence maps for that query. Thus the bindings of a variable shared between the two interval representations are ordered the same.

Algorithm 1. $\bowtie_{[]}(S_1, S_2)$

> **input** : Interval representations S_1 and S_2 with *disjoint* edge covers
> **output**: Interval representation res representing the join of the induced relations
> of the inputs

1 $\text{EC}_1 \leftarrow \text{edgeCover}(S_1)$; $\text{EC}_2 \leftarrow \text{edgeCover}(S_2)$;
2 $\text{AllVars} \leftarrow domS_1 \cup domS_2$;
3 $\text{SharedVars} \leftarrow domS_1 \cap domS_2$;
4 $\text{res} \leftarrow \emptyset$;

5 **foreach** $v \in \text{AllVars}$ **do**
6   **if** $v \notin domS_2$ **then** $\text{res} \leftarrow \text{res} \cup \{(v, S_1(v))\}$;
7   **else if** $v \notin domS_1$ **then** $\text{res} \leftarrow \text{res} \cup \{(v, S_2(v))\}$;
8   **else**   *// v is in both*
           // 1 is the primary (fallback if v is in neither edge cover)
9      $\text{iter} \leftarrow S_1(v)$; $\text{alt} \leftarrow S_2(v)$;
10   **if** $(v', v) \in \text{EC}_2$ *for some* v' **then**
           // v is sink in EC_2, thus the order and number of entries must be as in 2
           (it can not be sink in EC_1 as Q tree query and edge covers disjoint)
11       $\text{iter} \leftarrow S_2(v)$; $\text{alt} \leftarrow S_1(v)$;
12     $S \leftarrow \emptyset$; $i, j, k \leftarrow 1$;
13   **while** $i \leq |\text{iter}|$ **do**
14       $(n_1, i) \leftarrow \text{nextBinding}(S_1(v), i)$; $(n_2, j) \leftarrow \text{nextBinding}(S_2(v), j)$; **if** $n_1 = n_2$ **then**   *// Retain binding if same*
15         $S[k] = (n_1, \text{intervals}(\text{iter}[i]) \cup \text{intervals}(\text{alt}[j]))$;
16   i++; j++; k++;
17   **else if** $n_1 < n_2$ **then**   *// "bomb" if in iter but not in alt*
18         $S[k] = \frac{1}{2}$;
19   i++; k++;
20   **else**   *// skip binding if in alt but not in iter*
21   j++;
22
23     $\text{res} \leftarrow \text{res} \cup \{(v, S)\}$;
24

25 **return** res

These observations are exploited in Algorithm 2.5.1 to give a merge-join [73] style algorithm for the join of two interval representations with disjoint edge cover that has linear time complexity in the (combined) size of the inputs. Since the bindings are already in the same order, we can omit the sort phase of the merge join and immediately merge the two binding sequences. However, we need to ensure that not only the order but also the number of bindings (and the position of eventual failure markers, cf. lines 18–20) reflects that for the same variable v in the sequence map where v's incoming edge is in the edge cover (lines 9–11).

Algorithm 2. NextBinding(S, i)

---

input : Sequence S containing, possibly, failure markers and start index i

output: The next element in S at or after i that is not a failure marker and its
index or (∞, ∞) if no such binding exists

1 **for** $j \leftarrow i$ **to** $|S|$ **do**
2 ⎿ **if** $S[j]$ *is not a failure marker* **then break**;

3 **if** $j = |S|$ *and* $S[j]$ *failure marker* **then** **return** (∞, ∞) ;
4 **return** $(S[j], j)$

Theorem 5. *Algorithm 2.5.1 computes* $\bowtie_{[}](S_1, S_2)$ *for interval representations with disjoint edge cover and set of shared variables* **Shared** *in* $O(b_{Shared}^{total} \cdot i) \leq O(|\text{Shared}| \cdot n \cdot i)$ *time where* b_{Shared}^{total} *is the total number of bindings associated in either input with a variable in* **Shared** *and i is the maximum number of intervals associated with any such binding. For tree, forest, and* CIG *data* $i = 1$, *for arbitrary graph data* $i \leq n$.

Proof. Algorithm 2.5.1 computes $S = \bowtie_{[}](S_1, S_2)$: For any variable v, if a binding for v occurs in the induced relation of both interval representations, it occurs also in S due to lines 15–17. If v's incoming edge is in the edge cover of one of the interval representations S', lines 9–11 ensure that the sequence of bindings for v is the same (except that some bindings are "bombed") in S as in S'. For the parent v' of v, if a binding is retained the set of intervals from both interval representations are copied en block. There are only intervals in S' (as (v', v) is not in the edge cover of the other interval representation) and thus only those relations between bindings of v and v' as in the induced relation of S' are retained. This is proper as in the induced relation of the other interval representation *all* bindings of v are related to all bindings of v' by definition of the induced relation. Both input sequences may be inconsistent: The presence of failure markers in either sequence does not affect the correctness of the algorithm: failure markers in alt are skipped, failure markers in iter are retained (lines 15–17) as intended. Dangling bindings do not affect the algorithm.

Algorithm 2.5.1 loops over all shared variables of S_1 and S_2 and for each such variable it iterates over all bindings in the primary interval representation iter and corresponding bindings in alt, skipping, if necessary, bindings in alt not in iter. In the loop lines 13–22 i or j is incremented (possibly multiple times, if failure markers are skipped in NextBinding) until either $i > |\text{iter}|$. If j ever becomes $> |\text{alt}|$ subsequent calls of binding(()alt[j]) return, by definition, a value larger than all $n \in \text{Nodes}(D)$.

Thus the algorithm touches, for each shared variable, each entry in either interval representation at most once (and touches one proper (not a failure mark) entry in each step of the loop 13–22). Thus it runs in $O(b_{Shared}^{total} \cdot i)$ where b_{Shared}^{total} is the total number of bindings in both interval representations for a shared variable and i is the maximum number of intervals per binding. This is bound by $O(|\text{Shared}| \cdot n \cdot i)$ for any interval represenation (including those for arbitrary graphs).

It is worth pointing out, that a tree query any variable is shared at most once for each in- or outgoing edge and for each unary relation associated with the variable. Thus, even if there are $O(q)$ joins in the expression, the accumulated number of shared variables among all those joins is also only $O(q)$ an thus the total complexity for only those joins is bounded by $O(q \cdot n \cdot i)$.

Comparison. The above outlined algorithm for a join on interval representations can be easily extended to an algorithm for full tree queries, see [69]. In the following we compare our approach, called clQcAG, to previous approaches for tree query evaluation on XML (i.e., limited to *tree data*).

To keep the discussion focused we ignore index-based evaluation of XML. Though path indices such as the DataGuide [76] or IndexFabric [55] and more recent variants [51] can significantly speed up path queries they suffer from two anomalies: First, if a tree query contains many branching nodes (i.e., nodes with more than one children) they generally do not perform better than, e.g., the structural join approach below. Second, even though only path queries can be directly answered from the index, the index size can be significantly higher than the size of the original XML documents.

We can classify most of the remaining approaches to the evaluation of XML tree queries in four classes (the corresponding complexity for evaluation XPath (and similar) tree queries on tree data is summarized in Table 14):

1. *Structural joins:* The first class is most reminiscent of query evaluation for relational queries and arguable inspired by earlier research on acyclic conjunctive queries on relational databases [78]. Tree queries are decomposed into a series of (structural) joins. Each structural join enforces one of the structural properties of the given query, e.g., a child or descendant relation between nodes or a certain label. Proposed first in [8], structural joins have also been used to great effect for studying the complexity of XPath evaluation and proposing the first polynomial evaluation of full XPath [77]. Due to its similarity with relational query evaluation it has proved to be an ideal foundation for implementing XPath and XQuery on top of relational databases [80]. It turns out, however, that the use of standard joins is often not an ideal choice and structure- or tree-aware joins [22] (that take into consideration, e.g., that only nodes in the sub-tree routed at another node can be that nodes a-descendants) can significantly improve XPath and XQuery evaluation.

2. *Twig joins:* In sharp contrast, the second class employs a single (thus called *holistic*) operator for solving an entire tree query rather than decomposing it into structural joins. These approaches are commonly referred to as twig or stack join [30,49] and essentially operate by keeping one stack for each step in, e.g., an XPath query representing partial answers for the corresponding node-set. Theses stacks are organized hierarchically with (where possible, implicit) parent pointers connecting partial answers for upper stack entries to those of lowers. The approaches mostly vary in how the stacks are populated. In contrast to the other approaches, twig joins are limited to vertical, i.e., child and descendant, axes and have not been adapted for the full range of XPath axes. They also, like structure-aware joins [22], exploit the tree-shape of the data and can, at best, be adapted to DAGs [48].

3. *PDA-based:* Where twig joins assume one stream of nodes from the input document for each stack (and thus XPath step), the third class of approaches based on pushdown automata aims to evaluate XPath queries on a single input stream similar to a SAX event stream. SPEX, e.g., [119,120,118] also maintains a record of partial

Table 14. Approaches for XML Tree Query Evaluation. n: number of nodes in the data, d: depth of data; q: size of query. We assume constant membership test for all structural relations.

	time	space
Structural Joins, relational join	$O(q \cdot n \cdot \log n)$	$O(q \cdot n^2)$
——————, structure-aware join	$O(q \cdot n)$	$O(q \cdot n^2)$
Twig or Stack Joins	$O(q \cdot n)$	$O(q \cdot n + n \cdot d)$
PDA-based (here: SPEX)	$O(q \cdot n \cdot d)$	$O(q \cdot n)$
Interval-based (here: clQcAG)	$O(q \cdot n)$	$O(q \cdot n)$

answers for each XPath step, but minimizes used memory more efficiently and exploits the existential nature of most XPath steps by maintaining only generic conditions rather than actual pointers to elements from the XML stream (except for candidates of the actual results set, of course). Also it supports all XPath axes in contrast to the twig join approaches. The cost is a slightly more complex algorithm.

4. *Interval-based*: Finally, interval-based approaches are a combination of the tree awareness in twig joins and SPEX and the structural join approach: The query is decomposed into a series of structural relations, but each relation is organised in such a way that all elements related to one element of its parent step are in a single continuous interval. This allows both an efficient storage and join of intermediate answers. The first interval-based approach are the Complete Answer Aggregates (CAA) [115,114]. Here theclQcAG algebra is proposed which improves on the complexity of CAA (to the linear complexity given in Table 14) and covers, in contrast to CAA, arbitrary tree-shaped relations. It is also shown that interval-based approaches can be extended even to a large, efficiently detectable class of graph data (so called continuous-image graphs) that is not covered by any of the other linear time approaches discussed above.

Currently, extensions of the above algorithms for larger classes of graph data are investigated, e.g., in [48] and [69].

Conclusion. In this chapter, we present a *novel characterization* of a class of graphs that is a proper, non-trivial superclass of trees that still exhibits a *labeling scheme with constant time, constant per-node space* adjacency and reachability tests. Furthermore, we give a quadratic algorithm that computes, for an arbitrary graph, such a labeling if one exists.

Constant time membership test almost immediately yields linear time evaluation for existential acyclic conjunctive queries on tree data. However, nave approaches for n-ary universal queries take at least quadratic time in the graph size. We show how the above labeling scheme can be exploited to give an *algorithm for evaluating acyclic conjunctive queries* that is $O(n \cdot q)$ wrt. time and space complexity, i.e., linear in both data and program complexity. Furthermore, our algorithm guarantees iteration in the size of the related nodes rather than in all nodes.

2.5.2   Evaluating Rules: Subsumption under Rich Unification

In the preceeding sections, we have considered the efficient evaluation of Xcerpt queries by a translation to a relational normal form. This section deals with the subsumption relationship between Xcerpt query terms. Deciding subsumption has traditionally been an important means for optimizing multiple queries against the same set of data and can be used for improving termination of Xcerpt programs in a backward chaining evaluation engine.

Xcerpt query terms (Definition 2) are an answer to accessing Web data in a rule-based query language. Like most approaches to Web data (or semi-structured data, in general), they are distinguished from relational query languages such as SQL by a set of query constructs specifically attuned to the less rigid, often diverse, or even entirely schema-less nature of Web data. As Definitions 2 (Xcerpt Query Term) and 7 suggest, Xcerpt terms are similar to normalized forward XPath (see [121]) but extended with variables, deep-equal, a notion of injective match, regular expressions, and full negation. Thus, they achieve much of the expressiveness of XQuery without sacrificing the simplicity and pattern-structure of XPath.

When used in the context of Xcerpt, query terms serve a similar role to terms of first-order logic in logic languages. Therefore, the notion of unification has been adapted for Web data in [134], there called "simulation unification". Simulation is recapitulated in Definition 8. This form of unification is capable of handling all the extensions of query terms over first-order terms that are needed to support Web data: selecting terms at arbitrary depth (`desc`), distinguishing partial from total terms, regular expressions instead of plain labels, negated subterms (`without`), etc.

The notions of query term, simulation and substitution sets are exemplified in Section 2.3 and formally defined in 2.4. In this section, we consider query containment between two Xcerpt terms.

Subsumption or containment of two queries (or terms) is an established technique for optimizing query evaluation: a query q_1 is said to be *subsumed* by or *contained* in a query q_2 if every possible answer to q_1 against every possible data is also an answer to q_2. Thus, given all answers to q_2, we can evaluate q_1 only against those answers rather than against the whole database.

For first-order terms, subsumption is efficient and employed for guaranteeing termination in tabling (or memoization) approaches to backward chaining of logic [144,50]. However, when we move from first-order terms to Web queries, subsumption (or containment) becomes quickly less efficient or even intractable. Xcerpt query terms have, as pointed out above, some similarity with XPath queries. Containment for various fragments of XPath is surveyed in [139], both in absence and in presence of a DTD. Here, we focus on the first setting, where no additional information about the schema of the data is available. However, Xcerpt query terms are a strict super-set of (navigational) XPath as investigated in [139]. In particular, the Xcerpt query terms may contain (multiple occurrences of the same) variables. This brings them closer to *conjunctive queries* (with negation and deep-equal), as considered in [152] on general relations, and in [18] for tree data. Basic Xcerpt query terms can be reduced to (unions of) conjunctive queries with negation. However, the injectivity of Xcerpt query terms (no two siblings may match with the same data node) and the presence of deep-equal (two nodes are deep-equal iff they have the same structure) have no direct counterpart in conjunctive query containment. Though [100] shows how inequalities in general affect conjunctive query containment, the effect of injectivity (or all-distinct constraints) on query containment has not been studied previously. The same applies to deep-equal, though the results in [103] indicate that in *absence* of composition deep-equal has no effect on evaluation and thus likely on containment complexity.

For Xcerpt query terms, subsumption is, naturally, of interest for the design of a terminating, efficient Xcerpt engine. Beyond that, however, it is particularly relevant in a Web setting. Whenever we know that one query subsumes another, we do not need to access whatever data the two queries access twice, but rather can evaluate both queries with a single access to the basic data by evaluating the second query on the answers of the first one. This can be a key optimization also in the context of search engines, where answers to frequent queries can be memorized so as to avoid their repeated computation. Even though today's search engines are rather blind of the tree or graph structure of HTML, XML and RDF data, there is no doubt that some more or less limited form of structured queries will become more and more frequent in the future (see Google scholar's "search by author, date, etc."). Query subsumption, or containment, is key to a selection of queries, the answers to which are to be stored so as to allow as many queries as possible to be evaluated against that small set of data rather than against the entire search engine data. Thus, the notion of simulation subsumption proposed in this chapter can be seen as a building block of future, structure-aware search engines.

Therefore, we study in this section subsumption of Xcerpt query terms. The main building blocks of this section are the following.

– we introduce and formalize a notion of subsumption for Xcerpt query terms, called *simulation subsumption*, in Section 2.5.2. To the best of our knowledge, this is the first notion of subsumption for queries with injectivity of sibling nodes and deep-equal.
– we show, also in Section 2.5.2, that simulation on ground query terms is equivalent to simulation subsumption.[40] This also shows that ground query term simulation as introduced in [134] indeed captures the intuition that a query term that simulates into another query term subsumes that term.
– we define, in Section 2.5.2, a *rewriting system* that allows us to reduce the test for subsumption of q in q' to finding a sequence of syntactic transformations that can be applied to q to transform it into q'.
– we show, in Section 2.5.2, that this rewriting system gives rise to an algorithm for testing subsumption that is sound and complete and can determine whether q subsumes q' in time $O(n!^n)$. In particular, this shows that simulation subsumption is decidable.

Xcerpt Basics: Query Terms and Simulation. Query terms are an abstraction for queries that can be used to extract data from semi-structured trees. In contrast to XPath queries, they may contain (multiple occurrences of the same) variables and demand

---

[40] With small adaptions of the treatment of regular expressions and negated subterms in query term simulation.

an *injective mapping* of the child terms of each term. For example, the XPath query /a/b[c]/c demands that the document root has label a, and has a child term with label b that has itself a child term with label c. The subterm c that is given within the predicate of b can be mapped to the same node in the data as the child named c of b. Therefore, this XPath query would be equivalent to the query term $a\{\{b\{\{c\}\}\}\}$, but not to $a\{\{b\{\{c,c\}\}\}\}$. Simulation could be, however, easily modified to drop the injectivity requirement.

Simulation Subsumption. In this section, we first introduce simulation subsumption (Definition 28), then for several query terms we discuss whether one subsumes the other to give an intuition for the compositionality of the subsumption relationship. Subsequently, the transitivity of the subsumption relationship is proven (Lemma 4), some conclusions about the membership in the subsumption relationship of subterms, given the membership in the subsumption relationship of their parent terms are stated. These conclusions formalize the compositionality of simulation subsumption and are a necessary condition for the completeness of the rewriting system introduced in Section 2.5.2.

In tabled evaluation of logic programs, solutions to subgoals are saved in a solution table, such that for equivalent or subsumed subgoals, these sets do not have to be recomputed. As mentioned before, this avoidance of re-computation does not only save time, but can, in certain cases be crucial for the termination of a backward chaining evaluation of a program. In order to classify subgoal as solution or look-up goals, boolean subsumption as specified by Definition 28 must be decided. Although Xcerpt query terms may contain variables, n-ary subsumption as defined in [139] would be too strict for our purposes. To see this, consider the Xcerpt query terms $q_1 := a\{\{var\ X\}\}$ and $q_2 := a\{\{c\}\}$. Although all data terms that are relevant for q_2 can be found in the solutions for q_1, q_1 and q_2 cannot be compared by n-ary containment, because they differ in the number of their query variables.

Definition 28 (Simulation Subsumption). *A query term q_1 subsumes another query term q_2 if all data terms that q_2 simulates with are also simulated by q_1.*

Example 1 (Examples for the subsumption relationship). Let the query terms $q_1, \ldots q_5$ be given by:

- $q_1 := a\{\{\}\}$
- $q_2 := a\{\{desc\ b, desc\ c, d\}\}$
- $q_3 := a\{\{desc\ b, c, d\}\}$
- $q_4 := a\{\{without\ e\}\}$
- $q_5 := a\{\{without\ e\{\{without\ f\}\}\}\}$

Then the following subsumption relationships hold:

- q_2 subsumes q_3 because it requires less than q_3: While q_3 requires that the data has outermost label a, subterms c and d as well as a descendant subterm b, q_2 requires not that there is a direct subterm c, but only a descendant subterm. Since every descendant subterm is also a direct subterm, all data terms simulating with q_3 also simulate with q_2.

But the subsumption relationship can also be decided in terms of simulation: q_2 subsumes q_3, because there is a mapping π from the direct subterms $ChildT(q_2)$ of q_2 to the direct subterms $ChildT(q_3)$ of q_3, such that q_i subsumes $\pi(q_i)$ for all q_i in $ChildT(q_2)$.

- q_3 does not subsume q_2, since there are data terms that simulate with q_2, but not with q_3. One such data term is $d := a\{b, e\{c\}, d\}$.
  Again, the subsumption relationship between q_3 and q_2 (in this order) can be decided by simulation. There is no mapping π from the direct subterms of q_3 to the direct subterms of q_2, such that a simulates into $\pi(a)$.
- q_1 subsumes q_4 since it requires less than q_4. All data terms that simulate with q_4 also simulate with q_1.
- q_4 does not subsume q_1, since the data term $a\{\{e\}\}$ simulates with q_1, but does not simulate with q_4.
- q_5 subsumes q_4, but not the other way around.

Proposition 4. *The subsumption relationship between query terms is transitive, i.e. for arbitrary query terms q_1, q_2 and q_3 it holds that if q_1 subsumes q_2 and q_2 subsumes q_3, then q_1 subsumes q_3.*

Proposition 4 immediately follows from the transitivity of the subset relationship. Query term simulation and subsumption are defined in a way such that, given the simulation subsumption between two query terms, one can draw conclusions about subsumption relationships that must be fulfilled between pairs of subterms of the query terms. Lemma 1 formalizes these sets of conclusions.

Lemma 1 (Subterm Subsumption). *Let q_1 and q_2 be query terms such that q_1 subsumes q_2. Then there is an injective mapping π from $ChildT^+(q_1)$ to $ChildT^+(q_2)$ such that q_1^i subsumes $\pi(q_1^i)$ for all $q_1^i \in ChildT^+(q_1)$.*

Furthermore, if q_1 and q_2 are breadth-incomplete, then there is a (not necessarily injective) mapping σ from $ChildT^-(q_1)$ to $ChildT^-(q_2)$ such that $pos(\sigma(q_1^j))$ subsumes $pos(q_1^j)$ for all $q_1^j \in ChildT^-(q_1)$.

If q_1 is breadth-incomplete and q_2 is breadth-complete then there is no q_1^j in $ChildT^-(q_1)$ and $q_2^k \in ChildT^+(q_2) \setminus range(\pi)$ such that $pos(q_1^j) \leq q_2^k$.

Lemma 1 immediately follows from the equivalence of the subsumption relationship and the extended query term simulation (see Lemma 4 in the appendix of [35]).

Simulation Subsumption by Rewriting. In this section, we lay the foundations for a proof for the decidability of subsumption between query terms according to Definition 28 by introducing a rewriting system from one query term to another, which is later shown to be sound and complete. Furthermore, this rewriting system lays the foundation for the complexity analysis in Section 2.5.2.

The transformation of a query term q_1 into a subsumed query term q_2 is exemplified in Figure 2.5.2.

Definition 29 (Subsumption monotone query term transformations). *Let q be a query term. The following is a list of so-called* subsumption monotone *query term transformations.*

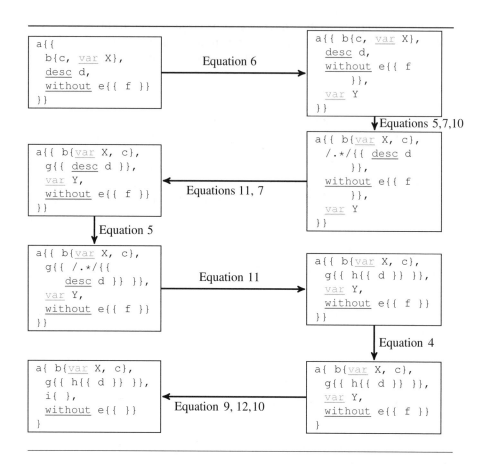

– *if q has incomplete subterm specification, it may be transformed to the analogous query term with complete subterm specification.*

$$\frac{a\{\{q_1,\ldots,q_n\}\}}{a\{q_1,\ldots,q_n\}},\tag{4}$$

– *if q is of the form desc q' then the descendant construct may be eliminated or it may be split into two descendant constructs separated by the regular expression* /.*/, *the inner descendant construct being wrapped in double curly braces.*

$$\frac{desc\ q}{q},\qquad\qquad \frac{desc\ q}{desc\ /.*/\{\{desc\ q\}\}}\tag{5}$$

– *if q has incomplete-unordered subterm specification, then a fresh variable X may be appended to the end of the subterm list. A* fresh *variable is a variable that does not occur in* q_1 *or* q_2 *and is not otherwise introduced by the rewriting system.*

$$X\ \text{fresh} \Rightarrow \frac{a\{\{q_1,\ldots,q_n\}\},}{a\{\{q_1,\ldots,q_n, var\ X\}\}}\tag{6}$$

– *if q has unordered subterm specification, then the subterms of q may be arbitrarily permuted.*

$$\pi \in Perms(\{1,\ldots,n\}) \Rightarrow \frac{a\{\{q_1, \quad \ldots, \quad q_n\}\}}{a\{\{q_{\pi(1)}, \quad \ldots, \quad q_{\pi(n)}\}\}} \tag{7}$$

$$\pi \in Perms(\{1,\ldots,n\}) \Rightarrow \frac{a\{q_1, \quad \ldots, \quad q_n\}}{a\{q_{\pi(1)}, \quad \ldots, \quad q_{\pi(n)}\}} \tag{8}$$

– *if q contains a variable* var X, *which occurs in q at least once in a positive context (i.e. not within the scope of a* without*) then all occurrences of* var X *may be substituted by another Xcerpt query term.*

$$X \in PV(q), t \in QTerms \Rightarrow \frac{q}{q\{X \mapsto t\}} \tag{9}$$

This rule may only be applied, if q contains all *occurrences of X in q_1. Furthermore, no further rewriting rules may be applied to the replacement term t.*
If a variable appears within q only in a negative context (i.e. within the scope of a without*), the variable cannot be substituted by an arbitrary term to yield a transformed term that is subsumed by q. The query terms* a{{ without var X }} *and* a{{ without b{ } }} *together with the data term* a{ c } *illustrate this characteristic of the subsumption relationship. For further discussion of substitution of variables in a negative context see Example 2.*
– *if q has a subterm q_i, then q_i may be transformed by any of the transformations in this list except for Equation 9 to the term $t(q_i)$, and this transformed version may be substituted at the place of q_i in q, as formalized by the following rule:* [41,42]

$$\frac{q_i}{t(q_i)} \Rightarrow \frac{a\{\{q_1,\ldots,q_n\}\}}{a\{\{q_1,\ldots,q_{i-1},t(q_i),q_{i+1},\ldots q_n\}\}} \tag{10}$$

– *if the label of q is a regular expression e, this regular expression may be replaced by any label that matches with e, or any other regular expression e' which is subsumed by e (see Definition 8 in the appendix of [35]).* [41]

$$e \in RE, e \text{ subsumes } e' \Rightarrow \frac{e\{\{q_1,\ldots,q_n\}\}}{e'\{\{q_1,\ldots,q_n\}\}} \tag{11}$$

– *if q contains a negated subterm q_i = without r and r' is a query term such that $t(r') = r$ (i.e. r' subsumes r) for some transformation step t, then q_i can be replaced by $q_i' :=$ without r'.*

$$(q_i = \text{without } r) \wedge \frac{r'}{r} \wedge (q_i' = \text{without } r') \Rightarrow \frac{a\{\{q_1,\ldots,q_i,\ldots,q_n\}\}}{a\{\{q_1,\ldots,q_i',\ldots q_n\}\}} \tag{12}$$

---

[41] The respective rules for complete-unordered subterm specification, incomplete-ordered subterm specification and complete-ordered subterm specification are omitted for the sake of brevity.

[42] The exclusion of Equation 9 ensures that variable substitutions are only applied to entire query terms and not to subterms. Otherwise the same variable might be substituted by different terms in different subterms.

Properties of the Rewriting System. In this section, we show that the rewriting system introduced in the previous section is sound (Section 2.5.2) and complete (Section 2.5.2). Furthermore, we study the structure of the search tree induced by the rewriting rules, show that it can be pruned without losing the completeness of the rewriting system and conclude that simulation subsumption is decidable. Finally we derive complexity results from the size of the search tree in Section 2.5.2.

Subsumption Monotonicity and Soundness

Lemma 2 (Monotonicity of the transformations in Definition 29). *All of the transformations given in Definition 29 are subsumption monotone, i.e. for any query term q and a transformation from Definition 29 which is applicable to q, q subsumes $t(q)$.*

The proof of Lemma 2 is straight-forward since each of the transformation steps can be shown independently of the others. For all of the transformations, inverse transformation steps t^{-1} can be defined, and obviously for any query term q it holds that $t^{-1}(q)$ subsumes q.

Lemma 3 (Transitivity of the subsumption relationship, monotonicity of a sequence of subsumption monotone query term transformations). *For a sequence of subsumption monotone query term transformations $t_1, \ldots, t_n$, and an arbitrary query term q, q subsumes $t_1 \circ \ldots \circ t_n(q_1)$.*

The transitivity of the subsumption relationship is immediate from its definition (Definition 28) which is based on the subset relationship, which is itself transitive.

As mentioned above, the substitution of a variable X in a negative context of a query term q by a query term t, which is not a variable, results in a query term $q' := q[X \mapsto t]$ which is in fact more general than q. In other words $q[X \mapsto t]$ subsumes q for any query term q if X only appears within a negative context in q. On the other hand, if X only appears in a positive context within q, then q' is less general – i.e. q subsumes q'. But what about the case of X appearing both in a positive and a negative context within q? Consider the following example:

Example 2. Let $q := $ a{{ var X, without b{{ var X }} }}. It may be tempting to think that substituting X by c[] to give q' makes the first subterm of q less general, but the second subterm of q more general. In fact, a subterm b[c] within a data term would cause the subterm without b{{ var X }} of q to fail, but the respective subterm of q' to succeed, suggesting that there is a data term that simulation unifies with q', but not with q, meaning that q does not subsume q'. However, there is no such data term, which is due to the fact that the second occurrence of X within q is only a consuming occurrence. When this part of the query term is evaluated, the variable X is already bound.

The normalized form for Xcerpt query terms is introduced, because for an unnormalized query term q_1 that subsumes a query term q_2 one cannot guarantee that there is a sequence of subsumption monotone query term transformations $t_1, \ldots, t_n$ such that $t_n \circ \ldots \circ t_1(q_1) = q_2$. To see this, consider example 3.

Example 3 (Impossibility of transforming an unnormalized query term). Consider $q_1 :=$ $a\{\{var\ X\ as\ b\{\{c\}\}, var\ X\ as\ b\{\{d\}\}\}\}$ and $q_2 := a\{\{b\{\{c,d\}\}, b\{\{c,d\}\}\}\}$. q_2 subsumes q_1, in fact both terms are even simulation equivalent. But there is no sequence of subsumption monotone query term transformations from q_2 to q_1, since one would have to omit one subterm from both the first subterm of q_2 and from the second one. But such a transformation would in general not be subsumption monotone.

Besides opening up the possibility of specifying restrictions on one subterm non-locally, duplicate restrictions for the same variable also allow the formulation of *unsatisfiable* query terms, as the following example shows:

Example 4 (Unsatisfiable query terms due to variable restrictions). Consider the query terms $q_1 := a\{\{var\ X\ as\ b,\ var\ X\ as\ c\}\}$ and $q_2 := b\{\{\}\}$. It is easy to see that q_1 is unsatisfiable, and thus q_2 subsumes q_1. However, there is no transformation sequence from q_2 to q_1.

Also single variable restrictions may in some cases be problematic, because they allow the specification of infinite, or at least graph structured data terms as example 5 shows:

Example 5 (Nested variable restrictions). Consider the query terms $q_1 := a\{\{var\ X\ as$ $b\{\{var\ X\}\}\}\}$ and $q_2 := a\{\{var\ Y\ as\ b\{\{b\{\{var\ Y\}\}\}\}\}\}$.

To overcome this issue, query terms are assumed to be in normalized form (Definition 30). In fact, almost all Xcerpt query terms can be transformed into normalized form.

Definition 30 (Query terms in normalized form). *A query term containing only a single variable restriction for each variable is a query term in* normalized form. *A query term which can be converted into an equivalent query term in normalized form is said to be* normalizable.

Unsatisfiability of query terms makes the decision procedure for subsumption more complex, and thus it is to be avoided whenever possible. Allowing the specification of unsatisfiable query terms does not add expressive power to a query language, and should thus be disallowed. Apart from the normal form, also *subterm injectivity* is a means for preventing the user of the Xcerpt query language from specifying unsatisfiable queries.

Example 6 (Unsatisfiability due to non-injectivity). In this example we use triple curly braces to state that the mapping from the siblings enclosed within the braces need not be injective. With this notation queries become less restrictive as the number of braces in the subterm specification increases. Let $q_1 := a\{\{\{b,\ without\ b\}\}\}$. Since q_1 both requires and forbids the presence of a subterm with label b, it is clearly unsatisfiable. Let $q_2 := b\{\{\}\}$. Although q_2 subsumes q_1, we cannot find a subsumption monotone transformation sequence from q_2 to q_1.

The above example shows that the the proof for the decidability of the subsumption relationship given in this section relies on the injectivity of the subterm mapping. Since there is no injectivity requirement for multiple consecutive predicates in XPath, the proof cannot be trivially used to show decidability of subsumption of XPath fragments.

Completeness

Theorem 6 (Subsumption by transformation). *Let q_1 and q_2 be two query terms in normalized form such that q_1 subsumes q_2. Then q_1 can be transformed into q_2 by a sequence of subsumption monotone query term transformations listed in Definition 29.*

Proof. We distinguish two cases:

- q_1 and q_2 are subsumption equivalent (i.e. they subsume each other)
- q_1 strictly subsumes q_2

The first case is the easier one. If q_1 and q_2 are subsumption equivalent, then there is no data term t, such that t simulates with one, but not the other. Hence q_1 and q_2 are merely syntactical variants of each other. Then q_1 can be transformed into q_2 by consistent renaming of variables (Equation 10), and by reordering sibling terms within subterms of q (Equation 7). This would not be true for unnormalized query terms as Example 3 shows.

The second is shown by structural induction on q_1.

For both the induction base and the induction step, we assume that q_1 subsumes q_2, but that the inverse is false. Then there is a data term d, such that q_1 simulates into d, but q_2 does not. In both the induction base and the induction step, we give a distinction of cases, enumerating all possible reasons for q_1 simulating into d but q_2 not. For each of these cases, a sequence of subsumption monotone transformations $t_1, \ldots t_n$ from Definition 29 is given, such that $q'_1 := t_n \circ t_{n-1} \circ \ldots \circ t_1(q_1)$ does *not* simulate into d. By Lemmas 2 and 3, q'_1 still subsumes q_2. Hence by considering d and by applying the transformations, q_1 is brought "closer" to q_2. If q'_1 is still more general than q_2, then one more dataterm d' can be found that simulates with q'_1, but not with q_2, and another sequence of transformations to be applied can be deduced from this theorem. This process can be repeated until q_1 has been transformed into a simulation equivalent version of q_2. For the proof, see the appendix of [35]. ☐

Decidability and Complexity. In the previous section, we establish that, for each pair of query terms q_1, q_2 such that q_1 subsumes q_2, there is a (possibly infinite) sequence of transformations $t_1, \ldots, t_k$ by one of the rules in Section 2.5.2 such that $t_k \circ \ldots \circ t_1(q) = q_2$.

However, if we reconsider the proof of Theorem 6, it is quite obvious that the sequence of transformations can in fact not be infinite: Intuitively, we transform at each step in the proof q_1 further towards q_2, guided by a data term that simulates in q_1 but not in q_2. In fact, the length of a transformation sequence is bounded by the sum of the sizes of the two query terms. As size of a query term we consider the total number of its subterms.

Proposition 5 (Length of Transformation Sequences). *Let q_1 and q_2 be two Xcerpt query terms such that q_1 subsumes q_2 and n the sum of the sizes of q_1 and q_2. Then, there is a sequence of transformations $t_1, \ldots, t_k$ such that $t_k \circ \ldots \circ t_1(q_1) = q_2$ and $k \in O(n)$.*

Proof. We show that the sequences of transformations created by the proof of Theorem 6 can be bounded by $O(n + m)$ if computed in a specific way: We maintain a mapping μ from subterms of q_1 to subterms of q_2 indicating how the query terms are mapped. μ is initialized with (q_1, q_2). In the following, we call a data term d *discriminating* between q_1 and q_2 if q_1 simulates in d but not q_2.

(1) For each pair (q, q') in μ, we first choose a discriminating data term that matches case 1 in the proof of Theorem 6. If there is such a data term, we apply Equation (11), label replacement, once to q obtaining $t(q)$ and update the pair in μ by $(t(q), q')$. This step is performed at most once for each pair as $(t(q), q')$ have the same label and thus there is no more discriminating data term that matches case 1.

(2) Otherwise, we next choose a discriminating data term that matches case 2.a.i or 2.b.i. In both cases, we apply Equation (6), variable insertion, to insert a new variable and update the pair in μ. This step is performed at most $|q_2| - |q_1| \le n$ times for each pair.

(3) Otherwise, we next choose a discriminating data term that matches case 2.a.ii and apply Equation (4), complete term specification and update the pair in μ. This step is performed at most once for each pair.

(4) Finally, the only type of discriminating data term that remains is one with the same number of positive child terms as q_2. We use an oracle to guess the right mapping σ from child terms of q_1 to child terms of q_2. Then we remove the pair from μ and add $(c, \sigma(c))$ to μ for each child term of q_1. This step is performed at most once for each pair in μ.

Since query subterms have a single parent, we add each subterm only once to μ in a pair. Except for case 2, we perform only a constant number of transformations to each pair. Case 2 allows up to n transformations for a single pair, but the total number of transformations (over all pairs) due to case 2 is bound by the size of q_2. Thus in total we perform at most $4 \cdot n$ transformations where n is the sum of the number of the sizes of q_1 and q_2.

Though we have established that the length of a transformation sequence is bound by $O(n)$, we also have to consider how to *find* such a transformation sequence. The proof of Proposition 5, already spells out an algorithm for finding such transformation sequences. However, it uses an oracle to guess the right mapping between child terms of two terms that are to be transformed. A naive deterministic algorithm needs to consider all possible such mappings whose number is bound by $O(n!)$. It is worth noting, however, that in most cases the actual number of such mappings is much smaller as most query terms have fairly low breadth and the possible mappings between their child terms are severely reduced just by considering only mappings where the labels of child terms simulate. However, in the worst case the $O(n!)$ complexity for finding the right mapping may be reached and thus we obtain:

Theorem 7 (Complexity of Subsumption by Rewriting). *Let q_1 and q_2 be two Xcerpt query terms. Then we can test whether q_1 subsumes q_2 in $O(n!^n)$ time.*

Proof. By proposition 5 we can find a $O(n)$ length transformation sequence in $O(n!^n)$ time and by Theorem 6 q_1 subsumes q_2 if and only if there is such a sequence.

Future Work in the area of Xcerpt Query Term Subsumption. Starting out from the problem of improving termination of logic programming based on rich kinds of simulation such as simulation unification, this section investigates the problem of deciding simulation subsumption between query terms. A rewriting system consisting of subsumption monotone query term transformations is introduced and shown to be sound and complete. By convenient pruning of the search tree defined by this rewriting system, the decidability of simulation subsumption is proven, and an upper bound for its complexity is identified.

Future work includes (a) a proof-of-concept implementation of the rewriting system, (b) the development of heuristics and their incorporation into the prototype to ensure fast termination of the algorithm in the cases when it is possible, (c) the study of the complexity of the problem in absence of subterm negation, descendant constructs, deep-equal, and/or injectivity, (d) the implementation of a backward chaining algorithm with tabling, which uses subsumption checking to avoid redundant computations and infinite branches in the resolution tree, and (e) the adaptation of the rewriting system to XPath in order to decide subsumption and to derive complexity results for the subsumption problem between XPath queries.

2.6 Conclusion

The Merriam-Webster dictionary defines versatile as "embracing a variety of subjects, fields or skills", as "turning with ease from one thing to another", and as "having many uses or applications". As shown in this chapter, the query language Xcerpt embraces both tree and graph-shaped data (in particular also relational data), Web and Semantic Web data, semantic data embedded in HTML as microformats and purely semantic data. It can be used to query data on a syntactic and on a semantic level, and it turns easily between the formats that it supports, allowing the transformation of one format to another within a single rule.

Having isolated the concept of simulation unification as a matching algorithm that can be adapted to any kind of semi-structured data, the single rule and multi-rule semantics of Xcerpt become versatile in the sense that new forms of simulations for new Web formats (e.g., topic maps) can be "plugged into" Xcerpt without having to adapt the semantics of single rules, and the semantics of negation as failure of possibly recursive multi-rule programs.

Datalog with negation and value invention can be used to precisely formulate the semantics of Xcerpt rules, no matter of the type of data being queried. Since it is a well-studied fragment of first order logic this provides an easy-to-understand semantics of Xcerpt for query authors that have some background knowledge in rule based formalisms. In particular, we use this translation for proving a number of computational properties of Xcerpt and some of its sub-languages.

Furthermore, the translation to Datalog with negation and value invention can serve as a basis for an implementation of Xcerpt in a relational database. For this aim, we also need a compact and efficient representation of both tree- and graph-shaped semi-structured data. Such a representation is discussed in Section 2.5.1. We showed that this representation allows constant time and constant per-node space reachability and adjacency test for all trees and many graphs.

While this article aims at giving an answer to the design questions for versatile web query languages, it has also raised a number of new questions and desires:

- Section 2.3 shows that Xcerpt is suitable for querying HTML, XML, RDF and microformat data. Xcerpt queries for extracting data from microformats, however, exhibit all the same underlying characteristics: excessive use of the *descendant* axis, ignoring XML element labels by using regular expressions, filtering elements according to the value of the `class` attribute. While in regular XML querying, the child axis is often more prevalent than the attribute axis, and element labels are more distinguishing than attribute values (except for `id`-attributes), these pairs have switched roles in microformat querying. With microformats becoming the de facto standard of the "lowercase semantic web" [97], query patterns specifically aimed at micro-formats and sharing the same characteristics as simulation are a valuable investigation. Alternatively a domain specific language for microformats only could be of use for Semantic Web programmers.
- As mentioned in Section 2.4, the idea of weak stratification could be carried over to rule based languages with a rich unficication algorithm such as simulation unification.
- Guaranteeing termination of backward chaining evaluation of possibly recursive multi-rule programs involving negation has received a large amount of attention in the past [144,133] [130]. Termination is even a bigger issue for recursive rule based languages with a rich unification algorithm, since there is a larger variety of infinite branches of subsumption monotone subgoals. A subsumption-aware resolution algorithm for rule based languages with rich unfication and negation as failure is currently being implemented by the authors.

References

1. Abiteboul, S., Buneman, P., Suciu, D.: Data on the Web: from relations to semistructured data and XML. Morgan Kaufmann Publishers Inc., San Francisco (2000)
2. Abiteboul, S., Hull, R., Vianu, V.: Foundations of Databases. Addison-Wesley Publishing Co., Boston (1995)
3. Abiteboul, S., Quass, D., McHugh, J., Widom, J., Wienerm, J.L.: The Lorel Query Language for Semistructured Data. Intl. Journal on Digital Libraries 1(1), 68–88 (1997)
4. Adida, B.: hGRDDL: Bridging microformats and RDFa. J. Web Sem. 6(1), 54–60 (2008)
5. Adida, B., Birbeck, M.: RDFa primer 1.0 embedding RDF in XHTML. W3c working draft, W3C (October 2007)
6. Agrawal, R., Borgida, A., Jagadish, H.V.: Efficient management of transitive relationships in large data and knowledge bases. In: Proc. ACM Symp. on Management of Data (SIGMOD), pp. 253–262. ACM, New York (1989)
7. Akhtar, W., Kopecky, J., Krennwallner, T., Polleres, A.: XSPARQL: Traveling between the XML and RDF worlds – and avoiding the XSLT pilgrimage. In: Hauswirth, M., Koubarakis, M., Bechhofer, S. (eds.) ESWC 2008. LNCS, vol. 5021, pp. 432–447. Springer, Heidelberg (2008)
8. Al-Khalifa, S., Jagadish, H.V., Koudas, N., Patel, J.M., Srivastava, D., Wu, Y.: Structural Joins: A Primitive for Efficient XML Query Pattern Matching. In: Proc. Int. Conf. on Data Engineering, Washington, DC, USA, p. 141. IEEE Computer Society, Los Alamitos (2002)

9. Apple Inc.: plist — Property List Format (2003)
10. Apt, K.R., Bol, R.N.: Logic programming and negation: A survey. J. Log. Program. 19(20), 9–71 (1994)
11. Assmann, U., Berger, S., Bry, F., Furche, T., Henriksson, J., Johannes, J.: Modular web queries — from rules to stores. In: Meersman, R., Tari, Z., Herrero, P. (eds.) OTM-WS 2007, Part I. LNCS, vol. 4805. Springer, Heidelberg (2007)
12. Augurusa, E., Braga, D., Campi, A., Ceri, S.: Design and implementation of a graphical interface to XQuery. In: SAC 2003: Proceedings of the 2003 ACM symposium on Applied computing, pp. 1163–1167. ACM, New York (2003)
13. Backett, D.: Turtle—Terse RDF Triple Language. Technical report, Institute for Learning and Research Technology, University of Bristol (2007)
14. Bailey, J., Bry, F., Furche, T., Schaffert, S.: Web and semantic web query languages: A survey. In: Eisinger, N., Małuszyński, J. (eds.) Reasoning Web 2005. LNCS, vol. 3564, pp. 35–133. Springer, Heidelberg (2005)
15. Beckett, D., McBride, B.: RDF/XML Syntax Specification (Revised). Recommendation, W3C (2004)
16. Benedikt, M., Koch, C.: Xpath leashed. ACM Computing Surveys (2007)
17. Berger, S., Bry, F., Bolzer, O., Furche, T., Schaffert, S., Wieser, C.: Xcerpt and visxcerpt: Twin query languages for the semantic web. In: McIlraith, S.A., Plexousakis, D., van Harmelen, F. (eds.) ISWC 2004. LNCS, vol. 3298. Springer, Heidelberg (2004)
18. Björklund, H., Martens, W., Schwentick, T.: Conjunctive query containment over trees. In: Arenas, M., Schwartzbach, M.I. (eds.) DBPL 2007. LNCS, vol. 4797, pp. 66–80. Springer, Heidelberg (2007)
19. Boag, S., Berglund, A., Chamberlin, D., Siméon, J., Kay, M., Robie, J., Fernández, M.F.: XML path language (XPath) 2.0. W3C recommendation, W3C (January 2007), http://www.w3.org/TR/2007/REC-xpath20-20070123/
20. Bolzer, O.: Towards Data-Integration on the Semantic Web: Querying RDF with Xcerpt. Diplomarbeit/diploma thesis, University of Munich (2005)
21. Bolzer, O.: Towards data-integration on the semantic web: Querying RDF with Xcerpt. Diplomarbeit/diploma thesis, Institute of Computer Science, LMU, Munich (2005)
22. Boncz, P., Grust, T., van Keulen, M., Manegold, S., Rittinger, J., Teubner, J.: MonetDB/XQuery: a fast XQuery Processor powered by a Relational Engine. In: Proc. ACM Symp. on Management of Data (SIGMOD), pp. 479–490. ACM Press, New York (2006)
23. Bonifati, A., Ceri, S.: Comparative analysis of five xml query languages. SIGMOD Rec. 29(1), 68–79 (2000)
24. Booth, K.S., Lueker, G.S.: Linear Algorithms to Recognize Interval Graphs and Test for the Consecutive Ones Property. In: Proc. of ACM Symposium on Theory of Computing, pp. 255–265. ACM Press, New York (1975)
25. Bray, T., Hollander, D., Layman, A., Tobin, R.: Namespaces in XML 1.0 (2nd edn.). W3C Rec. (August 16, 2006)
26. Bray, T., Hollander, D., Layman, A., Tobin, R.: Namespaces in XML (2nd edn.). Recommendation, W3C (2006)
27. Bray, T., Paoli, J., Sperberg-McQueen, C.M., Maler, E., Yergeau, F.: Extensible markup language (XML) 1.0 (4th edn.) (2006)
28. Bray, T., Paoli, J., Sperberg-McQueen, C.M., Maler, E., Yergeau, F.: Extensible Markup Language (XML) 1.0 (3rd edn.). Recommendation, W3C (2004)
29. Broekstra, Kampman, Harmelen: Sesame: A generic architecture for storing and querying RDF and RDF Schema (2003)
30. Bruno, N., Koudas, N., Srivastava, D.: Holistic Twig Joins: Optimal XML Pattern Matching. In: Proc. ACM SIGMOD Int. Conf. on Management of Data, pp. 310–321. ACM Press, New York (2002)

31. Bry, F., Eisinger, N., Eiter, T., Furche, T., Gottlob, G., Ley, C., Linse, B., Pichler, R., Wei, F.: Foundations of rule-based query answering. In: Antoniou, G., Aßmann, U., Baroglio, C., Decker, S., Henze, N., Patranjan, P.L., Tolksdorf, R. (eds.) Reasoning Web 2007. LNCS, vol. 4636, pp. 1–153. Springer, Heidelberg (2007)

32. Bry, F., Furche, T., Badea, L., Koch, C., Schaffert, S., Berger, S.: Querying the web reconsidered: Design principles for versatile web query languages. Journal of Semantic Web and Information Systems (IJSWIS) 1(2) (2005)

33. Bry, F., Furche, T., Ley, C., Linse, B.: RDFLog—taming existence - a logic-based query language for RDF (2007)

34. Bry, F., Furche, T., Ley, C., Linse, B., Marnette, B.: RDFLog: It's like datalog for RDF. In: Proceedings of 22nd Workshop on (Constraint) Logic Programming, Dresden, 30 September–1 October (2008)

35. Bry, F., Furche, T., Linse, B.: Simulation subsumption or déjà vu on the web (extended version). Technical Report PMS-FB-2008-01, University of Munich (2007)

36. Bry, F., Furche, T., Linse, B., Schroeder, A.: Efficient Evaluation of n-ary Conjunctive Queries over Trees and Graphs. In: Proc. ACM Int'l. Workshop on Web Information and Data Management (WIDM). ACM Press, New York (2006), 2 citations [Google Scholar]

37. Bry, F., Coskun, F., Durmaz, S., Furche, T., Olteanu, D., Spannagel, M.: The XML Stream Query Processor SPEX. In: Proc. Int'l. Conf. on Data Engineering (ICDE), pp. 1120–1121 (2005), 17 citations [Google Scholar]

38. Bry, F., Furche, T., Ley, C., Linse, B.: Rdflog: Filling in the blanks in rdf querying. Technical Report PMS-FB-2008-01, University of Munich (2007)

39. Bry, F., Furche, T., Ley, C., Linse, B., Marnette, B.: Taming existence in rdf querying. In: Calvanese, D., Lausen, G. (eds.) RR 2008. LNCS, vol. 5341, pp. 236–237. Springer, Heidelberg (2008)

40. Bry, F., Schaffert, S.: A Gentle Introduction into Xcerpt, a Rule-based Query and Transformation Language for XML. In: Proc. Intl. Workshop on Rule Markup Languages for Business Rules on the Semantic Web (2002)

41. Buneman, P., Fernandez, M.F., Suciu, D.: UnQL: a query language and algebra for semistructured data based on structural recursion. VLDB Journal: Very Large Data Bases 9(1), 76–110 (2000)

42. Bussche, J.V.D., Gucht, D.V., Andries, M., Gyssens, M.: On the completeness of object-creating database transformation languages. Journal of the ACM 44(2), 272–319 (1997)

43. Cabibbo, L.: The expressive power of stratified logic programs with value invention. Information and Computation 147(1), 22–56 (1998)

44. Carlos, J., Polleres, A., Polleres, A.: Sparql rules. Technical report, Universidad Rey Juan Carlos (2006)

45. Carroll, J.J., Bizer, C., Hayes, P., Stickler, P.: Named graphs, provenance and trust. In: WWW 2005: Proceedings of the 14th international conference on World Wide Web, pp. 613–622. ACM, New York (2005)

46. Ceri, S., Comai, S., Damiani, E., Fraternali, P., Paraboschi, S., Tanca, L.: XML-GL: a graphical language for querying and restructuring XML documents (1998)

47. Chamberlin, D.D., Robie, J., Florescu, D.: Quilt: An XML query language for heterogeneous data sources. In: Suciu, D., Vossen, G. (eds.) WebDB 2000. LNCS, vol. 1997, pp. 1–25. Springer, Heidelberg (2001)

48. Chen, L., Gupta, A., Kurul, M.E.: Stack-based algorithms for pattern matching on dags. In: Proc. Int'l. Conf. on Very Large Data Bases (VLDB), pp. 493–504. VLDB Endowment (2005)

49. Chen, T., Lu, J., Ling, T.W.: On Boosting Holism in XML Twig Pattern Matching using Structural Indexing Techniques. In: Proc. ACM SIGMOD Int. Conf. on Management of Data, pp. 455–466. ACM Press, New York (2005)

50. Chen, W., Warren, D.S.: Tabled evaluation with delaying for general logic programs. J. ACM 43(1), 20–74 (1996)
51. Chen, Z., Gehrke, J., Korn, F., Koudas, N., Shanmugasundaram, J., Srivastava, D.: Index structures for matching XML twigs using relational query processors. Data & Knowledge Engineering (DKE) 60(2), 283–302 (2007)
52. Cholak, P., Blair, H.A.: The complexity of local stratification. Fundam. Inform. 21(4), 333–344 (1994)
53. Cohen, E., Halperin, E., Kaplan, H., Zwick, U.: Reachability and Distance Queries via 2-hop Labels. In: Proc. ACM Symposium on Discrete Algorithms, Philadelphia, PA, USA, pp. 937–946. Society for Industrial and Applied Mathematics (2002)
54. Connolly, D.: Gleaning resource descriptions from dialects of languages (grddl). Recommendation, W3C (2007)
55. Cooper, B., Sample, N., Franklin, M.J., Hjaltason, G.R., Shadmon, M.: A Fast Index for Semistructured Data. In: Proc. Int. Conf. on Very Large Databases, pp. 341–350. Morgan Kaufmann Publishers Inc., San Francisco (2001)
56. Cowan, J., Tobin, R.: XML Information Set (2 edn.). Recommendation, W3C (2004)
57. Davis, I.: GRDDL primer (2006)
58. Deutsch, A., Fernández, M.F., Florescu, D., Levy, A.Y., Suciu, D.: XML-QL. In: QL (1998)
59. Dijkstra, E.W.: On the role of scientific thought (EWD447). In: Selected Writings on Computing: A Personal Perspective, pp. 60–66 (1982)
60. Droop, M., Flarer, M., Groppe, J., Groppe, S., Linnemann, V., Pinggera, J., Santner, F., Schier, M., Schöpf, F., Staffler, H., Zugal, S.: Translating xpath queries into sparql queries. In: On the Move (OTM 2007) Federated Conferences and Workshops (DOA, ODBASE, CoopIS, GADA, IS), 6th International Conference on Ontologies, DataBases, and Applications of Semantics (ODBASE 2007), pp. 9–10 (2007)
61. Eiter, T., Faber, W., Koch, C., Leone, N., Pfeifer, G.: DLV - a system for declarative problem solving. In: Proceedings of the 8th International Workshop on Non-Monotonic Reasoning (NMR 2000) (2000)
62. Eiter, T., Ianni, G., Krennwallner, T., Schindlauer, R.: Exploiting conjunctive queries in description logic programs. In: Choueiry, B.Y., Givan, B. (eds.) Informal Proceedings of the 10th International Symposium on Artificial Intelligence and Mathematics (ISAIM 2008), Ft. Lauderdale, Florida, January 2-4 (2008) (to appear) (invited paper)
63. Euzenat, J., Valtchev, P.: An integrative proximity measure for ontology alignment. In: Doan, A., Halevy, A., Noy, N. (eds.) Proceedings of the 1st Intl. Workshop on Semantic Integration. CEUR, vol. 82 (2003)
64. Euzenat, J., Valtchev, P.: Similarity-based ontology alignment in OWL-lite. In: de Mántaras, R.L., Saitta, L. (eds.) Proceedings of the 16th European Conference on Artificial Intelligence (ECAI 2004), pp. 333–337. IOS Press, Amsterdam (2004)
65. Fagin, R.: Multivalued dependencies and a new normal form for relational databases. ACM Transactions on Database Systems 2(3), 262–278 (1977)
66. Fallside, D.C., Walmsley, P.: XML Schema Part 0: Primer Second edn. Recommendation, W3C (2004)
67. Fernández, M., Malhotra, A., Marsh, J., Nagy, M., Walsh, N.: XQuery 1.0 and XPath 2.0 Data Model. Recommendation, W3C (2007)
68. Fulkerson, D.R., Gross, O.A.: Incidence Matrices and Interval Graphs. Pacific Journal of Mathematics 15(3), 835–855 (1965)
69. Furche, T.: Implementation of Web Query Language Reconsidered: Beyond Tree and Single-Language Algebras at (Almost) No Cost. Dissertation/doctoral thesis, Ludwig-Maxmilians University Munich (2008)

70. Furche, T., Linse, B., Bry, F., Plexousakis, D., Gottlob, G.: RDF querying: Language constructs and evaluation methods compared. In: Barahona, P., Bry, F., Franconi, E., Henze, N., Sattler, U. (eds.) Reasoning Web 2006. LNCS, vol. 4126, pp. 1–52. Springer, Heidelberg (2006)

71. Furche, T., Weinzierl, A., Bry, F.: Scalable, space-optimal implementation of xcerpt single rule programs—part 1: Data model, queries, and translation. Deliverable I4-D15a, REWERSE (2007)

72. Gandon, F.: GRDDL use cases: Scenarios of extracting RDF data from XML documents. W3c working group note 6 April 2007, W3C (2007)

73. Garcia-Molina, H., Ullman, J.D., Widom, J.: Database Systems: The Complete Book. Prentice Hall, Englewood Cliffs (2002)

74. Garshol, L.M., Moore, G.: ISO 13250-2: Topic Maps — Data Model. International standard, ISO/IEC (2006)

75. Gelfond, M., Lifschitz, V.: The stable model semantics for logic programming. In: Proceeding of the Fifth Logic Programming Symposium, pp. 1070–1080 (1988)

76. Goldman, R., Widom, J.: Dataguides: Enabling query formulation and optimization in semistructured databases. In: Proc. Int'l. Conf. on Very Large Data Bases (VLDB), pp. 436–445. Morgan Kaufmann Publishers Inc., San Francisco (1997)

77. Gottlob, G., Koch, C., Pichler, R.: Efficient Algorithms for Processing XPath Queries. ACM Transactions on Database Systems (2005)

78. Gottlob, G., Leone, N., Scarcello, F.: The Complexity of Acyclic Conjunctive Queries. Journal of the ACM 48(3), 431–498 (2001)

79. Groppe, S., Groppe, J., Linnemann, V., Kukulenz, D., Hoeller, N., Reinke, C.: Embedding sparql into xquery/xslt. In: SAC 2008: Proceedings of the 2008 ACM symposium on Applied computing, pp. 2271–2278. ACM, New York (2008)

80. Grust, T.: Accelerating XPath Location Steps. In: Proc. ACM Symp. on Management of Data (SIGMOD) (2002)

81. Grust, T., Keulen, M.V., Teubner, J.: Accelerating XPath Evaluation in any RDBMS. ACM Transactions on Database Systems 29(1), 91–131 (2004)

82. Grust, T., Teubner, J.: Relational Algebra: Mother Tongue - XQuery: Fluent. In: Proc. Twente Data Management Workshop on XML Databases and Information Retrieval (2004)

83. Grust, T., van Keulen, M., Teubner, J.: Staircase Join: Teach A Relational DBMS to Watch its (Axis) Steps. In: Proc. Int. Conf. on Very Large Databases (2003)

84. Habib, M., McConnell, R., Paul, C., Viennot, L.: Lex-BFS and Partition Refinement, with Applications to Transitive Orientation, Interval Graph Recognition and Consecutive Ones Testing. Theoretical Computer Science 234(1-2), 59–84 (2000)

85. Hayes, P., McBride, B.: Rdf semantics. Recommendation, W3C (2004)

86. Henzinger, M.R., Henzinger, T.A., Kopke, P.W.: Computing simulations on finite and infinite graphs. In: FOCS, pp. 453–462 (1995)

87. Hsu, W.L.: PC-Trees vs. PQ-Trees. In: Wang, J. (ed.) COCOON 2001. LNCS, vol. 2108, p. 207. Springer, Heidelberg (2001)

88. Hsu, W.L.: A Simple Test for the Consecutive Ones Property. Journal of Algorithms 43(1), 1–16 (2002)

89. Hull, R., Yoshikawa, M.: Ilog: Declarative creation and manipulation of object identifiers. In: Proc. Int'l. Conf. on Very Large Data Bases (VLDB), pp. 455–468. Morgan Kaufmann Publishers Inc., San Francisco (1990)

90. Jagadish, H.V., Lakshmanan, L.V.S., Srivastava, D., Thompson, K.: TAX: A Tree Algebra for XML. In: Ghelli, G., Grahne, G. (eds.) DBPL 2001. LNCS, vol. 2397, p. 149. Springer, Heidelberg (2002)

91. Jenner, B., Köbler, J., McKenzie, P., Torán, J.: Completeness results for graph isomorphism. Journal of Computer and System Sciences 66(3), 549–566 (2003)

92. Karvounarakis, G., Magkanaraki, A., Alexaki, S., Christophides, V., Plexousakis, D., Scholl, M., Tolle, K.: RQL: A functional query language for RDF. In: Gray, P.M.D., Kerschberg, L., King, P.J.H., Poulovassilis, A. (eds.) The Functional Approach to Data Management: Modelling, Analyzing and Integrating Heterogeneous Data. LNCS, pp. 435–465. Springer, Heidelberg (2004)

93. Kay, M.: Parsing in functional unification grammar. In: Dowty, D., Karttunen, L., Zwicky, A. (eds.) Natural Language Parsing: Psychological, Computational, and Theoretical Perspectives, pp. 251–278. Cambridge University Press, Cambridge (1985)

94. Kay, M.: Functional unification grammar: A formalism for machine translation. In: COLING 1984, Stanford, CA, pp. 75–78 (1984)

95. Kay, M.: XSL Transformations, Version 2.0. Recommendation, W3C (2007)

96. Kay, M.: XSL transformations (XSLT) version 2.0. W3C recommendation, W3C (January 2007), http://www.w3.org/TR/2007/REC-xslt20-20070123/

97. Khare, R.: Microformats: The next (small) thing on the semantic web? IEEE Internet Computing 10(1), 68–75 (2006)

98. Khare, R., Çelik, T.: Microformats: a pragmatic path to the semantic web. In: WWW 2006: Proceedings of the 15th international conference on World Wide Web, pp. 865–866. ACM Press, New York (2006)

99. Klaas, V.: Who's who in the world wide web: Approaches to name disambiguation. Diplomarbeit/diploma thesis, Institute of Computer Science, LMU, Munich (2007)

100. Klug, A.C.: On conjunctive queries containing inequalities. J. ACM 35(1), 146–160 (1988)

101. Klyne, G., Carroll, J.J., McBride, B.: Resource Description Framework (RDF): Concepts and Abstract Syntax. Recommendation, W3C (2004)

102. Knoblock, C.A., Minton, S., Ambite, J.L., Ashish, N., Modi, P.J., Muslea, I., Philpot, A., Tejada, S.: Modeling web sources for information integration. In: AAAI/IAAI, pp. 211–218 (1998)

103. Koch, C.: On the Complexity of Nonrecursive XQuery and Functional Query Languages on Complex Values. Tods 31(4) (2006)

104. Kochut, K., Janik, M.: SPARQLeR: Extended SPARQL for semantic association discovery. In: Franconi, E., Kifer, M., May, W. (eds.) ESWC 2007. LNCS, vol. 4519, pp. 145–159. Springer, Heidelberg (2007)

105. Kolaitis, P.G., Papadimitriou, C.H.: Why not negation by fixpoint? In: PODS, pp. 231–239. ACM, New York (1988)

106. Lenzerini, M.: Data integration: A theoretical perspective (2002)

107. Manola, F., Miller, E.: RDF primer, W3C recommendation. Technical report, W3C (2004)

108. Manola, F., Miller, E., McBride, B.: Rdf primer. Recommendation, W3C (2004)

109. Marsh, J.: XML Base. Recommendation, W3C (2001)

110. Martínez, J.M.: Mpeg-7 overview. Technical Report ISO/IEC JTC1/SC29/WG11N6828, International Organisation for Standardisation (ISO) (2004)

111. Marx, M.: Conditional XPath, the first order complete XPath dialect. In: Proceedings of the twenty-third ACM SIGMOD-SIGACT-SIGART symposium on Principles of database systems, pp. 13–22. ACM, New York (2004)

112. Marx, M.: Conditional XPath. ACM Transactions on Database Systems (TODS) 30(4), 929–959 (2005)

113. McBride, B.: Rdf vocabulary description language 1.0: Rdf schema (2004)

114. Meuss, H., Schulz, K.U.: Complete Answer Aggregates for Treelike Databases: A Novel Approach to Combine Querying and Navigation. ACM Transactions on Information Systems 19(2), 161–215 (2001)

115. Meuss, H., Schulz, K.U., Bry, F.: Towards Aggregated Answers for Semistructured Data. In: Proc. Intl. Conf. on Database Theory, pp. 346–360. Springer, Heidelberg (2001)

116. Milner, R.: An algebraic definition of simulation between programs. In: IJCAI, pp. 481–489 (1971)
117. Noy, N.F., Musen, M.A.: PROMPT: Algorithm and tool for automated ontology merging and alignment. In: AAAI/IAAI, pp. 450–455 (2000)
118. Olteanu, D.: SPEX: Streamed and Progressive Evaluation of XPath. IEEE Transactions on Knowledge and Data Engineering (2007)
119. Olteanu, D., Furche, T., Bry, F.: Evaluating Complex Queries against XML streams with Polynomial Combined Complexity. In: Williams, H., MacKinnon, L.M. (eds.) BNCOD 2004. LNCS, vol. 3112, pp. 31–44. Springer, Heidelberg (2004) 17 citations [Google Scholar]
120. Olteanu, D., Furche, T., Bry, F.: An Efficient Single-Pass Query Evaluator for XML Data Streams. In: Data Streams Track, Proc. ACM Symp. on Applied Computing (SAC), pp. 627–631 (2004) 17 citations [Google Scholar]
121. Olteanu, D., Meuss, H., Furche, T., Bry, F.: Xpath: Looking forward. In: Chaudhri, A.B., Unland, R., Djeraba, C., Lindner, W. (eds.) EDBT 2002, vol. 2490, pp. 109–127. Springer, Heidelberg (2002)
122. O'Neil, P., O'Neil, E., Pal, S., Cseri, I., Schaller, G., Westbury, N.: ORDPATHs: Insert-friendly XML Node Labels. In: Proc. ACM Symp. on Management of Data (SIGMOD), pp. 903–908. ACM Press, New York (2004)
123. Pepper, S.: The TAO of topic maps (2000)
124. Pérez, J., Arenas, M., Gutierrez, C.: Semantics and complexity of SPARQL. In: Cruz, I.F., Decker, S., Allemang, D., Preist, C., Schwabe, D., Mika, P., Uschold, M., Aroyo, L. (eds.) ISWC 2006. LNCS, vol. 4273, pp. 30–43. Springer, Heidelberg (2006)
125. Pérez, J., Arenas, M., Gutierrez, C.: nSPARQL: A navigational language for RDF. In: Sheth, A.P., Staab, S., Dean, M., Paolucci, M., Maynard, D., Finin, T.W., Thirunarayan, K. (eds.) ISWC 2008. LNCS, vol. 5318, pp. 66–81. Springer, Heidelberg (2008)
126. Pérez, J., Arenas, M., Gutierrez, C.: nSPARQL: A navigational language for rdf. In: Sheth, A.P., Staab, S., Dean, M., Paolucci, M., Maynard, D., Finin, T., Thirunarayan, K. (eds.) ISWC 2008. LNCS, vol. 5318, pp. 66–81. Springer, Heidelberg (2008)
127. Polleres, A.: From sparql to rules (and back). In: Williamson, C.L., Zurko, M.E., Patel-Schneider, P.F., Shenoy, P.J. (eds.) WWW, pp. 787–796. ACM, New York (2007)
128. Polleres, A., Krennwallner, T., Kopecky, J., Akhtar, W.: Xsparql: Traveling between the XML and rdf worlds – and avoiding the xslt pilgrimage. In: Bechhofer, S., Hauswirth, M., Hoffmann, J., Koubarakis, M. (eds.) ESWC 2008. LNCS, vol. 5021, pp. 432–447. Springer, Heidelberg (2008)
129. Przymusinska, H., Przymunsinski, T.C.: Weakly stratified logic programs. Fundam. Inf. 13(1), 51–65 (1990)
130. Przymusinski, T.C.: On the declarative semantics of deductive databases and logic programs. In: Foundations of Deductive Databases and Logic Programming, pp. 193–216. Morgan Kaufmann, San Francisco (1988)
131. Recordon, D., Reed, D.: OpenID 2.0: a platform for user-centric identity management. In: DIM 2006: Proceedings of the second ACM workshop on Digital identity management, pp. 11–16. ACM, New York (2006)
132. Ross, K.A.: Modular stratification and magic sets for DATALOG programs with negation. In: PODS, pp. 161–171. ACM Press, New York (1990)
133. Sagonas, K.F., Swift, T., Warren, D.S.: The XSB programming system. In: Workshop on Programming with Logic Databases (Informal Proceedings), ILPS, p. 164 (1993)
134. Schaffert, S.: Xcerpt: A Rule-Based Query and Transformation Language for the Web. PhD thesis, University of Munich (2004)
135. Schaffert, S.: Xcerpt: A Rule-Based Query and Transformation Language for the Web. Dissertation/doctoral thesis, University of Munich (2004)

136. Schenk, S., Staab, S.: Networked graphs: A declarative mechanism for sparql rules, sparql views and rdf data integration on the web. In: Proceedings of the 17th International World Wide Web Conference, Bejing, China (2008-04)

137. Schenkel, R., Theobald, A., Weikum, G.: HOPI: An Efficient Connection Index for Complex XML Document Collections. In: Bertino, E., Christodoulakis, S., Plexousakis, D., Christophides, V., Koubarakis, M., Böhm, K., Ferrari, E. (eds.) EDBT 2004. LNCS, vol. 2992, pp. 237–255. Springer, Heidelberg (2004)

138. Schneider, P.P., Simeon, J.: The yin/yang web: Xml syntax and rdf semantics. In: Proceedings of the eleventh international conference on World Wide Web, p. 11. ACM Press, New York (2002)

139. Schwentick, T.: Xpath query containment. SIGMOD Record 33(1), 101–109 (2004)

140. Seaborne, A., Manjunath, G., Bizer, C., Breslin, J., Das, S., Davis, I., Harris, S., Idehen, K., Corby, O., Kjernsmo, K., Nowack, B.: SPARQL/Update A language for updating RDF graphs. W3C Member Submission, W3C (July 2008),
http://www.w3.org/Submission/2008/04/

141. Seaborne, A., Prud'hommeaux, E.: SPARQL query language for RDF. W3C recommendation, W3C (January 2008),
http://www.w3.org/TR/2008/REC-rdf-sparql-query-20080115/

142. Siméon, J., Chamberlin, D., Florescu, D., Boag, S., Fernández, M.F., Robie, J.: XQuery 1.0: An XML query language. W3C recommendation, W3C (January 2007),
http://www.w3.org/TR/2007/REC-xquery-20070123/

143. Stickler, P.: Cbd - concise bounded description (2005)

144. Tamaki, H., Sato, T.: OLD resolution with tabulation. In: Shapiro, E.Y. (ed.) ICLP 1986. LNCS, vol. 225, pp. 84–98. Springer, Heidelberg (1986)

145. Trißl, S., Leser, U.: Fast and practical indexing and querying of very large graphs. In: Proc. ACM Symp. on Management of Data (SIGMOD), pp. 845–856. ACM, New York (2007)

146. Ullman, J.D.: Information integration using logical views. Theor. Comput. Sci. 239(2), 189–210 (2000)

147. van Gelder, A., Ross, K., Schlipf, J.: The well-founded semantics for general logic programs. Journal of the ACM 18, 620–650 (1991)

148. van Gelder, A., Ross, K.A., Schlipf, J.S.: The well-founded semantics for general logic programs. Journal of the ACM (1991)

149. W3C: Gleaning resource descriptions from dialects of languages (GRDDL). W3c recommendation, W3C (September 2007)

150. Walsh, N., Muellner, L.: DocBook: The Definitive Guide. O'Reilly, Sebastopol (1999)

151. Wang, H., He2, H., Yang, J., Yu, P.S., Yu, J.X.: Dual labeling: Answering graph reachability queries in constant time. In: Proc. Int'l. Conf. on Data Engineering (ICDE), Washington, DC, USA, p. 75. IEEE Computer Society, Los Alamitos (2006)

152. Wei, F., Lausen, G.: Containment of conjunctive queries with safe negation. In: Calvanese, D., Lenzerini, M., Motwani, R. (eds.) ICDT 2003. LNCS, vol. 2572, pp. 343–357. Springer, Heidelberg (2003)

153. Weigel, F., Schulz, K.U., Meuss, H.: The bird numbering scheme for xml and tree databases – deciding and reconstructing tree relations using efficient arithmetic operations. In: Bressan, S., Ceri, S., Hunt, E., Ives, Z.G., Bellahsène, Z., Rys, M., Unland, R. (eds.) XSym 2005. LNCS, vol. 3671, pp. 49–67. Springer, Heidelberg (2005)

Chapter 3
Evolution and Reactivity in the Semantic Web

José Júlio Alferes[1], Michael Eckert[2], and Wolfgang May[3]

[1] CENTRIA, Departamento de Informática, Faculdade de Ciências e Tecnologia,
Universidade Nova de Lisboa, Portugal
[2] Institut für Informatik, Ludwig-Maximilians-Universität München, Germany
[3] Institut für Informatik, Georg-August-Universität Göttingen, Germany

Abstract. Evolution and reactivity in the Semantic Web address the vision and concrete need for an active Web, where data sources evolve autonomously and perceive and react to events. In 2004, when the REWERSE project started, regarding work on Evolution and Reactivity in the Semantic Web there wasn't much more than a vision of such an active Web.

Materialising this vision requires the definition of a model, architecture, and also prototypical implementations capable of dealing with reactivity in the Semantic Web, including an ontology-based description of all concepts. This resulted in a general framework for reactive Event-Condition-Action rules in the Semantic Web over heterogeneous component languages.

Inasmuch as heterogeneity of languages is, in our view, an important aspect to take into consideration for dealing with the heterogeneity of sources and behaviour of the Semantic Web, concrete homogeneous languages targeting the specificity of reactive rules are of course also needed. This is especially the case for languages that can cope with the challenges posed by dealing with composite structures of events, or executing composite actions over Web data.

In this chapter we report on the advances made on this front, namely by describing the above-mentioned general heterogeneous framework, and by describing the concrete homogeneous language XChange.

3.1 Introduction

The Web and the Semantic Web, as we see it, can be understood as a "living organism" combining autonomously evolving data sources, each of them possibly reacting to events it perceives. The dynamic character of such a Web requires declarative languages and mechanisms for specifying the evolution of the data, and for specifying reactive behaviour on the Web.

Rather than a Web of data sources, we envisage a Web of information systems where each such system, besides being capable of gathering information (querying, both on persistent data, as well as on volatile data such as occurring events), can possibly update persistent data, communicate the changes, request changes of persistent data in other systems, and be able to react to requests from and changes on other systems. As a practical example, consider a set of data

F. Bry and J. Maluszynski (Eds.): Semantic Techniques for the Web, LNCS 5500, pp. 161–200, 2009.
© Springer-Verlag Berlin Heidelberg 2009

(re)sources in the Web of travel agencies, airline companies, train companies, etc. It should be possible to query the resources about timetables, availability of tickets, etc. But in such an evolving Semantic Web, it should also be possible for a train company to report on late trains, and travel agencies (and also individual clients) be able to detect such an event and react upon it, by rescheduling travel plans, notifying clients that in turn could have to cancel hotel reservations and book other hotels, or try alternatives to the late trains, etc.

ECA Rules

Some reactive languages have been proposed that allow for updating Web sources as the above ones, and are also capable of dealing-with/reacting-to some forms of events, evaluate conditions, and upon that act by updating data [12,11,4,61] (see Section 3.2). The common aspect of all of these languages is the use of declarative *Event-Condition-Action* (ECA) rules for specifying reactivity and evolution. Such kind of rules (also known as triggers, active rules, or reactive rules), that have been widely used in other fields (e.g. active databases [62,70]) have the general form:

on *event* **if** *condition* **do** *action*

They are intuitively easy to understand, and provide a well-understood semantics: when an event (atomic or composite) occurs, evaluate a condition, and if the condition (depending on the event, and possibly requiring further data) is satisfied then execute an action (or a sequence of actions, a program, a transaction, or even start a process).

In fact, we fully agree with the arguments exposed in the field of active databases for adopting ECA rules for dealing with evolution and reactivity in the Web (declarativity, modularity, maintainability, etc). Still, the existing languages fall short in various aspects, when aiming at the general view of an evolving Web as described above. In these languages, the events and actions are restricted to updates on the underlying data level; they do not provide for more composite events and actions. In a Semantic Web environment, actions are more than just simple updates to Web data (be it XML or RDF data), but application-level *actions*. As said above, besides that, actions can be notifications to other resources, or update requests of other resources, and they can be composed from simpler actions (like: do this, and then do that).

Moreover, events may in general be more than simple atomic events in Web data, as in the above languages. First, there are atomic events other than physical changes in Web data: events may be received messages, or even "happenings" in the global Web, which may require complex event detection mechanisms (e.g. (once) any train to Munich is delayed ...). Moreover, as in active databases [29,74], there may be *composite* events. For example, we may want a rule to be triggered when there is a flight cancellation and then the notification of a new reservation whose price is much higher than the previous (e.g. to complain to the airline company). In our view, a general language for reactivity in the Web should cater for such richer actions and events.

Such a general ECA language with richer actions and events, adapted to Web data, is not yet enough for fully materialising our view of an evolving Semantic Web. In fact, a main goal of the *Semantic Web* since its inception is to provide means for a unified view on the Web, which obviously includes to deal with the heterogeneity of data formats and languages. In this scenario, XML (as a format for exchanging data), RDF (as an abstract data model for states, sometimes stored natively, sometimes mapped to XML or a relational storage), OWL (as an additional framework for theory-based knowledge representation) provide the natural underlying concepts. The Semantic Web does not possess any central structure, neither topologically nor thematically, but is based on peer-to-peer communication between autonomous, and autonomously developing and evolving nodes. This *evolution* and *behaviour* depend on the cooperation of nodes. In the same way as the main driving force for the Semantic Web idea was the heterogeneity of the underlying data, the heterogeneity of concepts for expressing behaviour requires an appropriate handling on the semantic level. When considering evolution, the concepts and languages for describing and implementing behaviour will surely be diverse, albeit due to different needs, and it is unlikely that there will be a unique language for this throughout the Web. Since the contributing nodes are prospectively based on different data models and languages, it is important that *frameworks* for the Semantic Web are modular, and that their *concepts* are independent from the actual data models and languages, and allow for an integrated handling of these.

Our view is that a general framework for evolution and reactivity in the Semantic Web should be based on a general ECA language that allows for the usage of different event languages, condition languages, and action languages. Each of these different (sub)languages should adhere to some minimal requirements (e.g. dealing with variables), but it should be as free as possible.

Moreover, the ECA rules do not only operate on the Semantic Web, but are themselves also part of it. For that, the ECA rules themselves must be represented as data in the Semantic Web, based on an ontology of ECA rules and (sub)ontologies for events, conditions and actions, with rules specified in RDF. The ontology does not only cover the rules themselves but, for handling language heterogeneity, the rule components have to be related to actual languages, which in turn can be associated with actual processors. Moreover, for exchange of rules and parts of them, an XML Markup of ECA Rules, that is preferably closely related to the ontology, is needed.

In this chapter, after a brief overview of the state of the art, we present a general framework for evolution and reactivity in the Semantic Web, which caters for the just exposed requirements. The framework also provides a comprehensive set of concrete languages. We continue the chapter with a description of a concrete homogeneous language for reactivity and evolution, XChange.

Both the general framework and the XChange language have been developed (and implemented) in the REWERSE project. This work opened several possibilities of future applications and research areas, that are sketched in the last section of this chapter.

3.2   Starting Point and Related Work

As already mentioned before, the issue of reactivity, and even that of reactivity on the Web, had already been studied before the beginning of this work.

Reactivity in Databases. Reactivity has been extensively studied in the area of databases, e.g., in [62,70]. In these, notions of composition of events, as advocated above, have been proposed based on *event algebras* with their concise theory and semantics and well-understood detection mechanisms. A prominent representative of such approaches is e.g. the SNOOP algebra of the Sentinel system [28] for transactional rules and rule-driven business workflows. Also more recent approaches like RuleCore [9] use similar concepts more or less explicitly. Our work on composite events in the Semantic Web has its roots on this previous work.

Event Algebra expressions are formed by nesting operators and basic expressions that specify which atomic events are relevant. For this, every event algebra specifies a set of *operators*, e.g., "A and B", "A or B", "A and then B", "not C between A and B". From a declarative point of view, such an event algebra expression can be true or false over a given sequence of events. From the procedural point of view, a composite event is *detected* at the timepoint where it becomes true wrt. the sequence of events occurred up to that point. Event algebra terms are usually not evaluated like queries against the history (although their semantics is defined like that), but are detected *incrementally* against the stream of incoming events.

Process Definition Languages. Also for the specification of composite actions, work already existed on process algebras and other process definition languages. Well-known process algebras are *CCS (Calculus of Communicating Systems)* [59] or *CSP (Communicating Sequential Processes)* [46]; another prominent recent process specification language is *BPEL (Business Process Execution Language)* [60]. In these approaches, e.g., the following concepts can be specified:

– sequences of actions to be executed (as in simple ECA rules);
– processes that include "receiving" actions, like the corresponding actions a and $\bar{a}$ action in CCS that are used for modeling communication: $\bar{a}$ can only be executed together with a (sending) action a in another process. The semantics of $\bar{a}$ is thus similar to the event part of ECA rules on a if *condition* do *action* where the occurrence of a "wakes the rule up" and starts execution of the subsequent condition and action;
– guarded (i.e., conditional) execution alternatives;
– families of *communicating, concurrent processes*, and
– starting an iteration or even infinite processes.

Reasoning about Actions. Formalisms for representing and reasoning about actions and effects of actions have also long been studied in Artificial Intelligence. Action languages have been defined to account for just that [49,56,5,40,41,42,43, 44]. Central to this approach of formalizing actions is the concept of a transition system: a transition system is simply a labelled graph where the nodes are states and the arcs are labelled with actions or sets of actions. Usually the states are first-order structures, where the predicates are divided into static and dynamic ones, the latter called *fluents* (cf. [66]). Action programs are sets of sentences that define one such graph by specifing which dynamic predicates change in the environment after the execution of an action. Usual problems here are to predict the consequences of the execution of a sequence of (sets of) actions, or to determine a set of actions implying a desired conclusion in the future (planning).

Most of the above action languages are equipped with appropriate action query languages, that allow for querying such a transition system, going beyond the simple queries of knowing what is true after a given sequence of actions has been executed (allowing e.g. to query about which sets of actions lead to a state where some goal is true, which involves planning).

Web Update Languages. The above work sets up the foundation on which the definition of reactivity and evolution in the Semantic Web has been inspired. Furthermore, *reasoning* about such behaviour has its own specificity that requires a specific solution *after* the mechanisms have been defined. To start, there is the issue of how to update Web data, something that is much more concrete than e.g. the update of states in action languages. For this, as a starting point we could rely on a number of proposals such as XUpdate [72], the XQuery update extension of [69], XML-RL [51], XPathLog [53], and RUL [52].

XUpdate [72] makes use of XPath expressions for selecting nodes to be processed afterwards, in a way similar to XSLT. The XSLT-style syntax of the language makes the programming, and the understanding of complex update programs, very hard.

A proposal to extend XQuery with update capabilities was presented in [69]. In it XQuery is extended with a `FOR ... LET ... WHERE ... UPDATE ...` structure. The new `UPDATE` part contains specifications of update operations (i.e. delete, insert, rename, replace) that are to be executed in sequence. For ordered XML documents, two insertion operations are considered: insertion before a child element, and insertion after a child element. Using a nested `FOR...WHERE` clause in the `UPDATE` part, one might specify an iterative execution of updates for nodes selected using an XPath expression. Moreover, by nesting update operations, updates can be expressed at multiple levels within a XML structure.

The XML-RL Language [51] incorporates features of object-oriented databases and logic programming. The XML-RL Update Language extends XML-RL with update capabilities. Five kinds of update operations are supported by the XML-RL Update Language, viz. `insert before`, `insert after`, `insert into`, `delete`, and `replace with`. Using the built-in position function, new elements can be inserted at the specified position in the XML document (e.g. insert first, insert second). Also, complex updates at multiple levels in the document structure can be easily expressed.

XPathLog [53] is a rule-based logic-programming style language for querying, manipulating and integrating XML data. XPathLog can be seen as the migration from F-Logic [48], as a logic-programming style language, for semistructured data to XML. It uses XPath as the underlying selection mechanism and extends it with the Datalog-style variable concept. XPathLog uses rules to specify the manipulation and integration of data from XML resources. As usual for logic-programming style languages, the query and construction parts are strictly separated: XPath expressions in the rule body, extended with variables bindings, serve for selecting nodes of XML documents; the rule head specifies the desired update operations intensionally by another XPath expression with variables, using the bindings gathered in the rule body. As a logic-programming style language, XPathLog updates are insertions.

Reactive Web Languages. Also some reactive languages have been proposed, that do not only allow for updating Web data as the above ones, but are also capable of dealing-with/reacting-to some forms of events, evaluate conditions, and act by updating data. These are e.g. Active XQuery [11], the XML active rules of [12], the Event-Condition-Action (ECA) language for XML defined in [4], and the ECA reactive language RDFTL [61] for RDF data.

Active XQuery [11] expands XQuery with a trigger definition and the execution model of the SQL3 standard that specifies a syntax and execution model for ECA rules in relational databases (and using the same syntax for CREATE TRIGGER). It adapts the SQL3 notions of BEFORE vs. AFTER triggers and, moreover, the ROW vs. STATEMENT granularity levels to the hierarchical nature of XML data. The core issue here is to extend the notions from "flat" relational tuples to hierarchical XML data.

Another approach to ECA rules reacting on updates of standard XML documents, in the style of SQL3 triggers, is the one of [4]. It defines ECA rules, of the usual form on ... if ... do, where events can be of the form INSERT e or DELETE e, where e is an XPath expression that evaluates to a set of nodes; the nodes where the event occurs are bound to a system-defined variable $delta$ where they are available for use in condition and action parts. An extension for a replace operation is sketched. The condition part consists of a boolean combination of XPath expressions. The action part consists of a sequence of actions, where each action represents an insertion or a deletion in XML. For insertion operations, one can specify the position where the new elements are to be inserted using the BELOW, BEFORE, and AFTER constructors. This work has been extended to RDF data (serialised as XML data) in [61].

These approaches are "local", in that, as in SQL3, work on a local database, are defined inside the database by the database owner, and only consider local events and actions. On the contrary, the XML active rules of [12] establishes an infrastructure for user-defined ECA rules on XML data, where rules to be applied to one given repository can be defined by arbitrary users (using a predefined XML ECA rule markup), and can be submitted to that repository where they are then executed. The definition of events and conditions is up to the user (in terms of changes and a query to an XML instance). The actions are restricted to sending messages. This approach further implements a subscription system that enables users to be notified upon changes on a specified XML document somewhere on the Web. For this, the approach extends the server where the document is located by a rule processing engine. Users that are interested in being notified upon changes in the document submit suitable rules to this engine that manages all rules corresponding to documents on this server. Thus, evaluation of events and rules is local to the server, and notifications are "pushed" to the remote users. Note that the actions of the rules do not modify the information, but simply send a message.

None of these languages considers composite events in general. There is some preliminary work on composite events in the Web [8], but it only considers composition of events of modification of XML-data in a single document.

Besides being mostly limited to updates, and reaction on updates, on XML (or RDF) data, and with mostly no support for composite events or actions, none of these proposals tackles the issues of heterogeneity of behaviour and languages in the Semantic Web, of dealing with composite events in the Web, and dealing with composite actions, required for materialising our initial vision of an active Semantic Web, where reactivity, evolution and propagation of changes play a central role. Having all of these aspects combined in a single framework is the goal of the work presented in this chapter.

3.3 Conceptualization of ECA Rules and Their Components: A General Framework for ECA Rules

The idea of a *General Framework for ECA Rules* aims at covering (i) active concepts w.r.t. the domain ontologies and (ii) heterogeneity of domain-independent conceptualization of activity in a comprehensive way [55].

Active Concepts of Domain Ontologies. The general framework assumes actions and events to be first-class citizens of the domain ontologies. While static notions like classes, properties, and their instances are represented in the state of one or more nodes, events and actions are present as volatile entities. Events and actions are represented by XML (including RDF/XML) fragments that are exchanged (e.g., by HTTP) between nodes. For instance,

```
<travel:CanceledFlight travel:code="LH1234">
    <travel:reason>bad weather</travel:reason>
</travel:CanceledFlight>
```

is the representation of an event (raised e.g. by an airport).

Every domain ontology – e.g. for banking or travel-ing – defines *static notions* (classes, relationships) and *dynamic notions*, i.e., the types of possible events and actions as classes of the re-spective domain ontologies as shown in UML notation in Fig. 1.

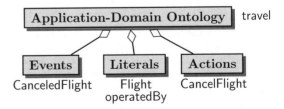

Fig. 1. Components of Domain Ontologies

Next, we identify the types of languages used in the rules to deal with state, events, and actions.

Domain-Independent Conceptualization of Activity. This aspect deals with the modeling and specification of active rules and rule components, i.e., the description of relevant events, including composite events, conditions, including queries against the state of domain nodes, and actions, including the specification of composite actions up to (even infinite) processes. While the atomic constituents are provided by the domain ontology, any kind of *composition* requires domain-independent notions. As mentioned above, multiple formalisms and languages for composite events, queries, and actions have been proposed. Each of them can be seen as an *ontology* of composite events, processes, etc.

The general framework covers this matter by following a modular approach for rule markup that considers the markup and ontology on the generic rule level separately from the markup and ontologies of the components.
Two different variants of the general idea have been implemented:

MARS – Modular Active Rules for the Semantic Web is an open architecture that allows for combining nearly arbitrary existing languages and services. For this, MARS includes a meta-level ontology of languages and services in general.

r^3 – Resourceful Reactive Rules follows an integrated design that is based on a toolbox for defining and implementing heterogeneous languages in a homogeneous programming environment.

In the following, we first present the common ideas underlying the general framework, and then point out the different design decisions in MARS and r^3. Services from both approaches can also interoperate.

3.3.1   The Rule Level

The core of the general framework is a model and architecture for ECA rules that use *heterogeneous* event, query, and action languages. The condition component is divided into queries to obtain additional information (from potentially different sources; the queries can be expressed in different languages) and a test component (that consists only of a boolean combination of generic comparison operators e.g., from XPath):

ON *event* AND *additional knowledge* IF *condition* THEN DO *something.*

The approach is parametric regarding the component languages. Users write their rules by using component languages of their choice. While the semantics of the ECA rules provides the global semantics, the components are handled by specific services that implement the respective languages. Fig. 2 (from [54]) illustrates the structure of the rules and the corresponding types of languages.

Markup. The markup of the rule level, i.e., the ECA-ML markup language, mirrors this structure as shown in Figure 3. It allows to embed the components as nested subexpressions in their own markup (using their own namespaces). The conceptual border between the ECA rule level and the particular concepts of the E, C and A components is manifested in language borders between the ECA level

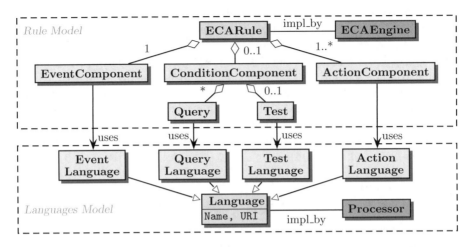

Fig. 2. ECA Rule Components and Corresponding Languages (from [54])

language and the languages of the nested components. The language borders are used at execution time to organize the cooperation between appropriate processors. The components are specified as nested subexpressions of the form

<eca:**Component-Type** xmlns:*lang*= "*embedded-lang-ns*" >
  embedded fragment in the embedded language's markup and namespace
</eca:**Component-Type**>
in arbitrary formalisms or languages.

As we shall see, analogous conceptual borders are found between the level of *composite expressions* and their *atomic subexpressions*.

Communication and Data Flow Between Components. The data flow throughout the rules, and between the ECA engine and the event, query, test, and action components is provided by *logical variables* in the style of deductive rules, or of

<eca:**Rule** xmlns:eca= "http://www.semwebtech.org/languages/2006/eca-ml#" >
  <eca:**Event**>
      <ns_1:el_1>*nested expression in event specification language markup* </ns_1:el_1>
  </eca:**Event**>
  <eca:**Query**" >
      <ns_2:el_2>*nested expression in query language markup* </ns_2:el_2>
  </eca:**Query**>
  ⋮
  <eca:**Query** > ... </eca:**Query**>
  <eca:**Test**>
      <ns_3:el_3>*test expression over obtained information* </ns_3:el_3>
  </eca:**Test**>
  <eca:**Action**>
      <ns_4:el_4>*nested expression in action language markup* </ns_4:el_4>
  </eca:**Action**>
</eca:**Rule**>

Fig. 3. ECA-ML Markup Pattern

production rules. The state of a rule evaluation, and the information sent and returned by service calls is always a *set of tuples of variable bindings*. Thus, all paradigms of query languages, following a functional style (such as XPath/X-Query), a logic style (such as Datalog or SPARQL [64]), or both (F-Logic [48]) can be used. The semantics of the event part (that is actually a "query" against an event stream that is evaluated incrementally) is –from that point of view– very similar to that of queries, and the action part takes variable bindings as input. Given this semantics, the ECA rule combines the evaluation of the components in the style of production rules (evaluated in forward-chaining mode "if body then head", cf. Figure 4):

$$action(X_1, \ldots, X_n, \ldots, X_k) \leftarrow$$
$$event(X_1, \ldots, X_n), \ query(X_1, \ldots, X_n, \ldots, X_k), \ test(X_1, \ldots, X_n, \ldots, X_k) \ .$$

The evaluation of the event component (i.e., the successful detection of a, possibly composite, event) binds variables to values that are then extended in the query component, possibly constrained in the test component, and propagated to the action component.

For the actual data exchange, an XML format has been defined. Alternatively, for local services, internal data structures can be exchanged as references, and also a shared database storage is provided.

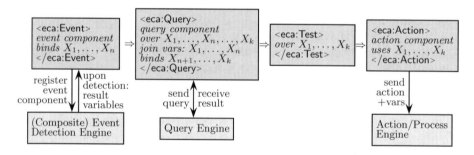

Fig. 4. Use of Variables in an ECA Rule

3.3.2   The Event Component

For the event component, two levels of specifications are combined (cf. Figure 5): The specification of the (algebraic) structure of the composite event is given as a *temporal combination* of atomic events, and the specification of the *contributing* atomic events as the leaf expressions is given by small "queries" against the actual events that check if an event matches the specification.

Atomic Event Specifications. As shown at the beginning of this section, actual events are volatile items, considered to be represented as XML fragments. Atomic event specifications (AESs) are used in the rules' event components for specifying which atomic events are relevant; they form the leaves of the event component tree. Their specification needs to consider the type and contents of atomic events,

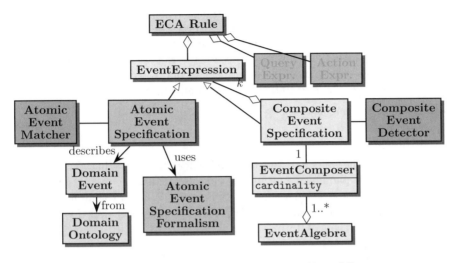

Fig. 5. Event Component: Languages (from [7])

e.g. reacting to an event "if a flight is canceled due to bad weather conditions then
...". In case of detecting the composite event "if a flight is first delayed and later
canceled then ...", the flight number must be *extracted* from the event to use it in
a *join condition* between *different* atomic events. The following fragment shows
an atomic event specification in an XML-QL-style [32] matching formalism, that
extracts the flight number and the reason from such an event:

```
<xmq:AtomicEvent
      xmlns:xmq="http://www.semwebtech.org/languages/2006/xmlql#" >
   <travel:CanceledFlight xmlns:travel="..." travel:flight="{$flightno}" >
     <travel:reason>{$reason}</travel:reason>
   </travel:CanceledFlight>
</xmq:AtomicEvent>
```

In an atomic event specification, there are always *two* languages involved as
shown in Figure 5: (i) a domain language (associated with the namespace of
the event; above: travel), and (ii) an atomic event description/matching/query
language (above: xmq) for *describing* what events should actually be matched.

Composite Event Specifications. Given an event algebra, the markup of event algebra expressions is straightforward, forming a tree term structure over atomic event
specifications. As a sample composite event language, a variant of the SNOOP
event algebra [29] extended with relational data flow has been developed [7].

3.3.3   The Condition Component

The condition component consists of one or more queries to obtain additional
information from domain nodes, and a test that is evaluated over the obtained
variable bindings.

As query languages, *opaque* components play an important role: most domain nodes are assumed to provide an existing query language according to their data model, e.g., SQL, XPath/XQuery, or SPARQL. While all these languages support variables, they are not in any XML markup[1], but queries are given as *strings* that have to be parsed at the respective nodes. For the ECA rules, these strings are opaque. Opaque embedded fragments are of the form

<eca:Opaque eca:language=*"lang-id"* eca:uri=*"uri"* >
query string
</eca:Opaque>

which indicates the language and the URI where the query has to be sent for being answered. Opaque code fragments can also be used in the action part.

Additionally, a query language for RDF and OWL data that has an RDF syntax (whose XML markup is its RDF/XML serialization), called OWLQ, has been developed in the MARS project.

3.3.4   The Action Component

The action component specifies the actual reaction to be taken. This again can be an atomic action, or a composite action, often called *process*. For its specification, process languages or process algebras can be used. Given a process language, the markup on the process level is again straightforward, forming a tree expression structure (note that BPEL [60] is originally defined as an XML language).

Atomic actions are those of the application domains, again represented as XML fragments, belonging to some domain namespace. Such atomic actions are then sent to the appropriate nodes to be executed. The specification of an action to be executed thus consists of the specification to generate an XML fragment which is then submitted to the corresponding domain nodes, or to a domain broker [6] that will in turn submit it to appropriate domain nodes (using the namespace identification). For that, also multiple languages exist.

Conditions, Queries and Events inside Process Specifications. The specification of a process, which e.g. includes branching or waiting for a response, can also require the specification of queries to be executed, and of events to be waited for. For that, we allow event specifications, queries and conditions as regular, executable components of a process:

- "executing" a query means to evaluate the query, to extend the variable bindings, and to continue.
- "executing" a condition means to evaluate it, and to continue for all tuples of variable bindings where the condition evaluates to "true". For instance, for a *conditional alternative* process $((c : a_1) + (\neg c : a_2))$, all variable bindings that satisfy c will be continued in the first branch with action a_1, and the others are continued with the second branch.
- "executing" an event specification means to wait for an occurrence of the respective event.

---

[1] With exceptions, such as XQueryX as an XML markup for XPath/Query

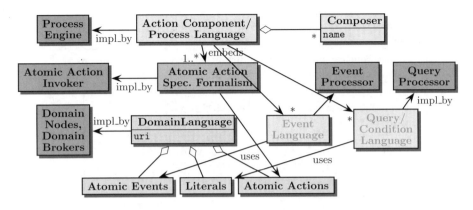

Fig. 6. Structure of the Action Component as an Algebraic Language (from [7])

Figure 6 shows the relationship between the process algebra language and the contributions of the event and test component languages, and those of the domain languages. As a sample process language, an enriched variant of CCS [59] has been defined [7, 47] that works on relational states (i.e., a set of tuples of variable bindings) that are manipulated by the atomic actions.

3.3.5 Languages and Language Borders

In the above design, rules and (sub)expressions are represented by XML trees. Nested fragments correspond to subtrees in different languages, corresponding to XML namespaces, as illustrated in Figure 7: the rule reacts on a composite event (specified in the extended SNOOP [29, 7] event algebra as a sequence) "if a flight is first delayed and later cancelled", and binds the flight number and the reason of the cancellation. Two expressions in an XML-QL-style matching formalism contribute the atomic event specifications. Note the occurrence of the travel domain namespace inside the atomic event specification.

Processing of an XML fragment in a given language, or more abstractly, executing some task for a fragment, is organized by using the namespace URI of the fragment's outermost element. Every processor (e.g., the one responsible for the crosshatched event part in the snoopy namespace) controls the processing of "his" level (the SNOOP event algebra), and whenever an embedded fragment (e.g., an atomic event specification) has to be processed, the appropriate processor (here, for XML-QL) is invoked.

The aim of the General Framework idea is to allow to embed *arbitrary* languages of appropriate types by only minimal restrictions on the languages. The cornerstones of the framework w.r.t. this issue are the following:

– the approach does only minimally constrain the component languages: the information flow between the ECA engine and the event, query, test, and

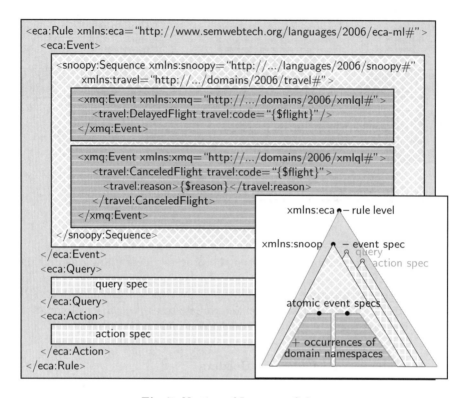

Fig. 7. Nesting of Language Subtrees

action components is provided by (i) XML language fragments, and (ii) current *variable bindings* (cf. Fig. 4),

- a comprehensive ontology of language types, service types and tasks (as described below),
- an open, service-oriented architecture,
- a *Language and Service Registry (LSR)* that holds information about actual services and how to do the actual communication with them.

The XML and RDF concept of *namespaces* provides a powerful and built-in mechanism to identify the language of an XML fragment: namespaces are the languages' URIs. The concrete languages are related to actual services, and the namespace information of a fragment to be processed is used to select and address an appropriate processor.

With that, all necessary information what to do with an embedded fragment of a "foreign language" is contained in (i) the language fragment (via the namespace of its root element), (ii) the local knowledge of the currently processing service (i.e., what it expects the fragment to be, and what it wants do to with it), and (iii) the LSR information.

3.3.6   Languages Types, Service Types, and Tasks

For every kind of language there is a specific type of service that provides a specific set of tasks – these are independent from the concrete language; only the actual implementation by a service is language-dependent.

The *Languages and Services Ontology* is shown in Figure 8; sample instances are denoted by dashed boxes. The ontology contains two levels: the level of *language types* and corresponding *service types*, and the concrete *languages* and *services*.

There are the following *language types*, with the corresponding *service types*:

Rule Languages, e.g. the ECA rule language, are handled by rule engines. There, e.g., rules can be registered.

Event Specification Languages (specifications of composite or atomic events): composite event specifications are processed by *Composite Event Detection Engines (CEDs)*; atomic event specifications are processed by *Atomic Event Matchers (AEMs)*. In both cases, event specifications can be registered at such services. Upon occurrence/detection of the event, the registrant will be notified (asynchronously).

Query Languages are handled by *query engines*. Queries can be sent there, and they are answered (synchronously or asynchronously).

Test Languages: tests (i.e., boolean queries) are also handled by query engines, or locally (as they involve rather simple comparisons). Tests can be submitted, and they are answered (synchronously or asynchronously).

Action Languages: Composite and atomic actions are processed by action services. Action specifications can be submitted there for execution (either processes, or atomic actions).

Domain Languages: Every domain defines a language that consists of the names of actions (understood to be executed), classes/properties/predicates (depending on the respective data model), and events (emitted by the domain services) as shown in Figure 1. Domain services support these names and carry out the real "businesses", e.g., airlines (in the travel domain) or universities. They are able to answer queries, to execute (atomic) actions of the domain, and they emit (atomic) events of the domain. *Domain Brokers* implement a portal functionality for a given domain.

For every kind of language there is intuitively a typical set of tasks (e.g., given a query language QL, one expects that a service that implements QL provides the task "answer-query"). In the *Languages and Services Ontology*, the tasks are not associated with the language, but with the corresponding service type (in programming languages terminology, a *service type* is an interface that is implemented by the concrete services; thus the provided tasks can be seen as (part of) its signature).

For a concrete language, e.g., SNOOP, there are one or more concrete services (of the appropriate service type) that implement it (here: snoopy). Each such service has a URI, and has to provide the characteristic tasks of the service type (in programming languages terminology: implement the signature of the

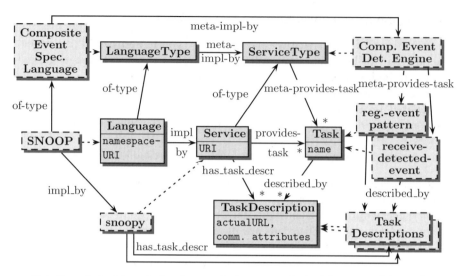

Notation: dashed boxed denote sample instances; dashed lines stand for MOF's ≪instance-of≫ and denote instanceship.

Fig. 8. Ontology of Languages and Services

interface). For each provided task, there is a *task description* that contains the information how to establish communication (described in detail at the end of this section).

The relationships on the meta level are called *meta-implemented-by* and *meta-provides-task*, while on the concrete level, they are called *implemented-by* and *provides-task*. The reason for not just overloading names is that the RDFS description then allows to distinguish domains and ranges.

Figure 9 shows the most important tasks for each service type; additionally, the actual communication flow is indicated: e.g., rule engines **p**rovide the task "register-rule", which in turn **c**alls the task "register-event-pattern" that is **p**rovided by event detectors. When the task "register-event-pattern" of a CED is called, it will in turn **c**all the task "register-event-pattern" for the embedded (atomic) subevents at some AEMs (for the respective AESLs). Domain nodes emit events that end up at the task "receive-event" that is **p**rovided by AEMs. If such an event matches a registered pattern, the AEM will **c**all the task "receive-detected-event" that is **p**rovided by the registrant (which is a CED or a rule engine). A more concrete example using the sample rule from Figure 7 will be given in Section 3.3.7.

Information about Concrete Services. The concrete information about available services for the concrete languages is managed by the *Language and Service Registry (LSR)*. For every such service, the LSR contains the URI, and for each provided task, there is a *task description* that contains information for establishing communication (cf. Figure 10 for a sample in RDF/XML markup):

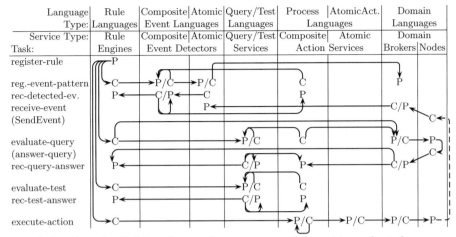

Language Type:	Rule Languages	Composite	Atomic Event Languages	Query/Test Languages	Process Languages	AtomicAct.	Domain Languages

Fig. 9. Services and Tasks

- the actual URL (as a service supports multiple tasks, each of them may have an own URL, which is not necessarily related to the service's URI),
- whether it supports to submit a set of tuples of variables, or only a single tuple at each time,
- information about the required message format:
  - send reply-to address and subject in the message header or in the body,
  - whether it requires a certain wrapper element for the message body,
- whether it will answer synchronously or asynchronously.

All MARS ontologies and an LSR snapshot in XML/RDF syntax can be found at http://www.dbis.informatik.uni-goettingen.de/MARS/#mars-ontologies
For processing a component or subexpression, a *language processor* for the indicated specification language is determined by asking an LSR for a processor for the *embedded-lang-ns* namespace and the task is submitted to that node according to the information given in the respective task description. The actual process of determining an appropriate service and organizing the communication is operationally performed by a *Generic Request Handler (GRH)*, that is used by all sample services. Details about the actual handling are described in [37].

In the current prototype, the LSR is implemented by a central RDF/XML file. In a fully operational MARS environment, the LSR would be realized as one or more LSR services where language services can register and deregister, and that are connected e.g. in a peer-to-peer way. By that, e.g., different LSRs can list their "friend" services, and only fall back to remote services if no local ones are available.

```
<mars:EventAlgebra rdf:about="http://.../languages/2006/snoop#" >
  <mars:is-implemented-by>
    <mars:CompositeEventDetectionEngine xml:base="http://.../services/2007/snoopy/"
        rdf:about="http://www.semwebtech.org/services/2007/snoopy" >
      <has-task-description> <TaskDescription>
        <describes-task rdf:resource="&mars;/ced#register-event-pattern" />
        <provided-at rdf:resource="register" /> <input>element register</input>
        <Reply-To>body</Reply-To> <Subject>body</Subject>
        <variables>*</variables>
      </TaskDescription> </has-task-description>
      :
    </mars:CompositeEventDetectionEngine>
  <mars:is-implemented-by>
</mars:EventAlgebra>
```

Fig. 10. MARS LSR: LSR entry with Service Description Fragment for SNOOP

3.3.7 Architecture and Processing: Cooperation between Resources

Imposed by the structure of the rule and the type of languages, each service plays a certain role when processing the parts it is responsible for. The basic pattern, according to the ECA structure is always the same, as illustrated in Figures 11 (global interaction) and 12 (more detailed view of the services and tasks that are involved in composite event detection including the prior registration of event specifications).

Consider again the example rule from Figure 7:

A client registers the rule (which deals with the travel domain) at the ECA engine (Step 1.1). Event processing is done in cooperation of an ECA engine, one or more *Composite Event Detection Engines (CEDs)* that implement the event algebras (in the example: SNOOP), and one or more *Atomic Event Matchers (AEMs)* that implement the *Atomic Event Specification Languages (AESLs)* (in the example: XML-QL). For this, the ECA engine submits the event component to the appropriate CED service (1.2), here, a SNOOP service. The SNOOP service inspects the namespaces of the embedded atomic event specifications and registers the atomic event specifications (for travel:DelayedFlight and travel:CanceledFlight) at the XML-QL AEM service (1.3). The AEM inspects the namespaces of the used domains, where in this case the travel ontology is relevant. It contacts a travel domain broker (1.4) to be informed about the relevant events (i.e., DelayedFlight and CanceledFlight).

The domain broker forwards relevant atomic events to the AEM (2.2; e.g., happening at Lufthansa (e.g., 2.1a: DelayedFlight(LH123), 2.1c: Canceled-Flight(LH123)) and Air France (2.1b: DelayedFlight(AF456))). The AEM matches them against the specifications and in case of a success reports the matched events and the extracted variable bindings to the SNOOP service (3). Only after detection of the registered composite event (after events 2.1a and 2.1c), SNOOP submits the result to the ECA engine (4).

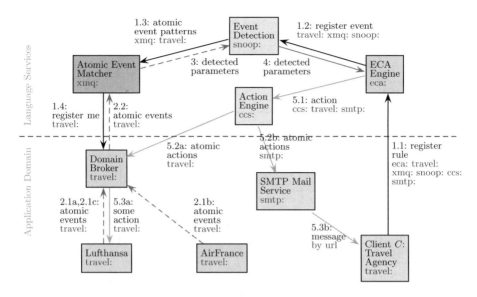

Fig. 11. Communication: Event Processing

This means the actual "firing" of the rule which then evaluates additional queries and tests (assumed to be empty here). Then, the action component is executed. Assume an action component (which is not given explicitly in Figure 7) that uses CCS and contains an atomic action concerning the corresponding airline node and an atomic action that sends a mail (using an SMTP action). The ECA engine inspects the used language namespace (ccs:) and forwards it to a CCS service (5.1). The CCS node forwards the action to the Lufthansa node via the domain broker (5.2a,b) and sends a mail (5.3a,b) via an SMTP service.

Embedding of Domain Languages. Domain languages and services are completely compatible with this approach (cf. Figure 9). For domain nodes, the tasks are register-for-event and execute-action. Domain brokers provide a portal functionality between language services and domain services.

3.3.8 The RDF Level: Language Elements and Their Instances as Resources

Rules on the semantic level, i.e., RDF or OWL, lift ECA functionality w.r.t. two (independent) aspects: first, the events, conditions and actions refer to the domain ontology level as described above. On an even higher level, the above rule ontology and event, condition, and action subontologies regard rules themselves as objects of the Semantic Web. Together with the languages and their processors, this leads directly to a resource-based approach: every rule, rule component, event, subevent etc. becomes a resource, which is related to a language which in turn is related to other resources.

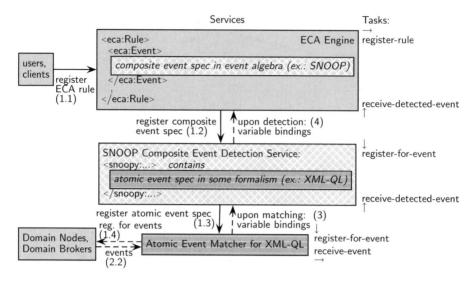

Fig. 12. Processing Event Components and Events

Describing rules in RDF provides an important base to be able to reason about rules. This will support several things:

- validation and support for execution,
- actual reasoning about the behaviour of a node, including correctness issues,
- expressing rules in abstract terms instead of w.r.t. concrete languages. The services can then e.g. choose which concrete languages support the expressiveness required by the rule's components.

For the RDF/OWL level, we assume that not only the data itself is in RDF, but also events and actions are given as XML/RDF fragments (using the same URIs for entities and properties as in the static data).

Based on the semantics of the component languages as algebraic structures, a representation in RDF is straightforward for each language. Actually, when designing a language having an RDF and an XML variant (such as developed for SNOOP and CCS), the XML markup is a stripped variant of a certain RD-F/XML serialization according to a target DTD of the RDF graph of the rule. The processing of rules given in RDF is actually done via transforming them to the XML syntax which is then executed as described above.

Figures 13 and 14 show an excerpt of the rule given in Figure 7 as RDF: "If a flight is first delayed and then canceled (note: use of a join variable), then ...". For atomic event matching, it uses the RDF-based OWLQ language.

3.3.9   MARS Implementation

Modular Active Rules for the Semantic Web – MARS – implements an open, service-oriented architecture exactly as described above. In MARS, every contributing service is completely autonomous. Making a language and a service

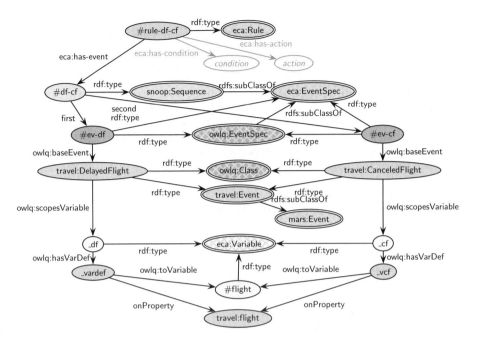

Fig. 13. Example Rule and Event Component as Resources

interoperable in MARS just consists of adding an appropriate language entry with a service description to the *Language and Service Registry (LSR)* and using it anywhere in a rule or process. Communication between services is always done via HTTP and the XML serialization of variable bindings.

In the MARS project, several sample languages on the XML and RDF level have been implemented.

XML Level. On the XML level, the focus is on having an XML markup for ECA rules using more or less well-known component languages that have been adapted to relational dataflow. XML is here just used as a markup format for rules and their components and subexpressions:

- Atomic Event Specifications: An XML-QL-style [32] pattern-based query mechanism,
- Composite Event Specifications: the SNOOP event algebra of the Sentinel system [28, 7],
- Queries: XPath and XQuery as *opaque* queries (i.e., non-markupped CDATA contents), XML-QL,
- Atomic Actions: An XML-QL-style pattern-based XML generation mechanism,
- Composite Actions: the *CCS – Calculus of Communicating Systems* process algebra [59, 7, 47].
  Process specifications (in CCS) as used in the *Action* part of ECA rules can also be defined and executed standalone.

```
<!DOCTYPE rdf:RDF [
  <!ENTITY rdf "http://www.w3.org/1999/02/22-rdf-syntax-ns#">
  <!ENTITY owlq "http://www.semwebtech.org/languages/2006/owlq#">
  <!ENTITY rdf "http://www.w3.org/1999/02/22-rdf-syntax-ns#">
  <!ENTITY travel "http://www.semwebtech.org/domains/2006/travel#"> ]>
<rdf:RDF xmlns:rdf="http://www.w3.org/1999/02/22-rdf-syntax-ns#"
  xmlns:eca="http://www.semwebtech.org/languages/2006/eca-ml#"
  xmlns:snoop="http://www.semwebtech.org/languages/2006/snoopy#"
  xmlns:owlq="http://www.semwebtech.org/languages/2006/owlq#"
  xmlns:travel="http://www.semwebtech.org/domains/2006/travel#"
  xml:base="foo:rule">

<eca:Rule rdf:ID="rule-df-cf">
 <eca:uses-variable rdf:resource="#flight"/>
 <eca:has-event>
  <snoop:Sequence>
   <snoop:first>
    <owlq:EventSpec rdf:ID="ev-df">
     <owlq:baseEvent rdf:resource="&travel;DelayedFlight"/>
     <owlq:scopesVariable>
      <owlq:Variable>
       <owlq:hasVariableDefinition>
        <owlq:VariableDefinition>
         <owlq:onProperty rdf:resource="&travel;flight"/>
         <owlq:toVariable rdf:resource="#flight"/>
        <owlq:VariableDefinition>
       </owlq:hasVariableDefinition>
      </owlq:Variable>
     </owlq:scopesVariable>
    </owlq:EventSpec>
   </snoop:first>
   <snoop:second>
    <owlq:EventSpec rdf:ID="ev-cf">
     <owlq:baseEvent rdf:resource="&travel;CanceledFlight"/>
     <owlq:scopesVariable>
      <owlq:Variable>
       <owlq:hasVariableDefinition>
        <owlq:VariableDefinition>
         <owlq:onProperty rdf:resource="&travel;flight"/>
         <owlq:toVariable rdf:resource="#flight"/>
        <owlq:VariableDefinition>
       </owlq:hasVariableDefinition>
      </owlq:Variable>
     </owlq:scopesVariable>
    </owlq:EventSpec>
   </snoop:second>
  </snoop:Sequence>
 </eca:has-event>
 <!-- ... query and action ... -->
</eca:Rule>
```

Fig. 14. Sample RDF Rule

RDF Level. On the RDF level, all language fragments are represented in RDF/XML. Here, new language proposals are embedded in MARS:

- SNOOP and CCS (in RDF version),
- OWLQ as query language and for atomic event specifications,
- RDF-CL (an OWLQ-style RDF generation language).

Domain Services. There is a prototypical domain broker (cf. [6]) and sample RDF data for demonstrating the rules; an active domain node with a demonstrator application is under development.

Openness. The MARS framework is open for foreign component languages and other sublanguages. Languages that have an XML markup smoothly integrate as shown above. For existing services, it is an easy task to implement a wrapper Web service that provides a suitable interface and to add the respective information to the LSR (the online MARS LSR and the demonstrator contain samples of foreign languages). Languages that do not have an XML markup but any other textual syntax can be integrated using the handling of *opaque* fragments (for details, see [2]), or also via an XML-based wrapper.

An online demonstrator of MARS is available at
http://www.semwebtech.org/mars/frontend/.

3.3.10 r^3 Implementation

Resourceful Reactive Rules – r^3 – is a prototype implementation of the general framework described above that, unlike MARS, follows an integrated design that is based on a toolbox for defining and implementing heterogeneous languages in a homogeneous programming environment.

r^3 is based on an OWL-DL foundational ontology [1], describing reactive rules and their components, and fully relies on the RDF Level. I.e., in r^3 rules, and their components are resources described in RDF according to the OWL-DL foundational ontology. As such, no concrete markup is expected, though a compatibility with the ECA-ML markup described above is provided.

The prototype is actually a network of r^3 engines that cooperate towards the evaluation of ECA rules. As in MARS, the communication between the r^3 engines is done via HTTP and XML serialization of variable bindings. However, making languages and services interoperable is not as simple as in MARS. The entry point is an r^3 main agent, providing operation for loading and removing ECA rules. This main engine then interfaces with r^3-aware language specific sub-engines, e.g. for detecting events, querying, testing conditions and performing actions. These r^3 sub-engines may either be language services or domain services.

For easing the construction of r^3 engines the prototype comes together with a toolbox, including a development library and a corresponding meta-model to describe component languages. This development library abstracts away communication protocols, bindings of variables, generation of alternative solutions, dealing with RDF models, etc. With this toolbox, building an r^3 engine for a language amounts to describing the language constructs in the meta-model, and implementing each of the constructs, or using already existing implementations. This provides a homogeneous programming environment for building r^3 engines.

In fact several r^3 engines have been built using this toolbox, for different languages and services:

- HTTP, providing functors for put, get, post and delete;
- Prova [50], allowing for querying prova (Prolog-like) rules bases, and for performing actions by using prova programs;
- XPath and XQuery for querying data in the Web;
- Xcerpt [63, 30] (see also Chapter 2), allowing for querying Web data using this language;
- SNOOP for specifying composite events;
- XChange [25] which allows for detecting events with XChange, raising XChange events, and performing XChange actions;
- Protune [10], allowing for query and acting upon policy knowledge bases, as defined in Chapter 4;
- Evolp [3] allowing to query and act in an updateable logic programming knowledge base.

Moreover, a broker has been implemented for allowing r^3 engines to access MARS, and an example domain service for the bio-informatics domains. All of this, plus the source code of r^3 and the toolbox, installation and usage manual, as well as an online demonstrator, can be found at `http://rewerse.net/I5/r3/`.

3.4   XChange – A Concrete Web-Based ECA Rule Language

XChange [25] is a reactive rule language addressing the need for evolution and reactivity on the Web, both local (at a single Web node) and global (distributed over several Web nodes) . As motivated in the beginning of this Chapter, it is based on ECA rules of the form "ON *event query* IF *Web query* DO *action*." When events answering the event query are received and the Web query is successful (i.e., has a non-empty result), the rule's action is executed.

In contrast to the ECA rule frameworks presented in the previous section, XChange aims at providing a single, homogenous, and elegant language that is tailored for working with Web data and that is easy to learn and use. A guiding idea of XChange is to build upon the existing Web query language Xcerpt [68,67]. (Xcerpt is discussed in Chapter 2 of this book; the core ideas as relevant for understanding XChange will also be presented shortly here.)

XChange builds on the pattern-based approach of Xcerpt for querying data, and additionally provides for pattern-based updating of Web data [63, 30]. Development of Xcerpt, XChange, and also of the complex event query language XChange^{EQ} [18] follows the vision of a stack of languages for performing common tasks on Web data such as querying, transforming, and updating static data, as well as reacting to changes, propagating updates, and querying complex

```
flights [                              <flights>
   flight {                               <flight>
      number { "UA917" },                    <number>UA917</number>
      from   { "FRA" },                      <from>FRA</from>
      to     { "IAD" }                       <to>IAD</to>
   },                                     </flight>

   flight {                               <flight>
      number { "LH3862" },                   <number>LH3862</number>
      from   { "MUC" },                      <from>MUC</from>
      to     { "FCO" }                       <to>FCO</to>
   },                                     </flight>

   flight {                               <flight>
      number { "LH3863" },                   <number>LH3863</number>
      from   { "FCO" },                      <from>FCO</from>
      to     { "MUC" }                       <to>MUC</to>
   }                                      </flight>
]                                      </flights>
```

       (a) Data term   (b) XML document

Fig. 15. An Xcerpt data term and its corresponding XML document

events. The result is a set of cooperating languages that provide, due to the pattern-based approach that is common to all of them, a homogenous look-and-feel. When a programmer has mastered the basics of querying Web data with Xcerpt's query terms, she can progress quickly and with smooth transitions to more advanced tasks.

3.4.1   Representing, Querying, and Constructing Web Data

Data terms. XML and other Web data is represented in XChange in the term syntax of Xcerpt that is arguably more concise and readable than the original formats. Further, data terms are the basis for query terms and construct terms, and the importance of conciseness and readability of the term syntax will become more pronounced when we introduce them shortly.

Figure 15(a) shows an Xcerpt data term for representing information about flights; its structure and contained information corresponds to the XML document shown in Figure 15(b). A data term is essentially a pre-order linearization of the document tree of an XML document. The element name, or *label*, of the root element is written first, then surrounded by square brackets or curly braces, the linearizations of its children as subterms separated by commas.

The term syntax provides two features that are not found in XML: First, child elements in XML are always ordered. The term syntax allows children to be specified as either ordered (indicated by square brackets [ ]) or unordered (indicated by curly braces { }). The latter brings no added expressivity to the data format (an unordered collection can always been given some arbitrary order) but is interesting for efficient storage based on reordering elements and for avoiding incorrect queries that attempt to make use of an order that should not exist. In the example of Figure 15(a), the order of the flight children of the flights element is indicated as relevant, whereas the order of the children of the flight elements is not.

Second, the data model of XML is that of a tree. Our terms are more general, supporting rooted graphs, which is necessary to transparently resolve links in XML documents (specified, e.g., with IDREFs [13,14] or with XLink [31,17,16]) and to support graph-based data formats such as RDF. However for understanding XChange in the scope of this article, this feature is not necessary and we therefore refer to [68,67] for more details.

Query Terms. A query term describes a pattern for data terms; when the pattern matches, it yields (a set of) bindings for the variables in the query term. Variable bindings are also called substitutions, and sets thereof (called substitution sets). The syntax of query terms resembles the syntax of data terms and extends it to accommodate variables, incompleteness, and further query constructs.

Variables in query terms are indicated by the keyword var. They serve as placeholders for arbitrary content and keep query results in the form of bindings. In the patterns of query terms, single brackets or braces indicate a *complete specification of subterms*. In order for such a pattern to match, there must be a one-to-one matching between subterms (or children) of the data term and the query term. Double brackets or braces in contrast indicate an *incomplete specification (w.r.t. to breadth)*: each subterm in the query term must find a match in the data term, but the data term may contain further subterms. As with data terms, square brackets indicate that the order of subterms is relevant to the query and curly braces that it is not. *Incompleteness in depth*, that is matching subterms that are not immediate children but descendants at arbitrary depth, is supported with the construct desc.

Query terms also cater for restricted variables, negated subterms, optional subterms, label variables, positional variables, regular expression matching, non-structural conditions such arithmetic comparisons, and more. Examples of query terms used in the ECA rule of Figure 16, which will be discussed in detail later, are the first term with root element xchange:event (following the keyword ON) and the term with root element flights.

Construct Terms. Construct terms are used to create new data terms using variable bindings obtained by a query. A construct term describes a pattern for the data terms that are to be constructed. The syntax of construct terms resembles the syntax of data terms and extends it to support variables and grouping.

In constructing new data, *variables* in construct terms are simply replaced by the bindings obtained from the query. The result is a new data term. If there are no grouping constructs, then a new data term is generated for each binding of the variables. For more complex restructuring of data, *groupings* can be expressed as subterms in a construct term of the form all c group by { var V }, where c is another construct term (which may of course contain further grouping constructs). Its effect is to generate a data term from the construct term c (as subterm for the overall construct term) for each distinct binding of the variable V. The group by part can also be left out; the default then is to group by the free variables immediately inside the construct term after all.

When grouping generates a list, the *order* of the generated subterms can be influenced with an `order by` clause. Grouping constructs can be nested.

Construct terms also cater for aggregation functions (e.g., max, count), groupings that are restricted to a fixed number of subterms, dealing with optional variables, and construction of graph rather than tree data. An examples of construct terms used in the ECA rule of Figure 16, which will be discussed in detail later, is the second term with root element `xchange:event` (following the keywords `DO` and `and`); note that this is a fairly simple construct term that does not make use of grouping constructs.

3.4.2 Event-Condition-Action (ECA) Rules

An XChange program consists of one or more reactive rules of the form `ON` *event query* `IF` *Web query* `DO` *action*.[2], with the intuitive meaning as described above. Both event query and Web query can extract data through variable bindings, which can then be used in the action. As we can see, both event and Web queries serve a double purpose of detecting *when* to react and influencing —through binding variables— *how* to react. For querying data, as well as for updating data, XChange embeds and extends the Web query language Xcerpt presented earlier.

Figure 16 shows an example of an XChange ECA rule, which will be used for our subsequent explanations. The individual parts of the rules employ Xcerpt and its pattern-based approach. Patterns are used for querying data in both the event and condition part, for constructing new event messages in the action part, and for specifying updates to Web data in the action part.

3.4.3 Events

Event messages. Events in XChange are represented and communicated as XML messages. The root element for all events is `xchange:event`, where the prefix `xchange` is bound to the XChange namespace. Events messages also carry some meta-data as children of the root element such as

- `raising-time` (i.e. the time of the event manager of the Web node raising the event),
- `reception-time` (i.e. the time at which a node receives the event),
- `sender` (i.e. the URI of the Web node where the event has been raised),
- `recipient` (i.e. the URI of the Web node where the event has been received), and
- `id` (i.e. a unique identifier given at the recipient Web node).

An example event that might represent the cancellation of a flight with number "UA917" for a passenger named "John Q Public" is shown in both XML and term syntax in Figure 17.

---

[2] In the course of the development of XChange, different keywords and orders for the rules have also been used. In particular, rules can also be written as `RAISE` *event raising action* `ON` *event query* `FROM` *Web query* or `TRANSACTION` *update action* `ON` *event query* `FROM` *Web query*.

```
ON
    xchange:event {{
       flight-cancellation {{
          flight-number { var N },
             passenger {{
                name { "John Q Public" }
    }} }} }}
IF
    in { resource { "http://www.example.com/flights.xml", "xml" },
       flights {{
          flight {{
             number { var N },
             from    { var F },
             to      { var T }
    }} }} }
DO
    and {
       xchange:event [
          xchange:recipient [ "http://sms-gateway.org/us/206-240-1087/" ],

          text-message [
             "Hi, John! Your flight ", var N,
             " from ", var F, " to ", var T, " has been canceled."
          ] ],

       in { resource { "http://shuttle.com/reservation.xml", "xml" },
          reservations {{
             delete shuttle-to-airport {{
                passenger { "John Q Public" },
                airport    { var F },
                flight     { var N }
    }} }} }
END
```

Fig. 16. An XChange ECA rule reacting to flight cancellations for passenger "John Q Public"

Simple ("atomic") event queries. The event part of a rule specifies a class of events that the rule reacts upon. This class of events is expressed as an event query. A simple (or atomic) event query is expressed as a single Xcerpt query term.

Event messages usually contain valuable information that will be needed in the condition and action part of a rule. By binding variables in the query term, information can flow from the event part to the other parts of a rule. Hence, event queries can be said to satisfy a dual purpose: (1) they specify classes of events the rule reacts upon and (2) they extract data from events for use in the condition and action part in the form of variable bindings.

An XChange program continually monitors the incoming event messages to check if they match the event part of one of its XChange rules. Each time an event that successfully matches the event query of a rule is received, the condition part of that rule is evaluated and, depending on the result of that, the action might be executed.

The event part of the ECA rule from Figure 16 would match the event message in Figure 17. In the condition and action part the variable N would then be bound to the flight number "UA917".

Event Composition Operators. To detect complex events, the original proposal of XChange supported composition operators such as **and** (unordered conjunction

```
<xc:event xmlns:xc="http://pms.ifi.lmu.de/xchange">
  <xc:sender>            http://airline.com   </xc:sender>
  <xc:recipient>         http://passenger.com</xc:recipient>
  <xc:raising-time>      2005-05-29T18:00     </xc:raising-time>
  <xc:reception-time>2005-05-29T18:01         </xc:reception-time>
  <xc:reception-id>      4711                 </xc:reception-id>

  <flight-cancellation>
    <flight-number>UA917</flight-number>
    <passenger>John Q Public</passenger>
  </flight-cancellation>
</xc:event>
```

(a) XML syntax

```
xchange:event [
        xchange:sender         ["http://airline.com"],
        xchange:recipient      ["http://passenger.com"],
        xchange:raising-time   ["2005-05-29T18:00"],
        xchange:reception-time ["2005-05-29T18:01"],
        xchange:reception-id   ["4711"],

        fligh-cancellation {
            flight-number { "UA917" },
            passenger { "John Q Public" }
        }
]
```

(b) Data term syntax

Fig. 17. Example of an event message

of events), andthen (ordered sequence of events), without (absence of events in a specified time window), etc. [33, 63, 24, 25]. This algebraic approach to query complex events with composition operators has been common in Active Database research [62, 29]; it is however not without problems and has weaknesses in terms of expressiveness and potential misinterpretations of operators [73, 38, 18, 21].

Querying Complex Events with XChange^{EQ}. Later work on the complex event query language XChange^{EQ} [18, 20] seeks to replace the original composition operators of XChange with an improved and radically different approach to querying complex events. The problems associated with composition operators can be attributed to a large extend to the operators mixing different aspects of querying (see [35] and [34] for an elaboration). For example in the case of a sequence operator (andthen), composition of events and their temporal order are mixed.

XChange^{EQ} is built on the idea that an expressive event query language must cover the following four orthogonal dimensions, and must treat them separately to gain ease-of-use and full expressiveness:

- **Data extraction:** Events contain data that is relevant to whether and how to react. For events that are received as SOAP messages (or in other XML formats), the data can be structured quite complex (semi-structured). The data of events must be extracted and provided (typically as bindings for variables) to construct new events or trigger reactions (e.g., database updates).

- **Event composition:** To support composite events, i.e., events that are made up out of several events, event queries must support composition constructs such as the conjunction and disjunction of events (or more precisely of event queries).
- **Temporal (and causal) relationships:** Time plays an important role in event-driven applications. Event queries must be able to express temporal conditions such as "events A and B happen within 1 hour, and A happens before B." For some applications, it is also interesting to look at causal relationships, e.g., to express queries such as "events A and B happen, and A has caused B."
- **Event accumulation:** Event queries must be able to accumulate events to support non-monotonic features such as negation (absence) of events, aggregation of data, or repetitive events. The reason for this is that the event stream is (in contrast to extensional data in a database) infinite; one therefore has to define a scope (e.g., a time interval) over which events are accumulated when aggregating data or querying the absence of events. Application examples where event accumulation is required are manifold. A business activity monitoring application might watch out for situations where "a customer's order has *not* been fulfilled within 2 days" (negation). A stock market application might require notification if "the *average* of the reported stock prices over the last hour raises by 5%" (aggregation).

XChange^{EQ} also adds support for deductive rules on events, relative temporal events (e.g., "five days longer than event i"), and enforces a clear separation between time specifications that are used as events (and waited for) or only as restrictions (conditions in the `where`-part).

The research on XChange^{EQ} also puts an emphasis on formal foundations for querying events [19]. Declarative semantics of XChange^{EQ} can be given as a model theory with accompanying fixpoint theory [18]. This is a well-understood approach for traditional (non-event) query and rule languages, and it is shown that with some important adaptations, this approach can be used for event query languages as well. Operational semantics for an efficient incremental evaluation of XChange^{EQ} programs are based on a tailored variant of relational algebra and finite differencing [19, 20]. The notion of temporal relevance is used in the operational semantics to garbage collect events that become irrelevant (to a given query) as time progresses during the evaluation [19, 20].

3.4.4   Conditions

Web queries. The condition part of XChange rules queries data from regular Web resources such as XML documents or RDF documents. It is a regular Xcerpt query, i.e., anything could come after the FROM part of an Xcerpt rule. Like event queries in the event part, Web queries in the condition part have a two-fold purpose: they (1) specify conditions that determine whether the rule's action is executed or not and (2) extract data from Web resources for use in the action part in the form of variable bindings.

The condition part in the rule from Figure 16 accesses a database of flights like the one from Figure 15 located at `http://www.example.com/flights.xml` (the resource is specified with a URI using the keyword `in`). It checks that the number (variable N) of the canceled flight exists in the database and extracts the flight's departure and destination airport (variables F and T, respectively).

Deductive rules. Web queries can facilitate Xcerpt rule chaining, that is, they can access not only extensional data (i.e., data in some Web resource) but also intensional data that has been constructed with deductive rules (i.e., results of these rules). For this, an XChange program can contain Xcerpt `CONSTRUCT-FROM` rules in addition to its ECA rules. Such rules are useful for example to mediate data from different Web resources. In our example we might want to access several flight databases instead of a single one and these might have different schemas. Deductive rules can then be used to transform the information from several databases into a common schema.

3.4.5 Actions

The action part of XChange rules has the following primitive actions: rasing new events (i.e., creating a new XML event message and sending it to one or more recipients) and executing simple updates to persistent data (such as deletion or insertion of XML elements). To specify more complex actions, compound actions can be constructed from these primitives.

Raising new events. Events to be raised are specified as a construct terms for the new event messages. The root element of the construct term must be labeled `xchange:event` and contain at least on child element `xchange:recipient` which specifies the recipient Web node's URI. Note that the recipient can be a variable bound in the event or condition part.

The action of the ECA rule in Figure 16 raises (together with performing another action) an event that is sent to an SMS gateway. The event will inform the passenger that his flight has been canceled. Note that the message contains variables bound in the event part (N) and condition part (F, T).

Updates. Updates to Web data are specified as so-called update terms. An update term is a (possibly incomplete) query pattern for the data to be updated, augmented with the desired update operations. There are three different types of update operations and they are all specified like subterms in an update term. An insertion operation `insert` c specifies a construct term c that is to be inserted. A deletion operation `delete` q specifies a query term q for deleting all data terms matching it. A replace operation `replace` q `by` c specifies a query term q to determine data items to be modified and a construct term c giving their new value. Note that update operations cannot be nested.

Together with raising a new event, the action of the ECA rule in Figure 16 modifies a Web resource containing shuttle reservations. It removes the reservation of our passenger's shuttle to the airport. The update specification employs variables bound in the event part (N) and condition part (F).

Due to the incompleteness in query patterns, the semantics of complicated update patterns (e.g., involving insertion and deletion in close proximity) might not always be easy to grasp. Issues related to precise formal semantics for updates that are still reasonably intuitive even for complicated update terms have been explored in [30]. So-called snapshot semantics are employed to reduce the semantics of an update term to the semantics of a query term.

Compound Actions. Actions can be combined with disjunctions and conjunctions. Disjunctions specify alternatives, only one of the specified actions is to be performed successfully. (Note that actions such as updates can be unsuccessful, i.e., fail.) The order in which alternatives are tried is non-deterministic and implementation dependent. Conjunctions in turn specify that all actions need to be performed. The combinations are indicated by the keywords or and and, followed by a list of the actions enclosed in braces or brackets.

The actions of the rule in Figure 16 are connected by and so that both actions, the sending of an SMS and the deletion of the shuttle reservation, are executed.

3.4.6   Applications

Due to its built-in support for updating Web data, an important application of XChange rules is local evolution, that is updating local Web data in reaction to events such as user input through an HTML form. Often, such changes must be mirrored in data on other Web nodes: updates need to be propagated to realize a global evolution. Reactive rules are well suited for realizing such a propagation of updates in distributed information portals.

A demonstration that shows how XChange can be applied to programming reactive Web sites where data evolves locally and, through mutual dependencies, globally has been developed in [45] and presented in [23, 22]. The demonstration considers a setting of several distributed Web sites of a fictitious scientific community of historians called the Eighteenth Century Studies Society (ECSS). ECSS is subdivided into participating universities, thematic working groups, and project management. Universities, working groups, and project management have each their own Web site, which is maintained and administered locally. The different Web sites are autonomous, but cooperate to evolve together and mirror relevant changes from other Web sites.

The different Web sites maintain XML data about members, publications, meetings, library books, and newsletters. Data is often shared, for example a member's personal data is present at his home university, at the management node, and in the working groups he participates in. Such shared data needs to be kept consistent among different nodes; this is realized by communicating changes as events between the different nodes using XChange ECA rules.

Events that occur in this community include changes in the personal data of members, keeping track of the inventory of the community-owned library, or simply announcing information from email newsletters to interested working groups. These events require reactions such as updates, deletion, alteration, or propagation of data, which are implemented using XChange rules. The rules run

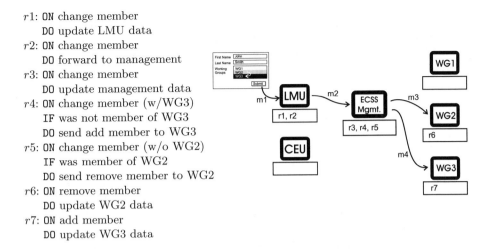

*r*1: ON change member
     DO update LMU data
*r*2: ON change member
     DO forward to management
*r*3: ON change member
     DO update management data
*r*4: ON change member (w/WG3)
     IF was not member of WG3
     DO send add member to WG3
*r*5: ON change member (w/o WG2)
     IF was member of WG2
     DO send remove member to WG2
*r*6: ON remove member
     DO update WG2 data
*r*7: ON add member
     DO update WG3 data

Fig. 18. Changing a member's personal data (including working group affiliation)

locally at the different Web nodes of the community, allowing for the processing of local and remote events.

For a concrete example, consider changing a member's personal data including his working group affiliation. The information flow is depicted in Figure 18. The initial change is entered by using a Web form at the member's home university LMU. The form generates event message *m*1. One ECA rule (*r*1) reacts to this event and locally updates the member's data at LMU accordingly. Another ECA rule (*r*2) forwards the change to the management node.

The management node has rules for updating its own local data about the member (*r*3) and for propagating the change to the affected working groups (*r*4 for adding, *r*5 for deleting a member). In the example, the member changes the working group affiliation from WG2 to WG3. Accordingly, event *m*4 is sent to WG3 by rule *r*4 and *m*3 is sent to WG2 by *r*5.

Finally, the working groups each have rules reacting to deletion and insertion events (*m*2 and *m*3) to perform the requested updates (here: *r*6 at WG2 and *r*7 at WG3).

In this description we have restricted ourselves for space reasons to this one example of changing member data. The demonstration realizes full member management of the community, a community-owned and distributed virtual library (e.g., lending books, monitions, reservations), meeting organization (e.g., scheduling panel moderators), and newsletter distribution. These other tasks are also implemented by ECA rules that are in place at the different nodes.

The full application logic of the distributed Web sites in the demonstration is realized in XChange ECA rules. While a similar behaviour as the one in the demo could be obtained with conventional programming languages, XChange provides an elegant and easy solution that also removes issues such as dealing low-level network communication protocols from the programmer's burden. Evolution of data and reactivity on the Web are easily arranged for by using readable and intuitive ECA rules. Moreover, by employing and extending Xcerpt as a query language, XChange integrates reactivity to events, querying of Web resources, and updating those resources in a single, easy-to-learn language. XChange ECA rules have also been investigated as way to realize workflows, e.g., in business processes. More details on this can be found in [65, 26].

3.5   Conclusions and Outlook

Reactivity in the Semantic Web was a quite untouched issue when the project started. The work developed in REWERSE, and described in this chapter, has established the basis for reactivity and evolution in the Semantic Web. It provides a proposal for a framework for active rules in the Semantic Web over heterogeneous component languages; the rule ontology and markup together with component languages have been developed. Moreover, a concrete homogeneous language, XChange, has also been defined, and integrated in the general framework. Both the language XChange and the general framework have been implemented, including the implementation of the integration of XChange in the r^3 implementation of the framework.

In other words, the work in REWERSE materialised the initial vision of an active Web, where reactivity, evolution and propagation of changes play a central role. Behaviour in the Semantic Web includes being able to draw conclusions based on knowledge in each Web node, but it also includes making updates on nodes and propagate these updates. Moreover, the specification of the behaviour must itself be part of this active Semantic Web, in as much as the specification of derivation rules must be part of the (static) Semantic Web. This requires an ontology of behaviour and rules, both derivation or reactive ones, to be formulated in this ontology, as well as concrete languages for detecting events in the Web, for querying the Web and testing conditions and for acting, including updates of Web data.

Despite all the advances made with the work described here, for having a (Semantic) Web with evolution and reactive behaviour, some issues remained untouched, and some new issues were raised, all of these calling for continuing the research in this area.

To start, taking more advantage of a semantical representation of behaviour rules is still pretty much an open issue. In our work an ontology for representing active rules semantically has been developed, and an execution framework has been realised. This semantical representation can also be used for working *on* rules and reasoning about the rules as objects themselves, e.g., doing rule analysis, verification, etc. Defining declarative semantics for reactive rules, along the lines of the existing languages in AI mentioned in the introduction, is certainly an interesting and important topic for further investigation when reasoning about the rules is desired. This work could also be seen as a generalisation for reactive rules, of the existing work of combining (derivation) nonmonotonic rules with ontologies, that is described in Chapter 7. Related with reasoning about rules is also the topic of model checking and verification methods for reactive rules in the Semantic Web. Preliminary studies have been made (cf. REX tool [36]), but much more is left to be done.

In the project we have developed specific languages for event querying in the Web and for update languages of Web data. Here again there is scope for further research, namely in detection of events at the semantic level, and on definition of updates on other data and meta-data formats on the Web such as RDF or TopicMaps. The extension of XChange (with its underlying Web query language Xcerpt) and the addition of new component languages to the general framework to deal with these data formats is an aspect of both practical relevance and research interest. Versatility [27], where data in different formats must be processed and reasoned with jointly to fulfill some task, becomes an important research issue with the inclusion of new data formats in reactive languages and frameworks. A further research issue is that Web formats such as RDF [71] (together with RDF Schema [15]) or OWL [57] can be considered more expressive than XML, allowing to specify inferences and more constraints on data. Updates on data in these formats may thus fail (because they violate constraints) or require additional, inferred updates. A related issue is also the integration of data formats and reasoning formalisms targeted for time and location, since time and location often play an important role in reactive applications.

Related to action languages, there is the whole issue of transactions in the Semantic Web which is a very important and by now almost untouched one. With an open environment as the (Semantic) Web, transactions following the ACID (Atomicity-Consistency-Isolation-Durability) properties as in databases are not desired, if at all possible. Surely *isolation* is something quite difficult to obtain in the Web, and independent nodes cannot wait, isolated, on actions being performed by other independent nodes. However, this does not rule out a relaxed notion of transaction. For instance, in our travel example, one may want to reserve both a flight and hotel room for a stay abroad in such a way that if one of these is not possible, then none should go ahead (i.e. if I cannot book the flight, then there is no point in keeping the hotel rooms, and vice-versa). Clearly, in such a case, some notion of *atomicity* is desired, even if *isolation* is not possible since one cannot expect the flights services to wait for the hotel reservation, nor vice-versa. This calls for defining a kind of long-running transactions, in the same spirit of those defined for heterogeneous databases [39], where *isolation* is only kept for (local) subtransactions, and irreversible actions on the global environment are associated to *compensation actions*, to account for a weaker form of atomicity. Though some preliminary work on transactions in the context of Web services exists [58], a lot remains to be done for having long-running transactions in the Semantic Web.

Another interesting new issue is that of automatic generation of ECA rules. ECA rules explicitly specify reactive behaviour, giving the events and conditions under which an action will be executed. Rather than authoring all rules manually, some applications may call for the automatic generation of ECA rules from higher level descriptions. Consider for example the distributed information portal with update propagation described in Section 3.4.6. Rather than writing ECA rules for all updates that are propagated manually, it may be conceivable to generate these rules automatically from a specification that takes the form of view definitions (e.g., over a global schema) that describe which nodes mirror which data. In a similar manner, the generation of ECA rules from process descriptions (e.g., in a language such as BPEL) is interesting [26].

Finally, for putting the whole work to usage in real practical applications, more work is needed regarding the efficiency of the systems, possibly fixing a smaller set of languages and services, and also on defining programming tools and methodologies for reactive rules in the Web. In fact, efficient execution of reactive rule sets has been given little consideration so far. The efficient execution of the individual parts of an ECA, i.e., event query evaluation, query evaluation and action processing, is well-understood. However, joint optimization of all parts of a rule as well as full rule sets has received little attention. It is conceivable for example to use multi-query optimization techniques to group together and jointly evaluate queries that are shared in multiple rules. Also, current reactive rule systems primarily work by evaluating all event queries first. It is also conceivable to use the evaluation of the condition part to enable or disable rules, thus saving the evaluation cost of the event query part when the condition part is not satisfied. Note however that this requires a mechanism where the condition part is re-checked whenever its underlying data changes.

References

1. Alferes, J.J., Amador, R.: r3 - a foundational ontology for reactive rules. In: Meersman, R., Tari, Z. (eds.) OTM 2007, Part I. LNCS, vol. 4803, pp. 933–952. Springer, Heidelberg (2007)
2. Alferes, J.J., Amador, R., Behrends, E., Fritzen, O., May, W., Schenk, F.: Prestandardization of the language. Technical Report I5-D10, REWERSE EU FP6 NoE (2008), http://www.rewerse.net
3. Alferes, J.J., Brogi, A., Leite, J.A., Pereira, L.M.: Evolving logic programs. In: Flesca, S., Greco, S., Leone, N., Ianni, G. (eds.) JELIA 2002. LNCS (LNAI), vol. 2424, pp. 50–61. Springer, Heidelberg (2002)
4. Bailey, J., Poulovassilis, A., Wood, P.T.: An event-condition-action language for XML. In: Proceedings of the 11th International Conference on World Wide Web (WWW 2002), pp. 486–495. ACM Press, New York (2002)
5. Baral, C., Gelfond, M., Provetti, A.: Representing actions: Laws, observations and hypotheses. Journal of Logic Programming 31(1-3), 201–243 (1997)
6. Behrends, E., Fritzen, O., Knabke, T., May, W., Schenk, F.: Based Active Domain Brokering for the Semantic Web. In: Marchiori, M., Pan, J.Z., Marie, C.d.S. (eds.) RR 2007. LNCS, vol. 4524, pp. 259–268. Springer, Heidelberg (2007)
7. Behrends, E., Fritzen, O., May, W., Schenk, F.: Embedding Event Algebras and Process Algebras in a Framework for ECA Rules for the Semantic Web. Fundamenta Informaticae 82, 237–263 (2008)
8. Bernauer, M., Kappel, G., Kramler, G.: Composite Events for XML. In: 13th Int. Conf. on World Wide Web (WWW 2004). ACM, New York (2004)
9. Berndtsson, M., Seiriö, M.: Design and Implementation of an ECA Rule Markup Language. In: Adi, A., Stoutenburg, S., Tabet, S. (eds.) RuleML 2005. LNCS, vol. 3791, pp. 98–112. Springer, Heidelberg (2005)

10. Bonatti, P.A., Olmedilla, D.: Driving and monitoring provisional trust negotiation with metapolicies. In: 6th IEEE International Workshop on Policies for Distributed Systems and Networks (POLICY 2005), Stockholm, Sweden, June 6-8, pp. 14–23. IEEE Computer Society, Los Alamitos (2005)
11. Bonifati, A., Braga, D., Campi, A., Ceri, S.: Active XQuery. In: Intl. Conference on Data Engineering (ICDE), San Jose, California, pp. 403–418 (2002)
12. Bonifati, A., Ceri, S., Paraboschi, S.: Pushing reactive services to XML repositories using active rules. In: Proceedings of the 10th International Conference on World Wide Web (WWW 2001), pp. 633–641. ACM Press, New York (2001)
13. Bray, T., et al.: Extensible markup language (XML) 1.0 (fourth edn.). W3C recommendation, World Wide Web Consortium (2006)
14. Bray, T., et al.: Extensible markup language (XML) 1.1 (2nd edn.). W3C recommendation, World Wide Web Consortium (2006)
15. Brickley, D., Guha, R.V. (eds.): Resource Description Framework Schema specification (RDFS) (2000), http://www.w3.org/TR/rdf-schema/
16. Bry, F., Eckert, M.: Processing link structures and linkbases in the Web's open world linking. In: Proc. ACM Conf. on Hypertext and Hypermedia, pp. 135–144. ACM, New York (2005)
17. Bry, F., Eckert, M.: Processing link structures and linkbases on the Web. In: Proc. Int. Conf. on World Wide Web, posters, pp. 1030–1031. ACM, New York (2005)
18. Bry, F., Eckert, M.: Rule-based composite event queries: The language XChange^{eq} and its semantics. In: Marchiori, M., Pan, J.Z., Marie, C.d.S. (eds.) RR 2007. LNCS, vol. 4524, pp. 16–30. Springer, Heidelberg (2007)
19. Bry, F., Eckert, M.: Towards formal foundations of event queries and rules. In: Proc. Int. Workshop on Event-Driven Architecture, Processing and Systems (2007)
20. Bry, F., Eckert, M.: On static determination of temporal relevance for incremental evaluation of complex event queries. In: Proc. Int. Conf. on Distributed Event-Based Systems, pp. 289–300. ACM, New York (2008)
21. Bry, F., Eckert, M.: Rules for making sense of events: Design issues for high-level event query and reasoning languages (position paper). In: Proc. AAAI Spring Symposium AI Meets Business Rules and Process Management. Number SS-08-01 in AAAI Technical Reports, pp. 12–16. AAAI Press, Menlo Park (2008)
22. Bry, F., Eckert, M., Grallert, H., Pătrânjan, P.L.: Evolution of distributed Web data: An application of the reactive language XChange. In: Proc. Int. Conf. on Data Engineering (Demonstrations) (2006)
23. Bry, F., Eckert, M., Grallert, H., Pătrânjan, P.L.: Reactive Web rules: A demonstration of XChange. In: Proc. Int. Conf. on Rules and Rule Markup Languages (RuleML) for the Semantic Web, Posters and Demonstrations (2006)
24. Bry, F., Eckert, M., Pătrânjan, P.L.: Querying composite events for reactivity on the Web. In: Shen, H.T., Li, J., Li, M., Ni, J., Wang, W. (eds.) APWeb Workshops 2006. LNCS, vol. 3842, pp. 38–47. Springer, Heidelberg (2006)
25. Bry, F., Eckert, M., Pătrânjan, P.L.: Reactivity on the Web: Paradigms and applications of the language XChange. J. of Web Engineering 5(1), 3–24 (2006)
26. Bry, F., Eckert, M., Pătrânjan, P.L., Romanenko, I.: Realizing business processes with ECA rules: Benefits, challenges, limits. In: Alferes, J.J., Bailey, J., May, W., Schwertel, U. (eds.) PPSWR 2006. LNCS, vol. 4187, pp. 48–62. Springer, Heidelberg (2006)
27. Bry, F., Furche, T., Badea, L., Koch, C., Schaffert, S., Berger, S.: Querying the Web reconsidered: Design principles for versatile Web query languages. Int. J. on Semantic Web and Information Systems 1(2), 1–21 (2005)

28. Chakravarthy, S., Krishnaprasad, V., Anwar, E., Kim, S.K.: Composite events for active databases: Semantics, contexts and detection. In: Proceedings of the 20th VLDB, pp. 606–617 (1994)
29. Chakravarthy, S., Mishra, D.: Snoop: An expressive event specification language for active databases. Data & Knowledge Engineering (DKE) 14, 1–26 (1994)
30. Coşkun, F.: Pattern-based updates for the Web: Refinement of syntax and semantics in XChange. Master's thesis (Diplomarbeit), Institute for Informatics, University of Munich (2007)
31. DeRose, S., Maler, E., Orchard, D.: XML linking language (XLink) version 1.0. W3C recommendation, World Wide Web Consortium (2001)
32. Deutsch, A., Fernandez, M., Florescu, D., Levy, A., Suciu, D.: XML-QL: A Query Language for XML. In: 8th. WWW Conference, W3C (1999), World Wide Web Consortium Technical Report, NOTE-xml-ql-19980819 (1999), http://www.w3.org/TR/NOTE-xml-ql
33. Eckert, M.: Reactivity on the Web: Event queries and composite event detection in XChange. Master's thesis (Diplomarbeit), Institute for Informatics, University of Munich (2005)
34. Eckert, M.: Complex Event Processing with XChange^{EQ}: Language Design, Formal Semantics, and Incremental Evaluation for Querying Events. PhD thesis, Institute for Informatics, University of Munich (2008), http://edoc.ub.uni-muenchen.de/9405/
35. Eckert, M., Bry, F.: Rule-based composite event queries: The language XChange^{eq} and its semantics. Int. Journal on Knowledge and Information Systems (to appear, 2009)
36. Ericsson, A., Berndtsson, M.: Rex, the rule and event explorer. In: Jacobsen, H.A., Mühl, G., Jaeger, M.A. (eds.) Proceedings of the 2007 Inaugural International Conference on Distributed Event-Based Systems – DEBS. ACM International Conference Proceeding Series, vol. 233, pp. 71–74. ACM, New York (2007)
37. Fritzen, O., May, W., Schenk, F.: Markup and Component Interoperability for Active Rules. In: Calvanese, D., Lausen, G. (eds.) RR 2008. LNCS, vol. 5341, pp. 197–204. Springer, Heidelberg (2008)
38. Galton, A., Augusto, J.C.: Two approaches to event definition. In: Hameurlain, A., Cicchetti, R., Traunmüller, R. (eds.) DEXA 2002. LNCS, vol. 2453, pp. 547–556. Springer, Heidelberg (2002)
39. Garcia-Molina, H., Salem, K.: Sagas. In: Dayal, U., Traiger, I.L. (eds.) Proceedings of the ACM Special Interest Group on Management of Data, pp. 249–259. ACM Press, New York (1987)
40. Gelfond, M., Lifschitz, V.: Representing actions and change by logic programs. Journal of Logic Programming 17, 301–322 (1993)
41. Gelfond, M., Lifschitz, V.: Action languages. Electronic Transactions on AI 3(16) (1998)
42. Giunchiglia, E., Lee, J., Lifschitz, V., Cain, N.M., Turner, H.: Representing actions in logic programs and default theories: a situation calculus approach. Journal of Logic Programming 31, 245–298 (1997)
43. Giunchiglia, E., Lee, J., Lifschitz, V., McCain, N., Turner, H.: Nonmonotonic causal theories. Artificial Intelligence 153, 49–104 (2004)
44. Giunchiglia, E., Lifschitz, V.: An action language based on causal explanation: Preliminary report. In: AAAI 1998, pp. 623–630 (1998)
45. Grallert, H.: Propagation of updates in distributed web data: A use case for the language XChange. Project thesis, Institute for Informatics, University of Munich (2006)

46. Hoare, C.: Communicating Sequential Processes. Prentice-Hall, Englewood Cliffs (1985)
47. Hornung, T., May, W., Lausen, G.: Process algebra-based query workflows. In: Conference on Advanced Information Systems Engineering (CAiSE) (to appear, 2009)
48. Kifer, M., Lausen, G.: F-Logic: A higher-order language for reasoning about objects, inheritance and scheme. In: Clifford, J., Lindsay, B., Maier, D. (eds.) ACM Intl. Conference on Management of Data (SIGMOD), Portland, pp. 134–146 (1989)
49. Kowalski, R.A., Sergot, M.: A logic-based calculus of events. New generation Computing 4, 67–95 (1986)
50. Kozlenkov, A., Schroeder, M.: PROVA: Rule-based Java-scripting for a bioinformatics semantic web. In: Rahm, E. (ed.) DILS 2004. LNCS (LNBI), vol. 2994, pp. 17–30. Springer, Heidelberg (2004)
51. Liu, M., Lu, L., Wang, G.: A Declarative XML-RL Update Language. In: Song, I.-Y., Liddle, S.W., Ling, T.-W., Scheuermann, P. (eds.) ER 2003. LNCS, vol. 2813, pp. 506–519. Springer, Heidelberg (2003)
52. Magiridou, M., Sahtouris, S., Christophides, V., Koubarakis, M.: Rul: A declarative update language for RDF. In: Gil, Y., Motta, E., Benjamins, V.R., Musen, M.A. (eds.) ISWC 2005. LNCS, vol. 3729, pp. 506–521. Springer, Heidelberg (2005)
53. May, W.: XPath-Logic and XPathLog: A logic-programming style XML data manipulation language. Theory and Practice of Logic Programming 4(3), 239–287 (2004)
54. May, W., Alferes, J.J., Amador, R.: Active rules in the semantic web: Dealing with language heterogeneity. In: Adi, A., Stoutenburg, S., Tabet, S. (eds.) RuleML 2005. LNCS, vol. 3791, pp. 30–44. Springer, Heidelberg (2005)
55. May, W., Alferes, J.J., Amador, R.: An ontology- and resources-based approach to evolution and reactivity in the semantic web. In: Meersman, R., Tari, Z. (eds.) OTM 2005. LNCS, vol. 3761, pp. 1553–1570. Springer, Heidelberg (2005)
56. McCarthy, J., Hayes, P.J.: Some philosophical problems from the standpoint of artificial intelligence. Machine Intelligence (4) (1969)
57. McGuinness, D., Harmelen, F. (eds.): OWL Web Ontology Language (2004), http://www.w3.org/TR/owl-features/
58. Mikalsen, T., Tai, S., Rouvellou, I.: Transactional attitudes: reliable composition of autonomous web services. In: Dependable Systems and Networks Conference (2002)
59. Milner, R.: Calculi for synchrony and asynchrony. Theoretical Computer Science, 267–310 (1983)
60. OASIS Web Services Business Process Execution Language Technical Committee: Business Process Execution Language (BPEL), http://docs.oasis-open.org/wsbpel/2.0/OS/wsbpel-v2.0-OS.html
61. Papamarkos, G., Poulovassilis, A., Wood, P.T.: RDFTL: An Event-Condition-Action Rule Language for RDF. In: Hellenic Data Management Symposium (HDMS 2004) (2004)
62. Paton, N.W. (ed.): Active Rules in Database Systems. Monographs in Computer Science. Springer, Heidelberg (1999)
63. Pătrânjan, P.L.: The Language XChange: A Declarative Approach to Reactivity on the Web. PhD thesis, Institute for Informatics, University of Munich (2005)
64. Prudhommeaux, E., Seaborne, A. (eds.): SPARQL Query Language for RDF (2006), http://www.w3.org/TR/rdf-sparql-query/

65. Romanenko, I.: Use cases for reactivity on the Web: Using ECA rules for business process modeling. Master's thesis (Diplomarbeit), Institute for Informatics, University of Munich (2006)
66. Sandewall, E.: Features and Fluents: A Systematic Approach to the Representation of Knowledge about Dynamical Systems. Oxford University Press, Oxford (1994)
67. Schaffert, S.: Xcerpt: A Rule-Based Query and Transformation Language for the Web. PhD thesis, Institute for Informatics, University of Munich (2004)
68. Schaffert, S., Bry, F.: Querying the Web reconsidered: A practical introduction to Xcerpt. In: Proc. Extreme Markup Languages (2004)
69. Tatarinov, I., Ives, Z.G., Halevy, A., Weld, D.: Updating XML. In: ACM Intl. Conference on Management of Data (SIGMOD), pp. 133–154 (2001)
70. Widom, J., Ceri, S. (eds.): Active Database Systems: Triggers and Rules for Advanced Database Processing. Morgan Kaufmann, San Francisco (1996)
71. World Wide Web Consortium: Resource Description Framework (RDF) (2000), http://www.w3.org/RDF
72. XML: DB: Xupdate - xml update language (2000), http://xmldb-org.sourceforge.net/xupdate/
73. Zhu, D., Sethi, A.S.: SEL, a new event pattern specification language for event correlation. In: Proc. Int. Conf. on Computer Communications and Networks, pp. 586–589. IEEE, Los Alamitos (2001)
74. Zimmer, D., Unland, R.: On the semantics of complex events in active database management systems. In: Intl. Conference on Data Engineering (ICDE), pp. 392–399 (1999)

Chapter 4
Rule-Based Policy Representations and Reasoning

Piero Andrea Bonatti[1], Juri Luca De Coi[2], Daniel Olmedilla[2,3],
and Luigi Sauro[1]

[1] Università di Napoli Federico II, Via Cinthia, 80126 Napoli, Italy
{bonatti,sauro}@na.infn.it
[2] Forschungszentrum L3S, Appelstr. 9a, 30167 Hannover, Germany
{decoi,olmedilla}@L3S.de
[3] Telefónica Research & Development, C/ Emilio Vargas 6, 28043 Madrid, Spain
danieloc@tid.es

Abstract. Trust and policies are going to play a crucial role in enabling the potential of many web applications. Policies are a well-known approach to protecting security and privacy of users in the context of the Semantic Web: in the last years a number of policy languages were proposed to address different application scenarios.

The first part of this chapter provides a broad overview of the research field by accounting for twelve relevant policy languages and comparing them on the strength of ten criteria which should be taken into account in designing every policy language. By comparing the choices designers made in addressing such criteria, useful conclusions can be drawn about strong points and weaknesses of each policy language.

The second part of this chapter is devoted to the description of the Protune framework, a system for specifying and cooperatively enforcing security and privacy policies on the Semantic Web developed within the network of excellence REWERSE. We describe the framework's functionalities, provide details about their implementation, and report the results of performance evaluation experiments.

4.1 Introduction

Trust is the top layer of the famous Semantic Web picture. It plays a crucial role in enabling the potential of the web. While security and privacy do not cover all the facets of trust, still they play a central role in raising the level of trust in web resources.

Security management is a foremost issue in large scale networks like the Semantic Web. In such a scenario, traditional assumptions for establishing and enforcing access control regulations do not hold anymore. In particular identity-based access control mechanisms have proved to be ineffective, since in decentralized and multicentric environments, the requester and the service provider are often unknown to each other.

F. Bry and J. Maluszynski (Eds.): Semantic Techniques for the Web, LNCS 5500, pp. 201–232, 2009.

Web Services obviously need some form of access control. Moreover, recent experiences with Facebook's "beacon" service[1] and Virgin's use of Flickr pictures[2] have shown that users are not willing to accept every possible use (or abuse) of their data.

Policies are a well-known approach to protecting security and privacy of users in the context of the Semantic Web: policies specify who is allowed to perform which action on which object depending on properties of the requester and of the object as well as parameters of the action and environmental factors (e.g., time). The application of suitable policies for protecting services and sensitive data may determine success or failure of a new service. In a near future, we might see Web Services compete with each other by improving and properly advertising their policies.

4.2   A Review of the State-of-the-Art in Policy Languages

The potential policies have proved to own is not fully exploited yet, since nowadays their usage is mainly restricted to specific application areas. On the one hand this depends on general lack of infrastructure services for such policies to truly function: for instance, there are no end user-oriented digital certification services (national digital ID providers are just appearing). On the other hand, lacking knowledge about currently available solutions is one of the main factors hindering widespread use of policies: in order to exploit a policy language the potential user needs to be provided with a clear picture of the advantages it provides in comparison with other solutions. Furthermore in the last years many policy languages were proposed, targeting different application scenarios and provided with different features and expressiveness: scope and properties of available languages have to be known to the user in order to help her in choosing the one most suitable to her needs.

In an attempt to help with these and other problems, comparisons among policy languages have been provided in the literature. However existing comparisons either do not consider a relevant number of available solutions or are mainly focused on the application scenarios the authors worked with (e.g., trust negotiation in [20] or ontology-based systems in [22]) Moreover policy-based security management is a rapidly evolving field and most of this comparison work is now out-of-date.

In this section we provide an extensive comparison covering twelve policy languages. Such a comparison will be carried out on the strength of ten criteria. Our analysis will hopefully have the side-effect of helping users in choosing the policy language mostly suiting their needs, as well as researchers currently investigating this area.

---

[1] http://www.washingtonpost.com/wp-dyn/content/article/2007/11/29/
AR2007112902503.html?hpid=topnews

[2] http://www.smh.com.au/news/technology/virgin-sued-for-using-teens-
photo/2007/09/21/1189881735928.html

This section is organized as follows. In section 4.2.1 related work is accounted for. Section 4.2.2 briefly sketches the evolution of the research field and introduces some concepts (e.g., role-based policy language as well as various kinds of policies) which will be massively exploited in the following. Sections 4.2.3 and 4.2.4 respectively introduce the languages which will be compared later on and the criteria according to which the comparison will be carried out. The actual comparison takes place in section 4.2.5, whereas section 4.2.6 presents overall results and draws some conclusions.

4.2.1   Related Work

The paper of Seamons et al. [20] is the basis of our comparison: some of the insights they suggested have proved to be still valuable right now and as such they are addressed in our work as well. Nevertheless in over six years the research field has considerably changed and nowadays many aspects of [20] are out of date: new languages have been developed and new design paradigms have been taken into account, what makes the comparison performed in [20] obsolete and many criteria according to which they were evaluated not suitable anymore.

The pioneer paper of Seamons et al. paved the way to future research on policy language comparisons like Tonti et al. [22], Anderson [2] and Duma et al. [14]: although [22] actually presents a comparison of two ontology-based languages (namely KAoS and Rei) with the object-oriented language Ponder, the work is rather an argument for ontology-based systems, since it clearly shows the advantages of ontologies.

Because of the impressive amount of details it provides, [2] restricts the comparison to only two (privacy) policy languages, namely EPAL and XACML, therefore a comprehensive overview of the research field is not provided, and features which neither EPAL nor XACML support are not taken into account at all among the comparison criteria.

Finally [14] provides a comparison specifically targeted to giving insights and suggestions to policy writers (*designers*): therefore the criteria, according to which the comparison is carried out, are mainly practical ones and scenario-oriented, whereas more abstract issues are considered out of scope and hence not addressed.

4.2.2   Background

In this section some concepts are introduced, which will help to smoothly understand the rest of the chapter. First an overall picture of the research field is provided by briefly outlining the historical evolution of policy languages, then the definitions of some policy types which will be used throughout the chapter are provided.

From uid/psw-based authentication to trust negotiation. Traditional access control mechanisms (like the ones exploited in traditional operating systems) make authorization decisions based on the identity of the requester: the user must provide a pair (*username, password*) and, if this pair matches with one of the entry in some static table kept by the system (e.g., the file /etc/pwd in Unix) the user is granted with some privileges. However, in decentralized or multicentric environments, peers are often unknown to each other, and access control based on identities may be ineffective. In order to address this scenario, role-based access control mechanisms were developed. In a role-based access control system a user is assigned with one or more roles, which are in turn exploited in order to take authorization decisions. Since the number of roles is typically much smaller than the number of users, role-based access control systems reduce the number of access control decisions. A thorough description of role-based access control can be found in [16].

In a role-based access control system the authorization process is split into two steps, namely assignment of one or more roles and check whether a member of the assigned role(s) is allowed to perform the requested action. The role-based languages we consider provide support only to one of the two steps: for instance, TPL (a role-assignment policy language) policies describe to which role the requester can be mapped; this role must then be fed as input to an existing role-based access control mechanism. A similar approach is taken by Cassandra and RT. On the other hand Ponder (authorization) policies are meant to support the second step, i.e., they allow to define which actions may be performed by a requester who has already been successfully authenticated.

Role-based authentication mechanisms require that the requester provides some information in order to map her to some role(s). In the easiest case this information can be once again a (uid, pwd) pair, but systems which need a stronger authentication usually exploit credentials, i.e., digital certificates representing statements certified by given entities (*certification authorities*) which can be used in establishing properties of their holder. More modern approaches (e.g., EPAL, WSPL and XACML) directly exploit the properties of the requester in order to make an authorization decision, i.e., they do not split the authorization process in two parts like role-based languages. Nevertheless they do not use credentials in order to certificate the properties of the requester.

Credentials, as well as declarations (i.e., not signed statements about properties of the holder) are however supported by PeerTrust, Protune and PSPL, which are languages designed to support the trust negotiation [24] vision. The notion of trust management was introduced by [7] as a new paradigm bringing together authentication and authorization in distributed systems. A scenario-based introduction to Trust Negotiation is provided in Section 4.3.2.

Policy types. Policies can be exploited in a number of fields and with different goals: security, management, conversation, quality-of-service, quality-of-protection, reliable messaging, reputation-based, provisional policies are just some examples of policies which are encountered in the literature. Here we focus on policy types which will be mentioned in the following, for instance because some language we consider has been explicitly designed to support that kind of policy.

Role-assignment policies. As the name suggests, role-assignment policies specify which conditions a requester must fulfill in order to belong to some server-defined role. Role-assignment policies are typically used in role-based policy languages like Cassandra, *RT* and TPL which postulate the existence of a back-end role-based access control mechanism to which the role will be fed in order to perform the actual authorization.

Access control policies. Access control is concerned with limiting the activities a user is allowed to perform. Consequently access control policies define the prerequisites the requester must fulfill in order to have the activity she asked for performed.

Privacy policies. Privacy policies are meant to protect the privacy of the user: they need to reflect current regulations and possibly promises made to the customers. Privacy policies arise further issues in comparison to access control policies, as they require a more sophisticated treatment of deny rules and conditions on context information; moreover privacy policy languages have to take into account the notion of "purpose", which is essential to privacy legislation. A subset of privacy policies are *enterprise* privacy policies which furthermore have to provide support to more restrictive enterprise-internal practices and may need to handle customer preferences. EPAL was especially designed in order to target enterprise privacy policies.

Obligation policies. Obligation policies specify the actions that must be performed when certain events occur, i.e., they are event-triggered condition-action rules. Obligation policies may be exploited, e.g., to specify which actions must be performed when security violations occur or under which circumstances auditing and logging activities have to be carried out. Obligation policies are supported, among others, by KAoS, Ponder and Rei.

4.2.3   Presentation of the Considered Policy Languages

To date a bunch of policy languages have been developed and are currently available: we have chosen those which at present seem to be the most popular ones, namely Cassandra [6], EPAL [3], [4], KAoS [23], PeerTrust [15], Ponder [13], Protune [8], [10], PSPL [9], Rei [17], *RT* [18], TPL [16], WSPL [1] and XACML [19], [21]. The information we will provide about the aforementioned languages is based on the referenced documents. Whenever a feature we are going to tackle is not addressed in the considered literature nor is it known to the authors in other way, the feature is supposed not to be provided by the language.

The number and variety of policy languages proposed so far is justified by the different requirements they had to accomplish and the different use cases they were designed to support. Ponder was meant to help local security policy specification and security management activities, therefore typical addressed application scenarios include registration of users or logging and audit events, whereas firewalls, operating systems and databases belong to the applications targeted by the language. WSPL's name itself (namely Web Services Policy Language) suggests its goal: supporting description and control of various aspects and features of a Web Service. Web Services are addressed by KAoS too, as well as general-purpose grid computing, although it was originally oriented to software agent applications (where dynamic runtime policy changes need to be supported). Rei's design was primarily concerned with support to pervasive computing applications (i.e. those in which people and devices are mobile and use wireless networking technologies to discover and access services and devices). EPAL (Enterprise Privacy Authorization Language) was proposed by IBM in order to support enterprise privacy policies. Some years before IBM had already introduced the pioneer role-based policy language TPL (Trust Policy Language), which paved the way to other role-assignment policy languages like Cassandra and *RT* (Role-based Trust-management framework), both of which aimed to address access control and authorization problems which arise in large-scale decentralized systems when independent organizations enter into coalitions whose membership and very existence change rapidly. The main goal of PSPL (Portfolio and Service Protection Language) was providing a uniform formal framework for regulating service access and information disclosure in an open, distributed network system like the web; support to negotiations and private policies were among the basic reasons which led to its definition. PeerTrust is a simple yet powerful language for trust negotiation on the Semantic Web based on a distributed query evaluation. Trust negotiation is addressed by Protune too, which supports a broad notion of "policy" and does not require shared knowledge besides evidences and a common vocabulary. Finally XACML (eXtensible Access Control Markup Language) was meant to be a standard general purpose access control policy language, ideally suitable to the needs of most authorization systems.

Given the multiplicity of available languages and the sometimes very specific contexts they fit into, one may argue that a meaningful comparison among them is impossible or, at least, meaningless. We claim that such a comparison is not only possible but even worth: to this aim we identified ten criteria which should be taken into account in designing every policy language. By comparing the choices designers made in addressing such criteria, useful conclusions can be drawn about strong points and weaknesses of each policy language.

4.2.4   Presentation of the Considered Criteria

We acknowledge the remark made by [14], according to which a comparison among policy languages on the basis of the criteria presented in [20] is only partially satisfactory for a designer, since general features do not help in understanding which kind of policies can be practically expressed with the constructs available in a language. Therefore in our comparison we selected a good deal of criteria having a concrete relevance (e.g., whether actions can be defined within a policy and executed during its evaluation, how the result of a request looks like, whether the language provides extensibility mechanisms and to which extent . . .). On the other hand, since we did not want to come short on theoretical issues, we selected four additional criteria, basically taken from [20] and somehow reworked and updated them. We called these more theoretical criteria *core policy properties* whereas more practical issues have been grouped under the common label *contextual properties*. In the following presentation core policy properties precede contextual properties.

Well-defined semantics. According to [20] we consider a policy language's semantics to be well-defined if the meaning of a policy written in that language is independent of the particular implementation of the language. Logic programs and Description logic knowledge bases have a mathematically defined semantics, therefore we assume policy languages based on either of the two formalisms to have well-defined semantics.

Monotonicity. In the sense of logic, a system is monotonic if the set of conclusions which can be drawn from the current knowledge base does not decrease by adding new information to the knowledge base. In the sense of [20] a policy language is considered to be monotonic if an accomplished request would also be accomplished if accompanied by additional disclosure of information by the peers: in other words, disclosure of additional evidences and policies should only result in the granting of additional privileges. Policy languages may be not monotonic in the sense of logic (as it happens with Logic programming-based languages) but still be monotonic in the sense of [20], like Protune.

Condition expressiveness. A policy language must allow to specify under which conditions the request of the user (e.g., for performing an action or for disclosing a credential) should be accomplished. Policy languages differ in the expressiveness of such conditions: some languages allow to set constraints on properties of the requester, but not on parameters of the requested action, moreover constraints on environmental factors (e.g., time) are not always supported. This criterion subsumes "credential combinations", "constraints on attribute values" and "inter-credential constraints" in [20].

Underlying formalism. A good deal of policy languages base on some well-known formalism. Knowledge about the formalism a language bases upon can be useful in order to understand some basic features of the language itself: e.g., the fact that a language is based on Logic programming with negation (as failure) entails consequences regarding the monotonicity of the language (in the sense of logic), whereas knowing that Description logic knowledge bases may contain contradictory statements could induce to infer that a Description logics-based language needs a way to deal with such contradictions.

Action execution. During the evaluation of a policy some actions may have to be performed: one may want to retrieve the current system time (e.g., in case authorization should be allowed only in a specific time frame), to send a query to a database or to record some information in a log file.

It is worth noticing that this criterion evaluates whether a language allows the *policy writer* to specify actions within a policy: during the evaluation of a policy the engine may carry out non-trivial actions on its own (e.g., both *RT* and TPL engines provide automatic resolution of credential chains) but such actions are not considered in our investigation.

Delegation. Delegation is often used in access control systems to cater for temporary transfer of access rights to agents acting on behalf of other ones (e.g., passing write rights to a printer spooler in order to print a file). The right of delegating is a right as well and as such can be delegated, too. Some languages provide a means for cascaded delegations up to a certain length, whereas others allow unbounded delegation chains.

In order to support delegation many languages provide a specific built-in construct, whereas others exploit more fine-grained features of the language in order to simulate high-level constructs. The latter approach allows to support more flexible delegation policies and is hence more suited for expressing the subtle but significant semantic differences which appear in real-world applications.

Evidences. The result of a policy's evaluation may depend on the identities or other properties of the peer who requested for its evaluation: a means needs hence to be provided in order for the peers to communicate such properties to each other. Such information is usually sent in the form of digital certificates signed by trusted entities (*certification authorities*) and called *credentials*. Credentials are not supported, among else, by languages not targeting authentication policies. PeerTrust, Protune and PSPL provide another kind of evidence, namely *declarations* which are non-signed statements about properties of the holder (e.g., credit-card numbers).

Negotiation support. [1] adopts a broad notion of "negotiation", namely a negotiation is supposed to happen between two peers whenever (i) both peers are allowed to define a policy and (ii) both policies are taken into account when processing a request. According to this definition, WSPL supports negotiations as well. In this chapter we adopt a narrower definition of negotiation by adding a third prerequisite stating that (iii) the evaluation of the request must be distributed, i.e., both peers must locally evaluate the request and either decide to terminate the negotiation or send a partial result to the other peer who will go on with the evaluation.

Whether the evaluation is local or distributed may be considered an implementation issue, as long as policies are freely disclosable. Distributed evaluation is required under a conceptual point of view as soon as the need for keeping policies private arises: indeed if policies were not private, simply merging the peers' policies would reveal possible compatibilities between them.

Policy engine decision. The result of the evaluation of a policy must be notified to the requester. The result sent back by the policy engine may carry information to different extents: in the easiest case a boolean answer may be sent (allowed vs. denied). Some languages support error messages. Protune is the only language providing enough informative content to let the user understand how the result was computed (and thereby why the query succeeded/failed).

Extensibility. Since experience shows that each system needs to be updated and extended with new features, a good programming practice requires to keep things as general as possible in order to support future extensions. Almost every language provides some support to extensibility: in the following we will provide a description of the mechanisms languages adopt in order to support extensibility.

4.2.5   Comparison

In this section the considered policy languages will be compared according to the criteria outlined in section 4.2.4. The overall results of the comparison are summarized in Table 1. Notice that Table 1 does not contain criterion "condition expressiveness" which can be hardly accounted for in a table.

Well-defined semantics. We assume policy languages based on Logic programming or Description logics to have well-defined semantics. Since the formalisms underlying the considered policy languages will be accounted for in the following, so far we restrict ourselves to list the languages provided with a well-defined semantics, namely, Cassandra, EPAL, KAoS, PeerTrust, Protune, PSPL, Rei and *RT*.

Monotonicity. In the sense of [20] a policy language is considered to be monotonic if disclosure of additional evidences and policies only results in the granting of additional privileges, therefore the concept of "monotonicity" does not apply to languages which do not provide support for credentials, namely EPAL, Ponder, WSPL and XACML. All other languages are monotonic, with the exception of TPL, which explicitly chose to support *negative certificates*, stating that a user can be assigned a role if there exists no credential of some type claiming something about it.

The authors of TPL acknowledge that it is almost impossible proving that there does not exist such a credential somewhere, therefore they interpret their statement in a restrictive way, i.e., they assume that such a credential does not exist if it is not present in the local repository. Despite this restrictive definition the language is not monotonic since, as soon as such a credential is released and stored in the repository, consequences which could be previously drawn cannot be drawn anymore.

Condition expressiveness. A role-based policy language maps requesters to roles. The assigned role is afterwards exploited in order (not) to authorize the requester to execute some actions. The mapping to a role may in principle be performed according to the identity or other properties of the requester (to be stated by some evidence) and eventually environmental factors (e.g., current time). Cassandra (equipped with a suitable constraint domain) supports both scenarios.

Environmental factors are not taken into account by TPL, where the mapping to a role is just performed according to the properties of the requester; such properties can be combined by using boolean operators, moreover a set of built-in operators (e.g., greater than, equal to) is provided in order to set constraints on their values.

Environmental factors are not taken into account by RT_0 either, where role membership is identity-based, meaning that a role must explicitly list its members; nevertheless since (i) roles are allowed to express sets of entities having a certain property and (ii) conjunctions and disjunctions can be applied to existing roles in order to create new ones, then role membership is finally based on properties of the requester.

RT_1 goes a step beyond and, by adding the notion of *parametrized role*, allows to set constraints not only on properties of the requester but even on the ones of the object, the requested action should be performed upon; the last feature makes the second step traditional role-based policy languages consist of unnecessary, therefore RT_1, as well as the other RT flavors basing on it, may be considered to lay on the border between role-based and non role-based policy languages.

A non role-based policy language does not split the authentication process in two different steps but directly provides an answer to the problem whether the requester should be allowed to execute some action. In this case the authorization decision can be made in principle not only depending on properties of the requester or the environment, but also according to the ones of the object the action would be performed upon as well as parameters of the action itself. EPAL introduces the further notion of "purpose" for which a request was sent and allows to set conditions on it.

Some non role-based languages make a distinction between conditions which must be fulfilled in order for the request to be taken into consideration (which we call *prerequisites*, according to the terminology introduced by [9]) and conditions which must be fulfilled in order for the request to be satisfied (*requisites* according to [9]); not always both kinds of conditions have the same expressiveness.

Let start checking whether and to which extent the non role-based policy languages we considered support prerequisites: WSPL and XACML allow only to use a simple set of criteria to determine a policy's applicability to a request, whereas Ponder provides a complete solution which allows to set prerequisites involving properties of requester, object, environment and parameters of the action. Prerequisites can be set in EPAL and PSPL as well; the expressiveness of PSPL prerequisites is the same as the one of its requisites, which we will discuss later.

With the exception of Ponder, which allows restrictions on the environment just for delegation policies, each other language supports requisites (Rei is even redundant in this respect): KAoS allows to set constraints on properties of the requester and the environment, Rei also on action parameters and Protune, PSPL, WSPL and XACML also on properties of the object. EPAL supports conditions on the purpose for which a request was sent but not on environmental factors. Attributes must be typed in EPAL, WSPL, XACML and typing can be considered a constraint on the values the attribute can assume, anyway the definition of the semantics of such attributes is outside WSPL's scope. Finally, in PeerTrust conditions can be expressed by setting guards on policies: each policy consists of a guard and a body, the body is not evaluated until the guard is satisfied.

Underlying formalism. The most part of languages provided with a well-defined semantics rely on some kind of Logic programming or Description logics. Logic programming is the semantic foundation of Protune and PSPL, whereas a subset of it, namely Constraint DATALOG, is the basis for Cassandra, PeerTrust and RT. KAoS relies on Description logics, whereas Rei combines features of Description logics (ontologies are used in order to define domain classes and properties

Table 1. Policy language comparison ("–" = not applicable)

	Cassandra	EPAL	KAoS	PeerTrust	Ponder	Protune	PSPL	Rei	RT	TPL	WSPL	XACML
Well-defined semantics	Yes	Yes	Yes	Yes	No	Yes	Yes	Yes	Yes	No	No	No
Monotonicity	Yes	–	Yes	Yes	–	Yes	Yes	Yes	Yes	No	–	–
Underlying formalism	Constraint DATALOG	Predicate logic without quantifiers	Description logics	Constraint DATALOG	Object-oriented paradigm	Logic programming	Logic programming	Deontic logic, Logic programming, Description logics	Constraint DATALOG	–	–	
Action execution	Yes (side-effect free)	Yes	No	Yes (only sending evidences)	Yes (access to system properties)	Yes	Yes (only sending evidences)	No	No	No	Yes	Yes
Delegation	Yes	No	No	Yes	Yes	Yes	No	Yes	Yes (RT^D)	No	No	No
Type of evaluation	Distributed policies, Local evaluation	Local	Local	Distributed	Local	Distributed	Distributed	Distributed policies, Local evaluation	Local	Local	Distributed policies, Local evaluation	Distributed policies, Local evaluation
Evidences	Credentials	No	No	Credentials, Declarations	–	Credentials, Declarations	Credentials, Declarations	–	Credentials	Credentials	No	No
Negotiation	Yes	No	No	Yes	No	Yes	Yes	No	No	Yes	No (policy matching supported)	No
Result format	A/D and a set of constraints	A/D, scope error, policy error	A/D	A/D	A/D	Explanations	A/D	A/D	A/D	A/D	A/D, not applicable, indeterminate	A/D, not applicable, indeterminate
Extensibility	Yes	Yes	Yes	Yes	Yes	Yes	No	Yes	No	No	No	Yes

associated with the classes), Logic programming (Rei policies are actually particular Logic programs) and Deontic logic (in order to express concepts like rights, prohibitions, obligations and dispensations). EPAL exploits Predicate logic without quantifiers. Finally, no formalisms underly Ponder (which only bases on the Object-oriented paradigm), TPL, WSPL and XACML.

Action execution. Ponder allows to access system properties (e.g., time) from within a policy, moreover it supports obligation policies, asserting which actions should be executed if some event happens: examples of such actions are printing a file, tracking some data in a log file and enabling/disabling user accounts.

XACML allows to specify actions within a policy; these actions are collected during the policy evaluation and executed before sending a response back to the requester. A similar mechanism is provided by EPAL and of course by WSPL, which is indeed a specific profile of XACML.

The only actions which the policy writer may specify in PeerTrust and PSPL are related to the sending of evidences, whereas Protune supports whatever kind of actions, not necessarily side-effect free, as long as a basic assumption holds, namely that action results do not interfere with each other (i.e., that actions are independent).

Cassandra (equipped with a suitable constraint domain) allows to call side-effect free functions (e.g., to access the current time).

It is worth noticing that languages allowing to specify actions within policies can to some extent simulate obligation policies, as long as the triggering event is the reception of a request, although the flexibility provided by Ponder is not met in such languages.

Finally, KAoS, Rei, *RT* and TPL do not support execution of actions.

Delegation. Ponder defines a specific kind of policies in order to deal with delegation: the field `valid` allows *positive* delegation policies to specify constraints (e.g., time restrictions) to limit the validity of the delegated access rights. Rei allows not only to define policy delegating rights but even policy delegating the right to delegate (some other right). Delegation is supported by RT^D ("D" stands indeed for "delegation"): being *RT* a role-based language, the right which can be delegated is the one of activating a role, i.e., the possibility of acting as a member of such a role.

Ponder delegation chains have length 1, whereas in *RT* delegation chains always have unbounded length. Cassandra and Protune provide a more flexible mechanism which allows to explicitly set the desired length of a delegation chain (as well as other properties of the delegation): in order to obtain such a flexibility the aforementioned languages do not provide high-level constructs to deal with delegation but simulate them by exploiting more fine-grained features of the language.

Delegation (of authority) can be expressed in PeerTrust by exploiting operator "@". Finally, EPAL, KAoS, PSPL, TPL, WSPL and XACML do not support delegation.

Type of evaluation. The most part of the considered languages require that all policies to be evaluated are collected in some place before starting the evaluation, which is hence performed locally: this is the way EPAL, KAoS, Ponder, *RT* and TPL work.

Other languages, namely Cassandra, Rei, WSPL and XACML, perform policy evaluation locally, nevertheless they provide some facility in order to collect policies (or policy fragments) which are spread over the net: e.g., in XACML combining algorithms define how to take results from multiple policies and derive a single result, whereas Cassandra allows policies to refer to policies of other entities, so that policy evaluation may trigger queries of remote policies (possibly the requester's one) over the network.

Policies can be collected into a single place if they are freely disclosable (assuming that the place they are collected into is not a trusted one), therefore the languages mentioned so far do not address the possibility that policies themselves may have to be kept private. Protection of sensitive policies can be obtained only by providing support to distributed policy evaluation, like the one carried out by PeerTrust, Protune or PSPL.

Evidences. Credentials are a key element in Cassandra, *RT* and TPL, whereas they are unnecessary in Ponder, whose policies are concerned with limiting the activity of users who have already been successfully authenticated.

The authors of PSPL were the first ones advocating for the need of exchanging non-signed statements (e.g., credit card numbers), which they called *declarations*; declarations are supported by PeerTrust and Protune as well.

Finally, EPAL, KAoS, Rei, WSPL and XACML do not support evidences.

Negotiation support. As stated above, we use a narrower definition of negotiation than the one provided in [1], into which WSPL does not fit, therefore only pretty few languages support negotiation in the sense we specified above, namely Cassandra, PeerTrust, Protune and PSPL.

Policy engine decision. The evaluation of a policy should end up with a result to be sent back to the requester. In the easiest case such result is a boolean stating whether the request was (not) accepted (and thereby accomplished): KAoS, PeerTrust, Ponder, PSPL, *RT* and TPL conform to this pattern.

Besides `permit` and `deny` WSPL and XACML provide two other result values to cater for particular situations: `not_applicable` is returned whenever no applicable policies or rules could be found, whereas `indeterminate` accounts for some error which occurred during the processing; in the latter case optional information is available to explain the error.

A boolean value, stating whether the request was (not) fulfilled, does not make sense in the case of an obligation policy, which simply describes the actions which must be executed as soon as an event (e.g., the reception of a request) happens, therefore besides the so-called *rulings* `allow` and `deny` EPAL defines a third value (`don't care`) to be returned by obligation policies; one of the elements an EPAL policy consists of is a global condition which is checked at the very beginning of the policy evaluation: not fulfilling such a condition is considered an error and a corresponding error message (`policy_error`) is returned; a further message (`scope_error`) is returned in case no applicable policies were found.

Cassandra's request format contains (among others) a set of constraints c belonging to some constraint domain; the response consists of a subset c' of c which satisfies the policy; in case $c' = c$ (resp. c' is the empty set) `true` (resp. `false`) is returned.

Protune allows for more advanced explanation capabilities: not only is it possible to ask why (part of) a request was (not) fulfilled (`Why` and `Why-not` queries respectively), but the requester is even allowed to ask since the beginning which steps she has to perform in order for her request to be accomplished (`How-to` and `What-if` queries).

A rudimentary form of `What-if` queries is supported also by Rei obligation policies: the requester can decide whether to complete the obligation by comparing the effects of meeting the obligation (`MetEffects`) and the effects of not meeting the obligation (`NotMetEffects`).

Extensibility. Extensibility is a fuzzy concept: almost all languages provide some extension points to let the user adapt the language to her current needs, nevertheless the extension mechanism greatly varies from language to language: here we will briefly summarize the means the various languages provide in order to address extensibility.

Extensibility is described as one of the criteria taken into account in designing Ponder: in order to provide smoothly support to new types of policies that may arise in the future, inheritance was considered a suitable solution and Ponder itself was therefore implemented as an object-oriented language.

XACML's support to extensibility is two-fold

- on the one hand new datatypes, as well as functions for dealing with them, may be defined in addition to the ones already provided by XACML. Datatypes and functions must be specified in XACML requests, which indeed consists of typed attributes associated with the requesting subjects, the resource acted upon, the action being performed and the environment
- as we mentioned above, XACML policies can consist of any number of distributed rules; XACML already provides a number of combining algorithms which define how to take results from multiple policies and derive a single result, nevertheless a standard extension mechanism is available to define new algorithms

Using non-standard user-defined datatypes would lead to wasting one of the strong points of WSPL, namely the standard algorithm for merging two policies, resulting in a single policy that satisfies the requirements of both (assuming that such a policy exists), since there can be no standard algorithm for merging policies exploiting user-defined attributes (except where the values of the attributes are exactly equal); use of non-standard algorithms would in turn mean that the policies could not be supported using a base standard policy engine. Being standardization the main goal of WSPL, no wonder that it comes short on the topic "extensibility", which is not necessarily a drawback, if the assertion of [1] holds: "most Web Services will probably use fairly simple policies in their service definitions".

Ontologies are the means to cater for extensibility in KAoS and Rei: the use of ontologies facilitates a dynamic adaptation of the policy framework by specifying the ontology of a given environment and linking it with the generic framework ontology; both KAoS and Rei define basic built-in ontologies, which are supposed to be further extended for a given application.

Extensibility was the main issue taken into account in the design of Cassandra: its authors realized that standard policy idioms (e.g., role hierarchy or role delegation) occur in real-world policies in many subtle variants: instead of embedding such variants in an *ad hoc* way, they decided to define a policy language able to express this variety of features smoothly; in order to achieve this goal, the key element is the notion of *constraint domain*, an independent module which is plugged into the policy evaluation engine in order to adjust the expressiveness of the language; the advantage of this approach is that the expressiveness (and hence the computational complexity) of the language can be chosen depending on the requirements of the application and can be easily changed without having to change the language semantics.

A standard interface to external packages is the means provided by Protune in order to support extensibility: functionalities of a component implementing such interface can be called from within a Protune policy.

Finally, PeerTrust, PSPL, *RT* and TPL do not provide extension mechanisms.

4.2.6 Discussion

In this section we review the comparison performed in section 4.2.5 and provide some general comments.

By carrying out the task of comparing a considerable amount of policy languages, we came to believe that they may be classified in two big groups collecting, so to say, *standard-oriented* and *research-oriented* languages respectively. EPAL, WSPL and XACML can be considered standard-oriented languages since they provide a well-defined but restricted set of features: although it is likely that this set will be extended as long as the standardization process proceeds, so far the burden of providing advanced features is charged on the user who need them; standard-oriented languages are hence a good choice for users who do not need advanced features but for whom compatibility with standards is a foremost issue.

Ponder, *RT* and TPL are somehow placed in between: on the one hand Ponder provides a complete authorization solution, which however takes place after a previously overcome authentication step, therefore Ponder cannot be applied to contexts (like pervasive environments) were users cannot be accurately identified; on the other hand *RT* and TPL do not provide a complete authorization solution, since they can only map requesters to roles and need to rely on some external component to perform the actual authentication (although parametrized roles available in RT_1 and the other *RT* flavors basing on it make the previous statement no longer true).

Finally research-oriented languages strive toward generality and extensibility and provide a number of more advanced features in comparison with standard-oriented languages (e.g., conflict harmonization in KAoS and Rei, negotiations in Cassandra, PeerTrust and PSPL or explanations in Protune); they should be hence the preferred choice for users who do not mind about standardization issues but require the advanced functionalities that research-oriented languages provide.

4.3   A Framework for Semantic Web Policies

The languages presented in Section 4.2 can be expected to be used by security experts or other computer scientists. Common users cannot profit for them, since almost no policy framework offers facilities or tools to meet the needs of users without a strong background in computer science. Yet *usability* is a major issue in moving toward a policy-aware web. It is well known that as protection increases, usability is affected by the extra steps required for authentication and other operations related to access control. The information collected for security and privacy purposes extends the amount of sensitive information released by users while navigating the web. Moreover, it is frequently not clear to a common user which policy is actually applied by a system, and which are its consequences (cf. Virgin's case mentioned in Section 4.1). Similarly, common users may find it difficult to formulate their own privacy requirements and compare them with whatever privacy policy is advertised by a Web Service.

The work on policies carried out within the network of excellence REWERSE has tackled these aspects by regarding *policies as semantic markup*. By regarding policies as pieces of machine understandable knowledge:

- it is possible to assist some of the operations related to access control and information release, thereby improving a user's navigation experience;
- it is easier to support attribute-based access control, that increases the level of privacy in on-line transactions;
- it is possible to create policy documentation automatically; in this way alignment is guaranteed between the policy enforced by the system and the policy documented in natural language for end users; moreover it is possible to specialize explanations to specific contexts (such as a particular transaction); this may help users to understand why a transaction fails (policy violation or technical problems?), how to get the permissions for obtaining a service, and so on;
- it is possible to create tools for verifying policies and more generally supporting policy authoring; other tools may help users to compare privacy policies and make (semi-)automated policy-aware service selections.

In this section we describe the policy framework *Protune*, designed and implemented within REWERSE to incarnate the above ideas. Protune is meant to support policy creation and advanced policy enforcement, providing not only traditional access control but also trust negotiation (to automate security checks and privacy-aware information release) and second generation explanation facilities (to improve user awareness about—and control on—policies).

In the next section we summarize the different semantic techniques applied in Protune. Section 4.3.2 introduces a possible reference scenario that inspires most of the examples used in the following sections. Then Section 4.3.3 recalls the policy language and the core functionalities of Protune, followed by a section devoted to the explanation facility Protune-X. When needed, we point out differences from the previous theoretical papers that set the foundations of Protune and Protune-X [11,12]. Sections 4.3.5 and 4.3.6 describe the implementation and the existing facilities for integrating Protune in a web application such as policy-driven personalized presentation of web content. In Section 4.3.7 we report the results of a preliminary experimental performance evaluation. The chapter is concluded by a section on further research perspectives.

4.3.1 Policies as Semantic Markup in Protune

Policies are semantic markup because they specify declaratively part of the semantics (in terms of behavior constraints and admissible usage) of the static or dynamic resources that policies are attached to. Accordingly, semantic techniques have several roles in Protune:

- Policies are formulated as sets of axioms and meta-axioms with a formal, processing-independent semantics; this is the basis for consistent treatment of policies for different tasks: enforcement, negotiation, explanations, validation, etc.;
- The aforementioned tasks involve different automated reasoning mechanisms, such as deduction for enforcement, abduction and partial evaluation for negotiation, pruning and natural language generation for explanations, etc.;
- The auxiliary concepts needed to formulate policies (such as what is a public resource, what is an accepted credit card, ...) and the link between such concepts and the evidence needed to prove their truth (e.g. which X.509 credentials are needed to prove authentication, or what forms need to be filled in) are defined by means of lightweight ontologies that may be included in the policy itself or referred to by means of suitable URIs; therefore, unlike XACML *contexts*, Protune's auxiliary concepts are machine understandable and allow agent interoperability.

4.3.2 Negotiations

In response to a resource request, a server may return its policy for accessing the resource. The policy may contain (a reference to) an auxiliary ontology, as explained in the previous section. In the simplest case, user agents may use such machine understandable information to check automatically whether the policy can be fulfilled and how, thereby (partially) automating the operations needed for (traditional) access control and facilitating navigation in the presence of articulated policies. In advanced scenarios, a user agent may reply with a counter-request in order to enforce the user's privacy policy, as explained below.

An example scenario. Bob's birthday is next week and Alice plans to use today's lunch break for buying on-line a novel of Bob's favourite writer. She finds out that an on-line bookshop she never heard about before sells the book at a very cheap price.

The bookshop groups its customers in different categories, according to personal data (country, age, profession ...) and purchase-related data (frequency, item, payment preferences ...). Different sale strategies are applied to different customer categories (e.g. prices discounted to different rates, no delivery fees, sending of promotional material, and so on).

By interacting with the bookshop's server Alice learns that she has to provide either her credit card number or a pair (*userId, password*) for a previously created account. This is just the bookshop's default policy, the custom-tailored policies described above are disclosed only after getting more information about the customer.

Alice does not want to create a new account on the fly, so releasing her credit card is the only option. However she is willing to give such information only to trusted on-line shops (let say, belonging to the Better Business Bureau – BBB), therefore she asks the bookshop to provide such information.

The bookshop belongs indeed to the BBB and is willing to disclose such credential to anyone. This satisfies Alice, who provides her credit card number.

After having interacted with the VISA server to check that the credit card is valid, the bookshop asks Alice for other information, in order to understand which customer category she belongs to and apply the corresponding sale strategy.

The lunch break is already over and Alice has no time left to provide the data requested, therefore she decides to abort the transaction.

Scenario revisited. Automated server *and client* policy processing may significantly speed up interactions like the above. As soon as Alice decides which book she wants to buy, a negotiation between her agent and the on-line bookshop's agent would be triggered. Since the bookshop is not willing to provide the book for free, it would answer by returning its (default) policy protecting the book. The returned policy would contain the description of the actions Alice has to perform (either sending a credit card number or providing log-in data): the use of shared ontologies to identify such actions would grant common understanding of their semantics. Alice's agent would then quickly check whether the server's policy can be fulfilled and how. Additionally, in the presence of a privacy policy, Alice's agent would reply with a counter-request asking the bookshop to provide a certificate. Again, the common vocabulary would allow the bookshop to understand the request, which would be accepted since the certificate is not protected by any policy, and as a consequence Alice would finally deliver her credit card number and have the book delivered.

The availability of a framework capable to enforce access control and negotiations automatically given the two policies has remarkable consequences on privacy as well as usability. On the one hand a direct intervention of the user in the decision process would be required less frequently, since the user's decision would be already embedded to some extent into the policies (s)he defines: therefore sensitive resources would (or would not) be disclosed without necessarily

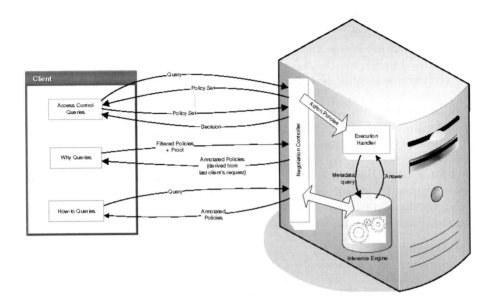

Fig. 1. Protune's architecture

asking the user each and every time. On the other hand, such usability improvement may encourage users to refine their policies by specifying articulated (eventually attribute-based) policies, thereby improving privacy guarantees.

4.3.3   Protune's Policy Language and Framework

In order to support assisted credential dislosure and handle negotiations (when needed), Protune's architecture comprises negotiation agents both on server side and on client side, as illustrated by Fig 1. Each agent reasons about access control or information disclosure policies to interpret the requests of the other peer and select possible negotiation actions.

Protune's policy language is a logic programming language enhanced with an object oriented syntax. For example, the rule that allows to buy a book by giving a credit card could be encoded with a set of rules including:

$allow(buy(Resource)) \leftarrow$
    $credential(C),\ valid\_credit\_card(C),\ accepted\_credit\_card(C).$

$valid\_credit\_card(C) \leftarrow$
    $C.expiration : Exp,\ date(Today),\ Exp > Today.$

where $C.expiration : Exp$ is an O.O. expression meaning that Exp is the value of C's attribute *expiration*. Protune policies may use and define different categories of predicates:

- *decision* predicates, used to specify a policy's outcome, such as *allow()* in the above example;[3]
- *provisional* predicates, that are meant to represent actions as described below;
- *abbreviation* predicates, defining useful abstractions such as *valid_credit_card*.

Protune supports two pre-defined provisional predicates: *credential* and *declaration*. An atom *credential(X)* is true when an object X representing an X.509 credential is stored in the current negotiation state. A peer may make *credential(X)* true on the other peer by sending the corresponding credential; this is the action attached to this particular provisional predicate. Predicate *declaration* is analogous but its argument is an unsigned semi-structured object similar to a web form that, for example, can be used to encode a traditional password-based authentication procedure as in:

authenticated ←
    declaration(D), valid_login_data(D.username, D.password).

When a set of rules like the above one is disclosed by a server in response to a client's request, the client—roughly speaking—works back from *allow(Request)* looking for the credentials and declarations in its portfolio that match the conditions listed in the rules' bodies. In logical terms, the selected credentials and declarations (represented as logical atoms) plus the policy rules should entail *allow(Request)*: this is called an *abduction problem* by the automated reasoning community. After receiving credential and declarations from a client, a server checks whether its policy is fulfilled by trying to prove *allow(Request)* using its own rules and the new atoms received from the client, as in a standard *deduction problem*.

When a client enforces a privacy policy and issues a counter-request as in Alice's scenario, the roles of the two peers are inverted: the client plays the role of the server and viceversa. For example, the client may publish rules governing credit card release such as:

allow(release(C)) ← *credit_card(C), bbb_member(Server),...*
bbb_member(Server) ← *credential(BBB), BBB.issuer =" BBB_CA",...*

Abbreviation predicates define in a machine understandable way the meaning of the conditions listed in rule bodies (unlike XACML *contexts*, which are black boxes). The rules defining such predicates constitute a lightweight, rule-based ontology. Abbreviation predicates are eventually defined on facts (e.g., listing the accepted credit cards, or the certification authorities recognized by a server) and/or on X.509 credentials and declarations, as in the rules for *authenticated* and *bbb_member*. Therefore, the ontologies associate each condition in a policy rule to the kind of *evidence* needed to fulfill the condition (specifying whether it should be signed or unsigned, issued by which certification authority, with what

---

[3] The specifications in [11] include also a predicate to sign and issue new credentials; this predicate is not yet implemented.

attributes, etc.) as well as the *actions* that need to be taken. In this way, negotiation agents can interoperate even if their policies use different abbreviation predicates.

So far, we have illustrated only information disclosure actions, such as those associated to *credential* and *declaration*. However, policies may require to execute actions that do not have negotiation purposes, such as logging some requests or notifying an administrator. New provisional predicates like these can be defined by means of *metapolicies* that specify the action associated to a predicate and the actor in charge of executing the action, for example:

$log(X) \rightarrow type : provisional.$
$log(X) \rightarrow action :' echo \$X > logfile'.$
$log(X) \rightarrow actor : self.$

where "$\rightarrow$" connects a metaterm to its metaproperties.

Rules and ontologies may be *sensitive*. For example, a server may want to publish which credit cards it accepts, but not the list of username and passwords encoded by predicate *valid_login_data*. As another example, in a social network scenario a rule such as

$allow(download(pictures)) \leftarrow best\_friend$

may have to be protected, because in case of a denial it may reveal to a friend that he or she is not considered as a *best* friend. The sensitivity level of predicates and rules is defined with metapolicies, e.g. by means of metafacts like

$valid\_login\_data(X, Y) \rightarrow sensitivity : private.$

Such metapolicies drive a *policy filtering* process that selects relevant rules (for efficiency), and removes sensitive parts if needed. The first definition of policy filtering [11] performed also partial evaluation w.r.t. the available facts. The current implementation does no partial evaluation anymore because (i) it may significantly increase the size of the messages exhanged during negotiations, and (ii) it destroys much of the structure of the policies thereby making the explanation facility (illustrated later on) much less effective.

Metapolicies are also used for other purposes, such as specifying atom verbalizations (see Section 4.3.4), controlling *when* actions are to be executed, and more generally driving negotiations in a declarative way. Metapolicies are an effective declarative way of adapting the framework to new application domains by means of activities much closer to configuration than general program encoding, thereby reducing deployment efforts and costs. For more details on the metapolicy language and its possible uses see [11] and REWERSE deliverable I2-D2 reachable from http://rewerse.net. Such documents illustrate also the facilities for integrating legacy software and data.

4.3.4 Explanations: Protune-X

Even with a policy with relatively few rules it could be hard for a common user—with neither a general training in Computer Science nor a specific knowledge of mechanisms and formats of the system—to understand what is actually required to access a certain service. Even more, a denial that results simply in a *no* does not help a user to see what has gone wrong during an acknowledgment process and hence may discourage new users from using a system. Therefore, a policy framework such as Protune would be effective only if it provides some explanation facilities that increase the user's awareness and control over a policy and provide a means to ask the system why a certain acknowledgment has been denied or granted.

Protune-X, the explanation facility of Protune, plays an essential role in improving user awareness about—and possibly control over—the policy enforced by a system. Protune-X is also a major element of Protune's *cooperative enforcement* strategy: the explanation system is meant to enrich the denials with information about how to obtain the permissions (if possible) for the requested service or resource.

For this purpose four kinds of queries are supported: *How-to* queries provide a description of a policy and may help a user in identifying the prerequisites needed for fulfilling the policy. How-to queries may also be used to verify a complex policy. *What-if* queries are meant to help users *foresee* the results of a hypothetical situation, which may be useful for validating a policy before its deployment. Finally, *why* and *why-not* queries explain the outcome of a concrete negotiation (i.e. provide a *context-specific* help). Why/why-not queries can be used both by end users who want to understand an unexpected response, and by policy administrators who want to diagnose a policy.

Some of the major desiderata that guided the design of Protune-X are:

- *Explanations should not increase significantly the computational load of the servers.* The explanation-related processes have not to be interwoven with the reasoning process of Protune. On the contrary, it would be desirable that the server simply provides the relevant piece of information (rules and facts) whenever an explanation is demanded and the client has the burden to produce it. For this reason explanations are produced in our approach by a distinct module, ProtuneX, which operates client-side.
- *Almost no further effort should be added to the policy instantiation phase.* This is achieved by exploiting *generic heuristics* as much as possible. For example *how-to* explanations exploit the *actor* meta-attributes defined in the metapolicy to distinguish automatically the prerequisites that should be satisfied by users from the conditions that are locally checked by the server. In most cases, the only extra effort needed for enabling explanations consists in writing verbalization metarules in order to specify how single, domain-specific atoms have to be rendered, e.g.:

  $passwd(X, Y) \rightarrow verbalization : Y$ & *"is the correct password of"* & X.

- *Explanation have to be presented in manageable pieces.* An acknowledgment process is essentially an attempt to show that some (state or provisional) facts satisfy/not satisfy a policy. This proof generally consists of an *AND-OR* tree where each node is a goal, *OR*-alternatives represent the different

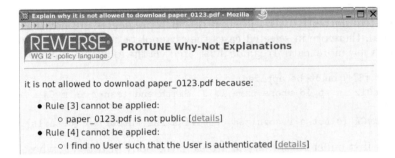

Fig. 2. A ProtuneX screenshot

rules that apply to that goal. Finally, *AND*-edges are the subgoals in the body of each rule. This structure cannot be easily captured all in a single view, therefore ProtuneX represents it by means of linked web-pages. Each web-page represents a view on a single goal and the rules that apply to it. Web-pages are linked in order to form a tree that reproduces the structure of the proof.

– *Explanations should be presented in a user-friendly format.* ProtuneX is meant to present explanations in natural language with the help of verbalization metarules.
– *Explanations should support so-called second generation features.* Such features include methods to highlight relevant information while pruning irrelevant parts and make easier to focus on the paths that do not match the expectations of the user.

In the following we illustrate some of these second-generation features by means of examples taken or adapted from the on-line demo. For a deeper discussion and a more complete description of Protune-X the reader is referred to [12].

A typical why-not explanation for a failed negotiation is an HTML hypertext whose first page may look like the screenshot in Fig. 2. The explanation may look different depending on the causes of failure. In Fig. 2 the negotiation fails because the paper is not public and the user released an invalid ID credential; if the ID credential were valid, then other conditions in the body of the rule corresponding to the second item would become relevant to explain the failure and would appear in the explanation, as in

```
– Rule [4] cannot be applied:
  • J. Smith is authenticated [details]
  but
  • There is no Subscription such that J. Smith subscribed the
    Subscription [details].
```

Note that the same rule can be explained in a completely different way depending on the context. This is an example of irrelevant information pruning, that results from another generic heuristic adopted by Protune-X. It exploits the metarules identifying so-called *blurred* predicates, that is, predicates whose definition is not communicated because the predicate is either sensitive or too large to be transmitted efficiently over the network. Such predicates are not (completely) evaluated, therefore in selected cases they cannot be responsible for success or failure. A few more features can be illustrated via the following explanation item:

```
– Rule [6] cannot be applied:
    • c012 is an id whose name is J. Smith and issuer is myCA
  but
    • myCA is not a recognized certification authority [details].
```

Here the first bullet covers *several* atoms in the body of Rule [6], whose internal format looks like

$$\ldots, credential(C),\ C.name{:}User,\ C.issuer{:}CA,\ \ldots$$

In this case the variables are bound to constants c012, 'J. Smith', myCA because there is a unique answer substitution for this group of atoms; this heuristics is called *unique answer propagation*. Moreover the group of atoms is verbalized in one phrase because the group constitutes a so-called *cluster*. These heuristics enhance readability by producing text about concrete entities (as opposed to variables with unspecified values) and referring to structured objects through their properties (name and issuer) rather than internal handlers (c012) that are meaningless to users.

However, if the user had provided a credential that had not been recognized as an identifier, the resulting explanation would have been

```
– Rule [6] cannot be applied:
    • I find no Credential such that the Credential is an id [details]
```

This is another example of pruning irrelevant information, if an object of a certain type is not found, as an id credential in this case, it is not relevant to report its properties.

The explanation hypertext can be navigated by clicking on the [details] links, that give more details about why the corresponding condition succeeds or fails. Note that this presentation technique combines local information (the rules that directly apply to a specific condition) with global information (which conditions eventually succeed, which of them fail, which answer substitutions are returned) that together describe a *set* of alternative (possibly incomplete or failed) proof *attempts*. For example, in case the types of subscription that allow to download the paper did not match the ones owned by J. Smith, we can obtain the following explanation.

```
– Rule [3 ] cannot be applied:
    • J. Smith is authenticated [details]
  but the following conditions cannot be simultaneously satisfied:
    • J. Smith subscribed some Subscription [Subscription = basic
      computer pubs] [Subscription = basic law pubs]
    • paper_0123.pdf is available for the Subscription [Subscription =
      gold subscription] [Subscription = complete computer pubs]
```

As you can see, the conditions *J. Smith subscribed some Subscription* and *paper_0123.pdf is available for the Subscription* do not have a common answer, therefore ProtuneX states that they cannot simultaneously satisfied. But, as they singularly succeed, ProtuneX provides a global view on the their possible results. So the user can more easily follow the paths that do not match the user expectations and focus more rapidly on the pages of interest.

4.3.5   The Engine

Protune can be entirely compiled onto Java bytecode. Network communications and the main flow of control for negotiations are implemented directly in Java, while reasoning (including filtering) is implemented in TuProlog, a standard Prolog that can be compiled onto Java bytecode.

Figure 3 shows the overall algorithm for a single negotiation step implemented within the Protune system.

– rfp ≡ Received filtered policy	1: add(rfp, s)
– s ≡ Negotiation state	2: add(rn, s)
– rn ≡ Received notifications	3: Action[] la = extractLocalActions(g, s)
– g ≡ Overall goal	4: **while**(la.length != 0)
– op ≡ Other peer	5: Notification[] ln = execute(la)
– ta ≡ Termination Algorithm	6: add(ln, s)
– ass ≡ Action Selection Strategy	7: la = extractLocalActions(g, s)
	8: **if**(prove(g, s))
	9: send(SUCCESS, op)
	10: **return**
	11: **if**(isNegotiationFinished(s, ta))
	12: send(FAILURE, op)
	13: **return**
	14: Action[] ua = extractUnlockedExternalActions(g, s)
	15: Action[] aa = selectActions(ass, ua, s)
	16: Notification[] sn = execute(aa)
	17: FilteredPolicy sfp = filter(g, s)
	18: add(sfp, s)
	19: add(sn, s)
	20: send(sfp, op)
	21: send(sn, op)

Fig. 3. Negotiation algorithm pseudocode

At each negotiation step a peer P_1 sends another peer P_2 a (potentially empty) filtered policy rfp and a (potentially empty) set of notifications rn, respectively stating the conditions to be fulfilled by P_2, and notifying the execution by P_1 of any actions it was asked for. As soon as P_2 receives this information, it adds it to its negotiation state.

Then P_2 processes its local policy in order to identify the local actions that can be performed taking into account the new information received. When such local actions are performed, other local actions may become ready for execution: this is the case e.g., if the instantiation of a variable is a prerequisite for the execution of an action and the instantiation of this variable is (part of) the result of another action's execution like in the following example, where the execution of *action1* makes *action2* ready for execution.

$$\ldots \leftarrow action1(X), action2(X).$$

$$action1(\_) \rightarrow actor : self.$$
$$action1(\_) \rightarrow execution : immediate.$$

$$action2(\_) \rightarrow actor : self.$$
$$action2(X) \rightarrow execution : immediate \leftarrow ground(X).$$

For this reason local action selection and execution are performed in a loop, until no more actions are ready to be executed. The need for iteration was overlooked in [11] and is documented in this chapter for the first time.

After having performed all possible local actions the local policy is processed in order to check whether the overall goal of the negotiation is fulfilled. If this is the case, a message is sent to P_1 telling that the negotiation can be successfully terminated. Otherwise the Termination Algorithm is consulted in order to decide whether the negotiation should continue or fail.

If the negotiation is not yet finished, then two processes have to be performed

- It is P_2's turn to filter its local policy and collect all items that have to be sent back to P_1;
- P_2 has to decide which of the actions whose execution has been requested by P_1 will be performed. Therefore, it processes its local policy and the (last) filtered policy received from P_1 in order to identify such actions. Notice that only actions such that the policies protecting them are fulfilled (*unlocked actions*) are collected.

Unlocked actions represent potential candidates to execution, i.e., those actions which can be performed according to P_2's local policy and its current negotiation state. However, just a subset of them will be actually performed, namely the one selected by the Action Selection Function. At each step of the negotiation, Protune builds an *AND-OR* tree with all the actions (e.g., information disclosure) that must be performed in order to advance the negotiation. This *AND-OR* tree is passed to a class implementing an Action Selection Function. Such a class can be custom and it just needs to follow an open API [4]. Protune provides out-of-the-box a "relevant" strategy that performs in parallel those actions required to advance the negotiation. We are also working on strategies based on preferences defined by the user between pairs of actions (e.g., *it is preferred to provide information related to my credit card than to my bank account*) and use them at run-time. "Good" negotiation strategies are discussed in [25,5].

Finally, the filtered policy and the notifications of the performed external actions are added to the negotiation state and sent to P_1.

---

[4] Cf. http://www.l3s.de/~olmedilla/policy/doc/javadoc/org/policy/strategy/ActionSelectionStrategy.html

4.3.6   Demo: Policy-Driven Protection and Personalization of Web Content

Open distributed environments like the World Wide Web offer easy sharing of information, but provide few options for the protection of sensitive information and other sensitive resources. Furthermore, many of the protected resources are not static, but rather generated dynamically, and sometimes the content of a dynamically generated web page might depend on the security level of the requester. Currently these scenarios are implemented directly in the scripts that build the dynamic web page. This typically means that the access control decisions that can be performed are either simple and inflexible, or rather expensive to develop and maintain. Moreover it is commonly accepted that access control and application logic should be kept separate, as witnessed by the design of policy standards such as XACML and the WS-* suite. Frameworks like Protune provide a flexible and expressive way of specifying access control requirements.

We have integrated Protune in a Web scenario capable of advanced decisions based on expressive conditions, including credential negotiation to establish enough trust to complete a transaction while obtaining some privacy guarantees on the information released [11]. We have developed a component that is easily deployable in web servers supporting servlet technology (we currently support Apache Tomcat), which adds support for negotiations and policy reasoning. It allows web developers to protect static resources by assigning policies to them. In addition to protection of static content, it also allows web developers to generate parts of dynamic documents based on the satisfaction of policies (possibly involving negotiations). We provide an extension to the web design tool Macromedia Dreamweaver in order to help web designers to easily and visually assign policies to their dynamic web pages[5].

A live demo is publicly available[6] as well as a screencast[7].

4.3.7   Experimental Evaluation

In order to evaluate the performance of Protune we first focused on its efficiency in carrying out negotiations. To this aim we measured the duration of each step of the negotiation algorithm described in Section 4.3.5 with a profiling tool we built exploiting the `log4j` [8] utility by the Apache foundation.

In the absence of large bodies of complex formalized policies, we further developed a module to automatically generate policies according to the following input parameters: number of negotiation steps, number of rules per predicate, number of literals per rule body.

---

[5] As described in `http://skydev.l3s.uni-hannover.de/gf/project/protune/wiki/admin/?pagename=Integration+with+Dreamweaver`

[6] `http://policy.l3s.uni-hannover.de/`

[7] `http://www.viddler.com/olmedilla/videos/1/`. We recommend viewing it in full screen.

[8] `http://logging.apache.org/log4j/`

Table 2. Overall reasoning and network time (msec)

			Reasoning time				Network time			
			Definitions				Definitions			
			1	2	3	4	1	2	3	4
3 steps	Literals	1	8 + 6 + 6	5 + 4 + 4	5 + 4 + 4	5 + 10 + 4	10	7	7	7
		2	5 + 4 + 5	5 + 4 + 4	5 + 4 + 10	5 + 3 + 4	9	7	7	6
		3	5 + 4 + 4	5 + 3 + 4	5 + 4 + 4	5 + 4 + 4	7	7	6	7
		4	5 + 10 + 5	5 + 3 + 5	5 + 3 + 4	5 + 4 + 4	7	7	7	7
			Definitions				Definitions			
			1	2	3	4	1	2	3	4
5 steps	Literals	1	16 + 20 + 34	34 + 54 + 50	81 + 93 + 111	199 + 173 + 208	14	17	18	18
		2	35 + 40 + 41	66 + 120 + 91	165 + 241 + 206	397 + 445 + 409	16	16	18	19
		3	42 + 78 + 56	116 + 212 + 161	291 + 481 + 347	646 + 867 + 719	17	17	19	19
		4	75 + 105 + 77	189 + 365 + 222	470 + 789 + 560	1012 + 1445 + 1173	17	17	20	21
			Definitions				Definitions			
			1	2	3	4	1	2	3	4
7 steps	Literals	1	65 + 37 + 47	196 + 100 + 63	1059 + 394 + 160	1922 + 893 + 230	19	12	20	14
		2	187 + 91 + 53	771 + 736 + 147	4423 + 3701 + 617	8526 + 12065 + 3030	12	12	14	14

Table 3. Realistic experiments for Protune core (left), and Protune-X performance (right)

Reasoning time	Network time
6 + 35 + 21	11
1 + 2 + 4	8
5 + 23 + 17	8
1 + 2 + 5	7
1 + 2 + 4	7

pol. size	output size	processing time	page rate	page squared rate
18	10	400 ± 70	40	4
35	20	1710 ± 60	85	4.3
71	22	2100 ± 50	95	4.3
42	31	3095 ± 31	99	3.2
40	32	3760 ± 40	117	3.7
42	35	3100 ± 40	88	2.5
40	41	6540 ± 130	159	3.9
39	41	6130 ± 30	150	3.6
59	42	5100 ± 60	121	2.9
83	46	6000 ± 60	130	2.8
57	50	8030 ± 90	160	3.2
109	63	20140 ± 110	319	5.0

Finally we assembled the components described above in a package which is freely available at http://skydev.l3s.uni-hannover.de/gf/project/protune/wiki/?pagename=Evaluation.

We ran a first set of experiments with realistic policies inspired by our reference scenarios. The results, reported in the leftmost part of Table 3, show that in these cases the system's performance is fully satisfactory. Then we tried the system on artificial policies that create large trees of dependencies: the root is the requested resource; its children (i.e., the 1st level of the tree) are the credentials needed to get the resource; the 2nd level is the set of counter-requests of the client that are needed to unlock the credentials in the 1st level, and so on. The artificial aspects in such examples consist in the exponential number of credentials involved (corresponding to tree nodes) and the chains of dependencies between them (usually shorter and sparser in realistic scenarios). Table 2 reports the results of these experiments, some of which are interrupted after 150sec. The frontier of terminating runs touches examples with hundreds or thousands of interrelated credentials, which explains the high values for reasoning time. Given the size of the examples involved, we conclude that this technology can scale up to policies and portfolios of credentials and declarations significantly larger than those applied today. This is interesting because the availability of frameworks like Protune may encourage the adoptions of policies more articulated and sophisticated than those deployed today.

A performance evaluation of the explanation facility ProtuneX has been done on a sample of 12 tests, including both realistic and artificial policies. We have used a ProtuneX implementation designed to be run through TuProlog, the Java-based Prolog adopted in the Protune framework. Each test has been run 20 times on a computer equipped with an 2GHz Intel dual-core duo and 2GB ram. Table 3 (on the right) shows the obtained results: the first column reports the size of the policy, that is, the number of its rules and metarules; the second column reports the number of generated web pages; the third column shows for each test the mean time (in msec) occurred to generate all the web pages and the relative mean squared error.

Tests are ordered according to the size of their policy and the reader can note that it is not easy to find out regularities between size and processing time. For example, tests 2 and 3 refer to policies of approximatively the same size, but the processing time of the latter is about 10 times longer than the former's one. However, if we consider test 7, whose policy size is notably larger than 3, the number of generated web-pages is a bit bigger and accordingly the processing time is. For this reason we have reported in column 4 the page rate, that is, the ratio between processing time and the output size, this value grows up linearly with respect to the output size, showing that the processing time is approximatively quadratic in the output size (cf. column 5).

Finally, we mention that there exists also a stand-alone implementation of ProtuneX, available at `http://cs.na.infn.it/rewerse/demos/protune-x/demo-protune-x.html`, that runs on XSB-Prolog, a Prolog engine written in C equipped with memoizing methods to improve performances and provide a more declarative semantics than standard Prolog. Even if a precise performance evaluation has not yet been carried out, its performance is remarkably better (>10 time faster) than the TuProlog counterpart.

The Java-based implementation is still appealing due to deployment ease (it is even possible to download the user agents as signed applets). However, the above performance estimates suggest that the explanation hypertext should rather be generated incrementally during navigation. Note that the computational load for the hypertext generation is essentially confined on the client; the server only needs to disclose verbalization metarules.

4.3.8 Discussion and Conclusions

We have illustrated the policy framework Protune and its implementation, reporting some positive, preliminary performance evaluation experiments. Protune is one of the most complete frameworks according to the desiderata laid out in the literature. It makes an essential use of semantic techniques to achieve its goals. More information about Protune and the vision behind it can be found on the web site of REWERSE's working group on Policies: `http://cs.na.infn.it/rewerse/`. There, on the software page, the interested reader may find links to Protune's software and some on-line demos and videos.

Unlike other applications of Semantic Web ideas, the main challenges for Protune are related to usability rather than tuple-crunching. Protune currently tackles usability issues by (partially or totally) automating the information exchange operations related to access control and information release control, and by supporting advanced, second generation explanation facilities for policies and negotiations.

We are planning to continue the development of Protune by adding new features and improving the prototype. In particular we plan to explore variants and enhancements of what-if queries to improve policy quality. Another interesting line of research concerns support for reliable forms of evidence not based on standard certification authorities, e.g. exploiting services such as OpenId and supporting user-centric credential creation (we can already support reputation-based policies via the external call predicates [11]). Support to obligation policies is another foreseen extension. Finally, we point the interested reader to Chapter 2, where the ACE front-end for Protune is discussed. Such front-end enables to exploit the controlled natural language ACE in order to define policies. As soon as (controlled) natural language is made Protune's standard user interface, usability evaluations will be carried out as well.

Another important line of research concerns standardization. We are investigating how Protune's policies and messages can be encoded by adapting and combining existing standards such as XACML (for decision rules), RuleML or RIF (for rule-based ontologies), WS-Security (for message exchange), and so on. Concerning W3C RIF initiative, our working group has contributed with a use case about policy and ontology sharing in trust negotiation.

References

1. Anderson, A.H.: An introduction to the web services policy language (wspl). In: 5th IEEE International Workshop on Policies for Distributed Systems and Networks (POLICY), pp. 189–192. IEEE Computer Society, Los Alamitos (2004)
2. Anderson, A.H.: A comparison of two privacy policy languages: Epal and xacml. In: Proceedings of the 3rd ACM workshop on Secure web services, pp. 53–60. ACM Press, New York (2006)
3. Ashley, P., Hada, S., Karjoth, G., Powers, C., Schunter, M.: Enterprise privacy authorization language (epal 1.2). Technical report, IBM (November 2003)
4. Backes, M., Karjoth, G., Bagga, W., Schunter, M.: Efficient comparison of enterprise privacy policies. In: Proceedings of the 2004 ACM symposium on Applied computing, pp. 375–382. ACM Press, New York (2004)
5. Baselice, S., Bonatti, P., Faella, M.: On interoperable trust negotiation strategies. In: IEEE POLICY 2007, pp. 39–50. IEEE Computer Society, Los Alamitos (2007)
6. Becker, M.Y., Sewell, P.: Cassandra: Distributed access control policies with tunable expressiveness. In: 5th IEEE International Workshop on Policies for Distributed Systems and Networks (POLICY 2004), Yorktown Heights, NY, USA, pp. 159–168. IEEE Computer Society, Los Alamitos (2004)
7. Blaze, M., Feigenbaum, J., Lacy, J.: Decentralized trust management. In: IEEE Symposium on Security and Privacy, pp. 164–173 (1996)
8. Bonatti, P., Olmedilla, D., Peer, J.: Advanced policy explanations. In: 17th European Conference on Artificial Intelligence (ECAI 2006), Riva del Garda, Italy. IOS Press, Amsterdam (2006)

9. Bonatti, P., Samarati, P.: Regulating service access and information release on the web. In: Proceedings of the 7th ACM conference on Computer and communications security, pp. 134–143. ACM Press, New York (2000)

10. Bonatti, P.A., Olmedilla, D.: Driving and monitoring provisional trust negotiation with metapolicies. In: 6th IEEE International Workshop on Policies for Distributed Systems and Networks (POLICY 2005), Stockholm, Sweden, pp. 14–23. IEEE Computer Society, Los Alamitos (2005)

11. Bonatti, P.A., Olmedilla, D.: Driving and monitoring provisional trust negotiation with metapolicies. In: 6th IEEE Policies for Distributed Systems and Networks (POLICY 2005), Stockholm, Sweden, pp. 14–23. IEEE Computer Society, Los Alamitos (2005)

12. Bonatti, P.A., Olmedilla, D., Peer, J.: Advanced policy explanations on the web. In: 17th European Conference on Artificial Intelligence (ECAI 2006), Riva del Garda, Italy, pp. 200–204. IOS Press, Amsterdam (2006)

13. Damianou, N., Dulay, N., Lupu, E., Sloman, M.: The ponder policy specification language. In: 2nd IEEE International Workshop on Policies for Distributed Systems and Networks (POLICY), pp. 18–38. Springer, Heidelberg (2004)

14. Duma, C., Herzog, A., Shahmehri, N.: Privacy in the semantic web: What policy languages have to offer. In: Eighth IEEE International Workshop on Policies for Distributed Systems and Networks-TOC (POLICY), pp. 5–8. IEEE Computer Society, Los Alamitos (2007)

15. Gavriloaie, R., Nejdl, W., Olmedilla, D., Seamons, K.E., Winslett, M.: No registration needed: How to use declarative policies and negotiation to access sensitive resources on the semantic web. In: Bussler, C.J., Davies, J., Fensel, D., Studer, R. (eds.) ESWS 2004. LNCS, vol. 3053, pp. 342–356. Springer, Heidelberg (2004)

16. Herzberg, A., Mass, Y., Michaeli, J., Ravid, Y., Naor, D.: Access control meets public key infrastructure, or: Assigning roles to strangers. In: 2000 IEEE Symposium on Security and Privacy, pp. 2–14. IEEE Computer Society, Los Alamitos (2000)

17. Kagal, L., Finin, T.W., Joshi, A.: A policy language for a pervasive computing environment. In: 4th IEEE International Workshop on Policies for Distributed Systems and Networks (POLICY), Lake Como, Italy, pp. 63–74. IEEE Computer Society, Los Alamitos (2003)

18. Li, N., Mitchell, J.C.: Rt: A role-based trust-management framework. In: Third DARPA Information Survivability Conference and Exposition (DISCEX III). IEEE Computer Society, Los Alamitos (2003)

19. Lorch, M., Proctor, S., Lepro, R., Kafura, D., Shah, S.: First experiences using xacml for access control in distributed systems. In: Proceedings of the 2003 ACM workshop on XML security, pp. 25–37. ACM Press, New York (2003)

20. Seamons, K.E., Winslett, M., Yu, T., Smith, B., Child, E., Jacobson, J., Mills, H., Yu, L.: Requirements for policy languages for trust negotiation. In: 3rd International Workshop on Policies for Distributed Systems and Networks (POLICY), Monterey, CA, USA, pp. 68–79. IEEE Computer Society, Los Alamitos (2002)

21. Simon Godik, T.M.: Oasis extensible access control markup language (xacml) version 1.0. Technical report, OASIS (February 2003)

22. Tonti, G., Bradshaw, J.M., Jeffers, R., Montanari, R., Suri, N., Uszok, A.: Semantic web languages for policy representation and reasoning: A comparison of kaos, rei, and ponder. In: Fensel, D., Sycara, K., Mylopoulos, J. (eds.) ISWC 2003. LNCS, vol. 2870, pp. 419–437. Springer, Heidelberg (2003)

23. Uszok, A., Bradshaw, J.M., Jeffers, R., Suri, N., Hayes, P.J., Breedy, M.R., Bunch, L., Johnson, M., Kulkarni, S., Lott, J.: Kaos policy and domain services: Toward a description-logic approach to policy representation, deconfliction, and enforcement. In: 4th IEEE International Workshop on Policies for Distributed Systems and Networks (POLICY), Lake Como, Italy, pp. 93–96. IEEE Computer Society, Los Alamitos (2003)

24. Winsborough, W., Seamons, K., Jones, V.: Automated trust negotiation. In: DARPA Information Survivability Conference and Exposition, DISCEX 2000. Proceedings, pp. 88–102. IEEE Computer Society, Los Alamitos (2000)

25. Yu, T., Winslett, M., Seamons, K.E.: Supporting structured credentials and sensitive policies through interoperable strategies for automated trust negotiation. ACM Trans. Inf. Syst. Secur. 6(1), 1–42 (2003)

Chapter 5
Component Models for Semantic Web Languages

Jakob Henriksson and Uwe Aßmann

Lehrstuhl Softwaretechnologie, Fakultät Informatik,
Technische Universität Dresden, Germany
{jakob.henriksson, uwe.assmann}@tu-dresden.de
http://st.inf.tu-dresden.de/

Abstract. Intelligent applications and agents on the Semantic Web typically need to be specified with, or interact with specifications written in, many different kinds of formal languages. Such languages include ontology languages, data and metadata query languages, as well as transformation languages. As learnt from years of experience in development of complex software systems, languages need to support some form of component-based development. Components enable higher software quality, better understanding and reusability of already developed artifacts. Any component approach contains an underlying component model, a description detailing what valid components are and how components can interact. With the multitude of languages developed for the Semantic Web, what are their underlying component models? Do we need to develop one for each language, or is a more general and reusable approach achievable? We present a language-driven component model specification approach. This means that a component model can be (automatically) generated from a given base language (actually, its specification, e.g. its grammar). As a consequence, we can provide components for different languages and simplify the development of software artifacts used on the Semantic Web.

Keywords: software engineering, composition, modularization, semantic web.

5.1 Introduction

The Semantic Web started out as a vision to enable "computers and people to work in cooperation" by creating an extension of the current Web "in which information is given well-defined meaning" [13]. The Semantic Web has since come to encompass a wide range of research areas and approaches; a very high-level overview of this diversity is provided by the Semantic Web Topic Hierarchy.[1] In particular, a large number of *languages* have been constructed to support the different ideas and approaches. This includes metadata languages (e.g. RDF [34]), query languages (e.g. XQuery [14], Xcerpt [45]), rule languages (e.g. RIF [1]) and ontology languages (e.g. OWL [44]), to only name a few. Most of these languages are developed by research groups, or developed through consortiums (such as the W3C[2]), where the focus of the language development is to cover certain use-cases that are considered important for the kind

---

[1] http://semanticweb.org/wiki/Semantic_Web_Topic_Hierarchy
[2] http://www.w3c.org

F. Bry and J. Maluszynski (Eds.): Semantic Techniques for the Web, LNCS 5500, pp. 233–275, 2009.

of problems the language is assumed to be able to address. Many of these languages have been developed over many years and are being adopted by a large number of users. More and more, they are considered mature enough to be deployed in production systems (XQuery and OWL are examples of such languages). Parallel with the development of the languages, development of associated tools is carried out, such as query engines, ontology reasoners and appropriate editors.

One commonality between most Semantic Web languages is that they are *domain-specific languages* (DSLs).[3] A DSL can been defined as a "programming language or executable specification language that offers, through appropriate notations and abstractions, expressive power focused on, and usually restricted to, a particular domain." [22, p. 26]. As such, DSLs stand in contrast to the existence of *general-purpose languages* (GPLs) in the field of software engineering. GPLs do not presume the kind of problems they will be used to solve, but can be used for many different purposes and solutions. However, only having a general-purpose language available can be sub-optimal for a very particular problem that is to be solved. For this reason—and the general usefulness of DSLs—there have been several approaches developed that enable the embedment of a DSL into a GPL, called the *host language* (e.g. [10,16,30,53]). Such DSLs are referred to as *embedded* DSLs (E-DSLs). The advantage of embedment is that both syntax and semantics from the host language can be reused for the DSL. For example, as a way to provide semantics for the embedded language, a translation into the host language can be specified. Thanks to the translational semantics, existing tooling for the general-purpose language may be reused. Examples of host languages are Java and Scala.[4] As an added benefit, the already existing abstraction and reuse constructs of the host language may be exploited, making it unnecessary to provide them in the DSL in the first place. This also holds for other useful language constructs, such as control-flow mechanisms (e.g. conditionals and loops).

Can these useful techniques also be applied to the DSLs used on the Semantic Web? Many Semantic Web DSLs are *non-embedded* DSLs (NE-DSLs, also called *standalone* or *external* [24]). This means that many Semantic Web languages are intended to be used as standalone languages, with their own syntax, well-defined semantics, and developed tools (e.g. query engines and ontology reasoners). As the authors of [22] point out in their work, the key to the definition of DSLs is the notion of *focused* expressive power. Thus, constructs and abstractions are provided specifically for formulating and solving problems related to the domain for which the language was developed. With such specialized constructs and appropriate domain-related abstractions at hand, programmers can concisely express what they want and need. However, DSLs often do not provide rich constructs that enable reuse and component-based development. Even though DSLs can be a tool to help cope with software complexity, unfortunately, they also introduce a new level of complexity that is not always initially foreseen. The new complexity arises because the DSL specifications themselves may grow in size. For example, XML query or transformation programs can easily grow large and become hard to manage and maintain. The same holds for ontology specifications which can contain thousands of concepts. As those specific parts of larger software systems grow,

---

[3] We say 'most' since we are obviously making a generalization, but a useful one.

[4] See http://java.sun.com/ and http://www.scala-lang.org, respectively.

there must be means in place to cope with that growth in order for DSLs to maintain their attractiveness. This also holds true for the NE-DSLs on the Semantic Web. While there exist general approaches to achieve abstraction and reuse constructs for DSLs in traditional software engineering through language embedment, no general approach exists to enable the same for NE-DSLs, in particular those used on the Semantic Web. With a large number of languages available, a general approach for the enablement of component-based development seems highly desirable.

In this chapter, we present such a general approach to address this problem; general in the sense that different languages and different component types (reuse constructs, or abstractions) can be realized. It should be mentioned that we only address the problem of how single programs/specifications can be decomposed and recomposed, and not the problem of how heterogenous languages/formalisms can be composed and integrated. We focus on the rule-based Web query language Xcerpt [45] (cf. Chapter 2) and the ontology language OWL [44]. Our approach is achieved by building on existing *invasive* software composition techniques: grammar-based modularization [36,37,39] and invasive software composition [5]. We follow the vision outlined in [4,6], where the employment of component-based software engineering techniques is proposed for several concerns of semantic applications. We have implemented our approach in the composition toolset and framework REUSEWARE.[5] A partial consequence of an invasive composition approach is that it is *static*. This means that the considered components must be composed at compile-time. Furthermore, all components are composed to valid instances of the addressed language (e.g. Xcerpt or OWL) before being executed or interpreted. This allows existing tools of the base language to be reused, which is essential for such a general approach.

This chapter is structured as follows. In Section 5.2 we briefly introduce the notion of composition systems, the query language Xcerpt and the ontology language OWL. In Section 5.3 we summarize the current state of the art wrt. modularization for query and ontology languages, and then invasive composition techniques in software engineering. In Section 5.4 we discuss use-cases for Web query and ontology components. In Section 5.5 we present our contribution of universal (invasive) component models and composition systems. In Section 5.6 we go into details of how component-based development with the query language Xcerpt can be realized in our approach. Finally, in Section 5.7, we summarize the chapter.

5.2   Background

In this section, we introduce three notions and languages that are further built upon in the remainder of the chapter: composition systems, the query language Xcerpt and the ontology language OWL.

5.2.1   Composition Systems and Component Models

The works of McIlroy [40] and Dennis [21] have, in the software engineering domain, introduced the notion of *components* (aka *modules*) and shown their usefulness for the

---

[5] http://www.reuseware.org

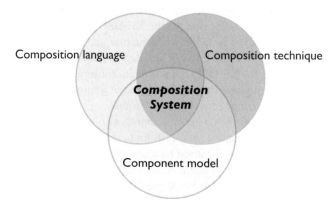

Fig. 1. A composition system consists of a *composition language, composition technique* and a *component model*

structuring of large software and the handling of product families [43]. Components hide volatile aspects under an interface, protect information, and serve for substitutability of parts. Many different component models have been developed; classical module systems a la Modula-2 are just a simple example. Recently, it has been proposed that component-based software engineering needs not only component models, but *composition systems* [5]. A *composition system* describes a particular compositional setting and is made up of three distinct parts: a composition language, a composition technique and a component model (see Figure 1). The *composition language* is used to specify exactly which components should be put together, and in what way. The composition language is thus used to write *composition programs*. The *composition technique* describes how components are joined, while the *component model* describes what kind of components may be defined (what they may look like) and how they are allowed to be accessed or transformed during composition (their interfaces).

A component model is essential to a composition system since it is the main instrument for controlling and restricting compositions. The exact restrictions posed by a component model often differ between composition systems, depending on their precise requirements.

Example 1 (Composition system of ASPECTJ*).* ASPECTJ is an extension of Java that enables the possibility of programming with *aspects* [32]. An aspect can contain a crosscutting concern of the overall system realization. It can be defined separately, and is then woven into the core system. We do not completely introduce ASPECTJ here, but use it as an example composition system since it is well-known, and see how it can be broken down into the three above-described components.

- *Composition language.* The composition language is a direct extension of Java, in which aspects can be defined, and how the aspects are to be woven into the core system.
- *Composition technique.* Aspects can be woven statically or dynamically. Static weaving changes the structure of programs without changing the program behavior,

while dynamic weaving does change the runtime behavior. The static weaving boils down to transforming the source code structure of Java programs. Hence, the composition technique can be seen as transforming Java source code. Dynamic weaving takes place at run time and involves technical details such as changing the underlying call stack.[6]

– *Component model.* The component model in ASPECTJ incorporates classes, aspects and the 'join point model.' The join point model defines all places in a program that are accessible for transformation, that is, the places where aspects can be woven in. In ASPECTJ, since only implicit interfaces are available (they are not marked by programmers), the component model mainly describes how components can be accessed, and less how they look.

5.2.2 Web Query Language Xcerpt

Xcerpt is a rule-based language for querying semi-structured data, for example XML or RDF (which has an XML serialization). The language follows, or is closely related to, the Logic Programming (LP) paradigm (see, for example, [41] for an introduction to LP). There are many publications on Xcerpt (see, e.g., [17,18,45] and in particular Chapter 2). Here we recall the basic constructs. An Xcerpt *program* consists of a finite set of Xcerpt *rules*. The rules of a program are used to define data, or to derive new data from existing data (i.e. the data being queried). In Xcerpt, two different kinds of rules are distinguished: *construct rules* and *goal rules*. Their syntax are given in Listings 1.1 and 1.2, respectively, where anything enclosed between angle brackets (< and >) will be explained later. We simply refer to (Xcerpt) *rules* when we do not distinguish between the two kinds of rules. Construct rules are used to produce intermediate results while

```
1 CONSTRUCT
2     <head>
3 FROM
4     <body>
5 END
```

```
1 GOAL
2     <head>
3 FROM
4     <body>
5 END
```

Listing 1.1. A *construct rule*   **Listing 1.2.** A *goal rule*

goal rules make up the output of programs. Rules have a *head* and optionally a *body*. Intuitively, rules are to be read: if *body* holds, then *head* holds. A rule lacking a *body* is interpreted as a *fact*, that is, the rule *head* always holds.

While Xcerpt works directly on XML data, it also has its own data format. Xcerpt *data terms* model XML data and there is a one-to-one correspondence between the two notions. While XML uses labeled "tags," Xcerpt data terms use a square bracket notation. The data term `book [ title [ "White Mughals" ] ]`, for example, corresponds to the `<book><title>White Mughals</title></book>` XML snippet. The data term syntax provides a more readable XML syntax to use in queries.

---

[6] Certain behavioral program modifications can also be achieved by changing program structure, but we do not elaborate on this here (see [33] for details).

```
1 GOAL
2     authors [ all author [ var X ] ]
3 FROM
4     book [[ author [ var X ] ]]
5 END
6
7 CONSTRUCT
8     book [ title [ "White Mughals" ], author [ "William Dalrymple" ] ]
9 END
10
11 CONSTRUCT
12     book [ title [ "Stanley" ], author [ "Tim Jeal" ] ]
13 END
```

Listing 1.3. The construct rules define data about books and their authors and the goal rule queries this data for authors

Formally, the *head* of a rule is a *construct term* and the *body* is a *query*. A query is a set of *query terms* joined by some logical connective (e.g. or or and). Query terms are used for querying data terms and intuitively describe patterns of data terms. Query terms are used with a pattern matching technique to match data terms.[7] Query terms can be configured to take partialness and/or ordering of the underlying data terms into account during matching. Square brackets are used in query terms when order is of importance, otherwise curly brackets may be used. E.g. the query term a [b [], c []] matches the data term a [b [], c []] while the query term a [c [], b []] does not. However, the query term a { c [], b [] } matches a [b [], c []] since ordering is said to be of no importance in the query term. Partialness of a query term can be expressed by using double, instead of single, brackets (i.e. [[...]] or {{ ... }}). Query terms may also contain logical variables (denoted by capitalized identifiers preceded by keyword var, for example, var X). If so, successful matching with data terms results in variable bindings used by rules for deriving new data terms. For example, matching the query term book [title [var X]] with the XML snippet above results in the variable binding {X / "White Mughals"}. *Construct terms* are essentially data terms with variables. The variable bindings produced by queries in the body of a rule can be applied to the construct term in the head of the rule in order to derive new data terms. In the rule head, construct terms including a variable can be prefixed with the keyword all to group the possible variable bindings around the specific variable.

An example Xcerpt program relating to books is shown in Listing 1.3. The last two rules are facts and define two books, each with a title and an author. The first rule—a goal rule—defines the output of the program. It queries authors of books, and constructs a list of all found authors. The program in Listing 1.3 would result in the following data term as output:

```
authors [ author [ "William Dalrymple" ], author [ "Tim Jeal" ] ]
```

---

[7] This technique is called *simulation unification*, please consult [46] for details.

```
1  GOAL
2      <head>
3  FROM
4      in { resource { "file:db.xml", "xml" },
5          <query>
6      }
7  END
```

Listing 1.4. A program with a single rule querying an external resource

Both authors are in the answer because of the grouping construct (all) used in the construct term of the goal rule. Furthermore, the query in the goal rule matches the two facts by not considering the book titles since the partialness construct is used ([[...]]).

A rule can also query an external resource, for example, a Web page or an XML database stored as a file. An example in given in Listing 1.4 where the XML file file:db.xml is being queried by a not further detailed query (*<query>*). The construct term of the rule is also omitted (*<head>*).

5.2.3   Description Logics and OWL

Description Logics (DLs) are a family of knowledge representation formalisms, where most members are sub-languages of first-order logics. DLs are used to capture the important *concepts* and relations (*roles* in DL parlance) between individuals of the modeled domain. Concepts and roles can be described by complex *concept* (resp. *role*) *descriptions* using the construction operators available in the particular DL.

The most widely used DL is the one underlying OWL DL [44]. To simplify the presentation, we do not cover datatypes here. An OWL DL *interpretation* is a tuple $I = (\Delta^I, \cdot^I)$ where the individual domain Δ^I is a nonempty set of individuals, and $\cdot^I$ is an individual interpretation function that maps (i) each individual name o to an element $o^I \in \Delta^I$, (ii) each concept name A to a subset $A^I \subseteq \Delta^I$, and (iii) each role name R to a binary relation $R^I \subseteq \Delta^I \times \Delta^I$. Valid OWL DL concept descriptions are defined by the DL syntax:

$$C ::= \top \mid \bot \mid A \mid \neg C \mid C \sqcap D \mid C \sqcup D \mid \{o\} \mid \exists R.C \mid \forall R.C \mid \geqslant mR \mid \leqslant mR$$

The interpretation function $\cdot^I$ is extended to interpret $\top^I = \Delta^I$ and $\bot^I = \emptyset$. The concept $\top$ ($\bot$) is called *owl:Thing* (*owl:Nothing*) in OWL. The interpretation function can further be extended to give semantics to the remaining concept and role descriptions (see [44] for details).

An OWL DL ontology consists of a set of *axioms*, including concept axioms, role axioms and individual axioms.[8] A DL knowledge base consists of a TBox, an RBox and an ABox. A *TBox* is a finite set of concept inclusion axioms of the form $C \sqsubseteq D$, where C, D are concept descriptions. An interpretation I satisfies $C \sqsubseteq D$ if $C^I \subseteq D^I$. An *RBox* is a finite set of role axioms, such as role inclusion axioms ($R \sqsubseteq S$). The

---

[8] Individual axioms are called *facts* in OWL.

kinds of role axioms that can appear in an RBox depend on the expressiveness of the ontology language. An interpretation I satisfies $R \sqsubseteq S$ if $R^I \subseteq S^I$. An *ABox* is a finite set of individual axioms of the form $a : C$, called *concept assertions*, or $\langle a, b \rangle : R$, called *role assertions*. An interpretation I satisfies $a : C$ if $a^I \in C^I$, and it satisfies $\langle a, b \rangle : R$ if $\langle a^I, b^I \rangle \in R^I$.

Let C, D be concept descriptions, C is *satisfiable* wrt. a TBox T iff there exist an interpretation I of T such that $C^I \neq \emptyset$; C subsumes D wrt. T iff for every interpretation I of T we have $C^I \subseteq D^I$. A knowledge base Σ is *consistent* (*inconsistent*) iff there exists (does not exist) an interpretation I that satisfies all axioms in Σ.

Human-readable syntax – Manchester OWL Syntax. OWL has several syntaxes, but OWL ontologies are most commonly represented by XML serializations. Such serializations are machine readable, which is good for tooling and interoperability, but less appealing to end-users and ontology designers. Many end-users prefer to use the Manchester OWL syntax [31], which is more user friendly for non-logicians, and also supported by ontology editors such as Protégé.[9] In short, the Manchester syntax "tries to minimize syntactic constructs that are difficult to enter or understand" [31, p. 3]. For example, the conjunction (disjunction) of concepts C and D, rather than using the mathematical symbol $\sqcap$ ($\sqcup$), can be written: C **and** D (C **or** D). Other concept constructors have similar intuitive English words that can be used. Ontology axioms can also be represented. The axiom defining concept C as a sub-concept of D ($C \sqsubseteq D$) can be written: **Class:** C **SubClassOf** D. There are other Manchester OWL constructs not detailed here, but they are intuitive to understand when seen in an example. More detail on the Manchester OWL syntax can be found in [31].

5.3   State of the Art: Semantic Web Components and Invasive Software Component Models

This section consists of three parts. In Sections 5.3.1–5.3.2 we study modularization (or component-based development) techniques for existing Semantic Web languages. Our main focus is on query and ontology languages. For ontology languages we mainly focus on OWL due to its wide adoption as a standard, but focus also on modularization techniques for its underlying logic – Description Logics. Then, in Section 5.3.3, we study existing invasive composition approaches.

5.3.1   Query Modularization

- *Xcerpt.* The only programming abstraction provided by Xcerpt is the *rule*. A rule can query an external resource or data constructed by another rule. An Xcerpt program is thus a set of rules with certain implicit dependencies. The programmer has the freedom of splitting the overall query task into any number of rules. During evaluation of a query program the same rule can be used several times and is in this sense reused for that particular query evaluation. Xcerpt does not, however, provide

---

[9] http://protege.stanford.edu/

a way of reusing rules across programs. Nor does Xcerpt provide means to reuse larger query tasks (sets of collaborating and related rules).

– *XQuery.* XQuery is a functional XML query language standardized by W3C [14]. XQuery provides a relatively rich set of constructs for reuse and component-based development, the main one being the *function.* A function takes a number of arguments and returns either an atomic value (e.g. an integer), or an element (document node). The function body consists of a single expression that can make use of the passed parameters, or other user-defined functions. Users can also define *modules.* Every application is a module, the *main module,* but so-called *library modules* can also be defined. Every module has an associated namespace. Library modules can be imported into other modules using the `import module` construct, by referring to their namespaces. Importing a module gives access to its declared functions and global variables. Imports do not cascade, that is, access is only given to modules that are imported directly. Cyclic imports are not allowed (wrt. different namespaces, since modules may share namespaces). The main advantage of XQuery functions (and modules) is the ability to hide schema complexity for users, and to enable recursive queries in a convenient way.

– *Extensible Stylesheet Language Transformations (XSLT).* XSLT is an XML transformation language [19]. XSLT provides two methods for combining *stylesheets* (transformation specifications): inclusion and importing. Including stylesheets retains the semantics of the combination, while importing stylesheets override each other depending on a precedence value (not described here, cf. [19]). XSLT is a rule-based language, but uses precedence rather than the standard union semantics for multiple applicable rules. Rule precedence is the dominating issue for XSLT's module system (stylesheet importing) which provides intricate mechanisms for determining the precedence of rules from different modules. The module system is cascading, and has limited parameterization via its `apply-imports` construct.

5.3.2   Ontology Modularization

There have been many approaches suggested for modularizing OWL ontologies, or its underlying logic, Description Logic. Below we give an overview of the main approaches, separated into two categories: ontology mapping and linking, and ontology importing.

Ontology Mapping and Linking. These approaches address the problem of how well-defined links can be established between ontologies. In such case, each component ontology can be seen as a module.

– *E-Connections.* An *E*-Connection is a set of "connected" ontologies, where each separate ontology is intended to cover a single topic [12]. It is an underlying assumption that the different involved ontologies have disjoint signatures (sets of non-logical symbols). An *E-Connected* ontology is a standard ontology, but contains special kinds of properties, *link properties.* Link properties are binary relations that relate elements from different (disjoint) domains (ontologies). Link properties are defined in a "source ontology," and relate some of its elements to elements in a "target" ontology. The elements in the target ontology are considered *foreign*. Link properties cannot be transitive or symmetric; each link property hence only connects two ontologies in a given E-Connection. E-Connected ontologies can be used both to integrate different ontologies (by adding knowledge of how the ontologies relate using link properties), and to decompose existing ontologies. The core idea is to keep ontologies small and disjoint and describe their relations with E-Connections, as seen fit by the modeler. E-Connections are not suitable for combining ontologies with overlapping domains.

– *Distributed Description Logics (D-DL).* D-DL is a formalism for combining different DL knowledge bases [15]. Each component ontology retains its independence and identity. The coupling between the different component ontologies is established by allowing a new set of inter-ontology axioms, *bridge rules*. Bridge rules take one of two forms [15]:

$$ C \overset{\sqsubseteq}{\longrightarrow} D \ \text{("into" bridge rule)} \quad \text{and} \quad C \overset{\sqsupseteq}{\longrightarrow} D \ \text{("onto" bridge rule)} $$

where C and D are concepts defined in different ontologies that are being related. The "into" bridge rule specifies that C-individuals in one ontology only correspond to D-individuals in the second ontology. The "onto" bridge rule specifies that every D-individual has a corresponding pre-image in concept C of the first ontology. Hence, D-DL can be used to relate ontologies developed independently. The main goal of D-DL is to connect existing ontologies, rather than to support collaborative development where different designers work on their separate modules.

Ontology Importing. These approaches address the problem of how an ontology or ontology module can be imported and hence reused.

– *owl:imports.* OWL natively provides some facilities for reusing ontologies and ontology parts. First, a feature inherited from RDF [34] (upon which OWL is layered) is *linking*—loosely referencing distributed Web content and other ontologies using URIs. Second, OWL provides an *owl:imports* construct which syntactically includes the complete referenced ontology into the importing ontology. The linking mechanism is convenient from a modeling perspective, but is semantically not well-defined—there is no guarantee that the referenced ontology or Web content exists. Furthermore, the component (usually an ontology class) is small and often hard to detach from the surrounding ontology in a semantically well-defined way. Usually, a full ontology import is required since it is unclear which other classes the referenced class depends on. The *owl:imports* construct can only handle complete ontologies and does not allow for partial reuse.

– *Semantic Import.* Semantic import differs from *owl:imports* (referred to as syntactic import) by allowing to import partial ontologies and by additionally enforcing the existence of any referred external ontologies and ontology elements by the notion of *ontology spaces* [42]. The goal in this work is controlled partial reuse of ontologies; the reuse units are concepts, properties or individuals.

– *Extracting ontology modules.* In [20], the authors propose an approach to extract *modules* from ontologies. A module is a (preferably minimal) set of axioms that define the concepts that are being reused. Importing a module M (from a larger ontology O_M) into an ontology B does not affect the modeling of B. Given a signature S (a set of non-logical symbols), an algorithm is presented for extracting *S-modules* from an existing ontology. An *S*-module is defined as [20]:

Let $Q_1 \subseteq Q$ be two ontologies and S a signature. Q_1 is an *S-module* in Q (wrt. a language L) if for every ontology P and every axiom α expressed in L with signature $Sig(P \cup \{\alpha\}) \cap Sig(Q) \subseteq S$, we have $P \cup Q \models \alpha$ iff $P \cup Q_1 \models \alpha$.

Thus, the approach can be used for extracting small and suitable modules from already developed monolithic ontologies. Extracted modules can then be imported into ontologies using *owl:imports*. Another approach in this direction is presented in [50].

- *Package-based Description Logics.* The work in [9] proposes package-based Description Logics (P-DLs) for collaborative ontology construction, sharing and reuse. A *package* is a fragment of an ontology. Each term (concept name, property name, or individual name) belongs to a particular package, its *home package*. A package can use terms defined in another package. A term that appears in package P, but has a different home package Q is called a foreign term in P (in this case, P *imports* Q). A *package-based ontology* (P-DL ontology) consists of multiple packages, each expressed in DLs. Packages may be organized in a hierarchical manner (e.g. a package can be defined as a sub-package of another package). Packages are defined by a *local semantics*, which is not influenced by foreign terms (simply treated as symbols; the same term can be interpreted differently in two different packages). Each package must be locally consistent. To ensure local semantics and knowledge hiding, it is possible to provide *scope limitation modifiers* to terms, regulating their visibility wrt. other packages. Terms and axioms can be visible only to their home packages (*private*), to their home packages and descendant packages in the package hierarchy (*protected*), or globally visible (*public*). The authors also define a *global* semantics, which gives the interpretation of all involved packages. Finally, there is also a *distributed* semantics, which gives interpretation to *some* of the involved packages. P-DL was developed to be used in a highly distributed environment such as the Web, and to support collaborative ontology development. P-DL is more expressive than D-DL. Bridge rules in D-DL only connect atomic concepts, while P-DL can express more complex relations, even involving concept expressions where terms belong to different modules (packages).

Ontology modularization is still largely an open issue. There are different approaches and motives for modularization. Most works have been focused on ontology design and reasoning in distributed environments. Less work has been done on component-based ontology development, that is, how to develop a single monolithic ontology in a flexible, modularized, and reusable fashion.

5.3.3   Invasive Component Models in Software Engineering

In this section we discuss the state of the art in invasive software component models. A component model in a composition approach can be said to support *invasive composition* if the involved components can be adapted internally to fit the particular reuse context. This stands in contrast to traditional black-box component models where the component interfaces consists of data flow between the components.[10]

Grammar-Based Modularization (GBM). BETA is an object-oriented programming language. One of the side projects developed around BETA was the Mjølner BETA system. The Mjølner system provides a *fragment system* (or *fragment language*) aimed for modularization of BETA program text. Essentially, any snippet of BETA source code— a *fragment*—can be a module. By putting such fragments together, a complete and executable program can be constructed. The technique used by the Mjølner fragment system is called *grammar-based modularization* (GBM) in the literature [36,37,39]. The technique is 'grammar-based' since the underlying language grammar dictates what are considered valid and deployable fragments for the modularization process.

The fragment system provided in the Mjølner system is directly connected to the BETA language itself. To simplify the presentation we do not introduce the BETA language and its syntax, but instead exemplify the technique using a Datalog-like language (see e.g. [41] for an introduction to Datalog). Modules, which here are equalled to fragments, are syntactical structures of the considered language and are called *forms*. Forms must belong to some syntactic category of the underlying grammar, and hence be derivable from some of its nonterminals. A form derived from nonterminal $\langle A \rangle$ is called an *A-form*. Forms can in principle be any sequence of terminal and nonterminal symbols of the considered grammar. Hence, forms are essentially sentential forms of a particular syntactic category of the grammar. Imagine that we have a grammar Datalog specifying the rule language Datalog (containing nonterminals such as $\langle Rule \rangle$, $\langle Atom \rangle$, $\langle Variable \rangle$, $\langle Num \rangle$ etc.).[11] Then the sentential form in (1) can be seen as a *rule*-form of the Datalog grammar with one $\langle Num \rangle$ and one $\langle Atom \rangle$ nonterminal (not yet derived to terminal symbols).

$$\text{bonus}(X, \langle Num \rangle) \text{ :- employee}(X), \langle Atom \rangle. \tag{1}$$

To be able to refer to nonterminals in forms, they are given names. Nonterminals meant to be replaced by the fragment system are called *slots* and have the following syntax:[12]

$$\text{«SLOT } T{:}A\text{»} \tag{2}$$

where T is the name of the slot and A is its syntactic category. The sentential form from (1) can thus be written as in (3), which contains a slot named value of syntactic category $\langle Num \rangle$ and a slot named condition of syntactic category $\langle Atom \rangle$ (when using nonterminals in slots we do away with the angle brackets). These slots describe where change can take place and are called *slot declarations*.

$$\text{bonus}(X, \text{«SLOT value:}Num\text{»}) \text{ :- employee}(X),$$
$$\text{«SLOT condition:}Atom\text{» }. \tag{3}$$

---

[10] Invasive components are referred to be supporting *gray-box* reuse abstractions [5].

[11] We do not present the grammar here, but it can easily be imagined.

[12] This syntax was originally chosen for its suitability wrt. the BETA language, and we use the same here.

When defining forms in the fragment system, they must be given a name and a syntactic category, and are then called *fragment-forms*. Following the style of [39], we use a graphical syntax for defining fragment-forms. The table in (4) demonstrates the graphical syntax (gray table rows indicate 'meta' information about forms, while white rows contain concrete forms).[13]

$$
\boxed{\begin{array}{|l|}
\hline
\texttt{F:}A \\
\hline
\texttt{ff} \\
\hline
\end{array}}
\tag{4}
$$

In (4) F is the name of the fragment-form, A is its syntactic category and ff is the form (derivable from nonterminal $\langle A \rangle$). The Mjølner system also introduces the notion of *fragment groups*, which are sets of fragment-forms associated by a name using the name construct (illustrated below). A fragment group containing a single fragment-form, corresponding to (3), is shown in (5).

name 'RuleGroup'
myRule:*Rule*
bonus(X, «SLOT value:*Num*») :- employee(X),
«SLOT condition:*Atom*» .

(5)

Complete programs are assembled by binding fragment-forms to declared slots. The *origin* construct can be used for this purpose. The *origin* construct takes the fragment group being operated on as an argument. The fragment-forms appearing in a fragment group with an *origin* construct are called *slot applications*.

name 'Rules'
origin 'RuleGroup'
value:*Num*
200
condition:*Atom*
efficient(X)

(6)

By matching the names of the fragment-forms in (6) (slot applications) with the slot names in the fragment group indicated by the *origin* construct (slot declarations), the fragment-form in (7) is constructed.

myRule:*Rule*
bonus(X, 200) :- employee(X), efficient(X).

(7)

Notice that the form in (7) is a valid Datalog sentence, stating that "efficient employees receive a bonus of 200." As such it is a useful entity constructed from its smaller fragment parts. The above has demonstrated the main idea of the Mjølner fragment system, but using the Datalog language rather than BETA itself. We have discussed the main features of the fragment system here. For more details we direct the reader to [39, Chapter 17].

---

[13] There is also a textual syntax available in the Mjølner BETA system, but is not further discussed here.

```
1 public void setTimeStamp() {
2     this.time = new java.util.Date().getTime();
3     System.out.println("Time set at: " + this.time);
4 }
```

Listing 1.5. A Java method box defined by an assignment and a print statement

```
1 public class Contract extends BankEntity {
2     // attributes ...
3     // methods ...
4     Contract() { ... }
5 }
```

Listing 1.6. A Java class box for a bank contract (attributes and methods not shown)

Invasive Software Composition in COMPOST. Invasive Software Composition (ISC) is a static composition approach where pieces of source code (fragments) are transformed into usable programs, the composition results [5]. The ISC demonstrator system is called COMPOST [52]. The entities being composed are programs, or partial programs, of a particular language. Such entities are, in ISC terminology, called *fragment boxes* (or simply *boxes*). As an example, a box containing a Java method—a Java "method box"—is shown in Listing 1.5. This particular method (setTimeStamp()), when invoked, assigns the class variable time the current time value and prints an informative message to standard output. The method cannot be used by itself (e.g. compiled by a Java compiler), but composed into a larger program it can provide certain functionality and can be reused.

Other kind of boxes can also be defined: perhaps larger entities such as elaborate Java class boxes. Listing 1.6 shows a (potentially complex) Java class box with class name Contract (attributes and methods are intensionally left out). But also simpler boxes can be defined. As an example of a less elaborate box, Listing 1.7 shows a Java attribute box defining an attribute named time of primitive type long.

To reuse boxes, it must be possible to adapt the environment (context) where they will be reused. ISC boxes—like any software components—need *composition interfaces* that can be exploited during reuse adaptation.

While many existing composition techniques mainly rely on only one kind of composition interface, ISC amalgamates two different kinds of interfaces: *explicit* and *implicit* interfaces. The possible *implicit interfaces* for boxes directly depend on the underlying language in which the boxes are defined. For a language having methods (e.g. Java), we can imagine the possibility of implicitly inserting debugging statements into such

```
1 public long time;
```

Listing 1.7. A Java attribute box defining an attribute named time, of type long

```
1  public class Contract extends genericSupertypeSuperClass {
2      // attributes ...
3      // methods ...
4      Contract() { ... }
5  }
```

Listing 1.8. A Java contract class box with unspecified super-class (a hook in ISC)

```
1  public class CompositionProgram {
2      public static void main(String argv[]) {
3          // load a classbox, methodbox and an attributebox
4          ClassBox cBox = new ClassBox("Contract");
5          MethodBox mBox = new MethodBox("setTimeStamp");
6          AttributeBox aBox = new AttributeBox("time");
7
8          // bind super—type hook
9          cBox.findGenericSuperClass("Supertype").bind("CarRentalEntity");
10         // extend class attribute list with attribute
11         cBox.findHook("Contract.members").extend(aBox);
12         // extend class method list with method
13         cBox.findHook("Contract.members").extend(mBox);
14     }
15 }
```

Listing 1.9. Composition specification for composing a contract class with time-stamping capabilities using ISC

methods. For a language lacking a method concept (e.g. Datalog), this cannot be done. For each language, the implicit interfaces are different, and directly depend on that language. However, as in GBM, boxes can also be adapted to new contexts using *explicit interfaces*, called (explicit) *hooks* in ISC lingo. The explicit interfaces of boxes make plain to their users which points can, or must, be modified before reuse.

For example, we might realize that the Java class in Listing 1.6 can be used for different kinds of contracts, not only bank contracts (e.g. car rental contracts). Say that we want to specify the same contract class, but without having to, a priori, commit to a specific super-type entity. However, we want to signal users (or systems supporting the users) that there needs to be a super-class specified when the class is reused. Hence, we want to make the super-class an explicit hook. To make super-classes recognizable as hooks in ISC, their names need to conform to the predefined naming convention for super-class hooks. In COMPOST this convention is generic + [hook name] + SuperClass. The super-class specified in Listing 1.8, for example, is a super-class hook with name Supertype.

One of the results from work on ISC was the distillation of two simple, yet fundamental, composition operators for boxes: *bind()* and *extend()*. These two operators comply with the observation in software composition and reuse in general of two pivotal composition and reuse styles: *parameterization* and *extension*. Hence, these two operators

```
 1 public class Contract extends CarRentalEntity {
 2     // attributes ...
 3     public long time;
 4     // methods ...
 5     Contract() { ... }
 6     public void setTimeStamp() {
 7         this.time = new java.util.Date().getTime();
 8         System.out.println("Time set at: " + this.time);
 9     }
10 }
```

Listing 1.10. Composed class for car rental contracts with time stamping functionality

correspond to composition phenomena observable in almost any language, and are as such very general. When executed, these composition operators work by transforming the abstract syntax trees (ASTs) of the fragment boxes they are applied to.

Example 2. As an example, we will use the above-mentioned Java fragments to compose a usable and compilable Java class. Say we want to use the generic contract class (Listing 1.8) to model car rental contracts. Furthermore, we want to be able to time stamp such contracts. We could implement this functionality directly in the contract class, but separating the two also allows to reuse the time stamping functionality in other applications. To achieve this, the method box of Listing 1.8 can be modified for reuse using both its implicit and explicit interface. The Java program in Listing 1.9, the language of choice in COMPOST [5] for describing compositions, details the steps needed to achieve the result.[14] First the fragment boxes are declared such that they are accessible (Lines 4–6). On Line 9 the super-class is bound, on Line 11 the class member list is extended with the attribute box (Listing 1.7), and on Line 13 the same member list is extended with the time-stamping method (Listing 1.5). The result of this composition can be found in Listing 1.10. The resulting class now sub-classes CarRentalEntity, contains an attribute holding the time stamp and a method to set it.

The above was a simple example, but demonstrated the use of both implicit and explicit interfaces, as well as the primitive operators *bind()* and *extend()*, which are the cornerstones of ISC. Based on these techniques, ISC can be used to realize aspect-oriented programming, hyperspace programming, collaboration-based design and other composition techniques [5].

The main drawback with the current realization of ISC (as demonstrated by COMPOST) is its hand-coded component models. For example, COMPOST has been tailored for Java and cannot easily be adapted to support another language.

Aspect-Oriented Programming (AOP) in ASPECTJ. The goal of AOP is to capture and separate *crosscutting concerns* of systems [32]. The most prominent implementation of AOP concepts can be found in ASPECTJ [2]. Crosscutting concerns can be both *scattered* (spread over the system) and *entangled* (mixed together with other concerns).

---

[14] Certain details of the specification have been left out for space and comprehensibility reasons.

The separation of crosscutting concerns is achieved by defining *aspects* that can be woven into the system, at the appropriate places such that the desired functionality is made available. The allowed points in programs where aspects can be woven in is governed by a *join point model*. The selection of a set of join points is called a *point cut*.

Aspects and AOP can be classified along different dimensions. In relating to ISC we are mainly interested in two dimensions: static vs. dynamic crosscuts, and basic vs. advanced dynamic crosscuts. Static crosscutting only changes the structure of a program, while dynamic crosscutting changes the execution behavior of the program. When using a basic dynamic crosscut, the join points can be determined statically. In contrast, an advanced dynamic crosscut can only be determined during runtime. This means that a purely static approach such as ISC can handle static and basic dynamic crosscuts, but not advanced dynamic crosscuts (since a static approach is not active during runtime, to e.g. inspect and possibly modify the call stack to change program behavior). However, as it turns out, advanced dynamic crosscuts are rarely used in practice [3].

Both ASPECTJ and COMPOST support composition approaches with component models that dictate valid interfaces. In ASPECTJ the component model is made up of classes, aspects, and the join point model. ASPECTJ only supports implicit interfaces, due to its reliance on *obliviousness* [23]. In contrast, COMPOST supports both implicit and explicit interfaces (hooks, inspired by the GBM notion of slots). The join point model of ASPECTJ is predefined, and changing it is not straight-forward.

Collaboration-Based Design. A composition approach the predates AOP is *Collaboration-Based Design* (CBD). Intuitively, a collaboration describes a set of 'collaborating' classes that together implement a functionality. Collaborations are interesting reuse units since they can be large (contain many classes) while still be general enough to be reusable. A CBD approach is presented in [47], where the collaborations are called *mixin layers*. In this approach, the collaborations are defined in a strict hierarchy. This layering requirement can be problematic for supporting unanticipated changes. Other approaches exists that can be used to realize CBD (e.g. [11]), but are not further described here. CBD play an important role in object-oriented programming and design and have received much attention in the last years. They are predominately invasive in their approach.

5.4   Use-Cases: Components on the Web

We believe that component-based development on the Semantic Web will play a more and more important role. It will be necessary for developers to be able to reuse already developed parts, and to have a choice of different abstractions. An abstraction and a component (or reusability through a component) are closely related issues (see for example [38]). In this section we introduce desirable abstractions, or components types, for two Semantic Web languages that have previously not been studied (apart from our work reported on in this chapter). First we discuss *modules* for Xcerpt, which is a declarative rule-based languages without explicit data flow control. Then we discuss *role models* for ontology languages, such as OWL. For further details, see [8] and [28]. Then, in Section 5.5, we will make the connection between the desirable component types and their languages, with the invasive composition techniques surveyed in the previous section.

5.4.1 Modular Xcerpt

For reuse of Xcerpt rules or query tasks, we introduce the notion of Xcerpt modules. An *Xcerpt module* is a set of rules that can be imported and reused across programs. A module defines *interfaces* dictating how the module may successfully be used. The interfaces are defined by adorning construct terms or queries of the module's rules. Adorned query terms are part of the *required* interface and adorned construct terms are part of the *provided* interface. Modules can contain both construct and goal rules, but construct terms of goal rules cannot be part of module interfaces since goal rules only result in program output.

Definition 1 (Xcerpt module). *Let Q represent a query, C a construct term in a construct rule, and G a construct term in a goal rule. We denote $C \leftarrow Q$ a construct rule, and $G \leftarrow Q$ a goal rule. Then the following is an* Xcerpt module *consisting of n rules:*

$$\widehat{C_1} \leftarrow Q_1, \ldots, G_k \leftarrow Q_k, \ldots, C_n \leftarrow \widehat{Q_n}$$

where each C_i or Q_j adorned with a $\widehat{}$ (hat) is part of the module interface.[15] *The following properties hold for a module: (i) No Q_i or $\widehat{Q_j}$ will match any $\widehat{C_k}$, and (ii) No $\widehat{Q_i}$ will match any C_j or $\widehat{C_k}$. That is, no rule in the module depends on a rule with an adorned construct term, and adorned queries can only match rules outside of the module.*

In general a module can have several input and output interfaces. Most of our examples will have one output interface, and possibly one input interface. Below we define and discuss concrete constructs needed to define modules and for making use of them in programs. It should be noted that it is also possible for modules to make use of other modules, called module *nesting*.

1. **Defining modules – constructs for module programmers.** Module programmers need constructs for defining sets of rules as modules and ways of declaring their interfaces.

   (a) *Module definition.* We can group sets of rules into modules and give such a set a mnemonic identifier using the *module* construct.

      $\langle module \rangle ::= \texttt{MODULE} \ \langle module\text{-}id \rangle \ \langle import \rangle * \ \langle rule \rangle *$

      The $\langle module\text{-}id \rangle$ construct is a simple string identifier, the $\langle import \rangle$ construct is defined below and the $\langle rule \rangle$ construct is the rule construct of Xcerpt. The *import* constructs inform us that a module can in turn import any number of other modules. The *module* construct is assumed, along with the *program*, to be a fundamental *unit* formulable by programmers.

   (b) *Module interfaces.* A module is considered to have a *required* interface if any of its rules are meant to query data produced by rules outside of the module. This can be allowed by adorning a top-level query with the `public` keyword.

---

[15] A single rule may have both its construct term and query adorned.

⟨*interface-in*⟩ ::= public ⟨*top-level-query*⟩

The ⟨*top-level-query*⟩ construct is defined in Xcerpt and represents a query that is either a query contained directly in the rule body, or as the top-most query term inside a complex query (conjunction or disjunction). Similarly, a module will require an *provided* interface if the data produced by the module is intended to be further processed. To achieve this, the public keyword may adorn a top-level construct term.

⟨*interface-out*⟩ ::= public ⟨*top-level-construct-term*⟩

The ⟨*top-level-construct-term*⟩ construct is again defined by Xcerpt, and is a construct term directly contained in a rule head. Both the ⟨*interface-in*⟩ and the ⟨*interface-out*⟩ constructs are assumed to be valid alternatives for the constructs they encompass. That is, where a ⟨*top-level-query*⟩ can be programmed, an ⟨*interface-in*⟩ construct can be placed. The equivalent holds for ⟨*interface-out*⟩.

Thus, a module programmer defines a set of rules, gives them a suitable name, and possibly defines the input and output interfaces of the module, all depending on the programmer's intension with the module.

2. **Deploying modules – constructs for module users.** Module users need to be able to (a) declare which modules they want to use in a program, to (b) query those declared modules, and to (c) provide data to the same modules, if required.

   (a) *Module import.* We can import modules into other modules or programs. This is done using the IMPORT-AS construct, defined by:

   ⟨*import*⟩ ::= IMPORT ⟨*module-ref*⟩ AS ⟨*alias-id*⟩

   The ⟨*module-ref*⟩ is the location or unique identifier of the module, while the ⟨*alias-id*⟩ is a string identifier. The ⟨*alias-id*⟩ can be used in the same program to refer to the declared module. The IMPORT-AS construct can be used before the rules of the module (or program) being defined.

   (b) *Module querying.* We can query the data produced by a module using the IN-MODULE construct:

   ⟨*in-module*⟩ ::= IN ⟨*alias-id*⟩ (⟨*query*⟩)

   The ⟨*alias-id*⟩ construct represents the precise module to query and the ⟨*query*⟩ represents the actual Xcerpt query. The query can only match against data produced by *provided* interfaces of the referred module. The IN-MODULE construct can be used where an Xcerpt ⟨*query*⟩ construct is allowed.

   (c) *Module provision.* We can feed (provision) data to a module with the TO-MODULE construct:

   ⟨*to-module*⟩ ::= TO ⟨*alias-id*⟩ (⟨*top-level-construct-term*⟩)

```
 1 MODULE participants
 2 IMPORT file:student.mx AS stud
 3
 4 CONSTRUCT
 5  public
 6   participants [
 7     all name [ var N ] ]
 8 FROM
 9  IN stud (
10   students [[
11    name [ var N ]
12   ]]
13  )
14 END
```

```
 1 MODULE student
 2
 3 CONSTRUCT
 4  public students [
 5    name [ "John Rowlands" ],
 6    name [ "Henry Stanley" ],
 7    name [ "Edmund Morel" ],
 8    name [ "Roger Casement" ] ]
 9 END
10
11 CONSTRUCT
12   students [
13    name [ "William Sheppard" ] ]
14 END
```

Listing 1.11. Module A: Participants module in file file:particip.mx

Listing 1.12. Module B: Student data module in file file:student.mx

The ⟨*alias-id*⟩ construct represents the precise module to feed data into. The data produced by the TO-MODULE construct can only be matched by rules in the referred module that are part of its *required* interface, that is, rules with they keyword public used in its body. The TO-MODULE construct can be used where *top-level-construct-term*s are allowed.

Below we present a simple example making use of the above introduced constructs, and briefly study the consequences in terms of module encapsulation.

Example 3 (Simple Xcerpt modules and their usage). This example deals with two modules and a main program. Module A (Listing 1.11) imports module B (Listing 1.12) and is itself imported into the main program P (Listing 1.13). We thus have the following dependency between the modules and the program (where ⟶ denotes the dependency relation):

$$P \longrightarrow A \longrightarrow B$$

Module B defines data about students, their names in particular. Some of the data is declared to be part of the module interface, namely, where the construct term is adorned with the public keyword. Module A imports module B and queries it for student names using the IN-MODULE construct. Furthermore, module A "exports" the matched names, but in a different format. Again, this is the case since the construct term is adorned with the public keyword. The result of executing the main query program P is shown in Listing 1.14 (in Xcerpt's data term format).

The simple modules and query program in this example essentially passes the *public* data declared in module B into the main program P, via module A, as can be seen in the query result in Listing 1.14. Notice that the name "William Sheppard" is not part of the result since this data is not declared to be part of the interface of module B.

The programs in Listings 1.15 and 1.16 are constructed to test the encapsulation capabilities of the module system. Both the programs in Listings 1.15 and 1.16 return

```
1  IMPORT file:particip.mx AS part
2
3  GOAL
4    results [ all name [ var N ] ]
5  FROM
6    IN part (
7      participants [[
8        name [ var N ] ]]
9    )
10 END
```

Listing 1.13. Program P: The main query program.

```
1  results [
2    name [
3      "John Rowlands" ],
4    name [
5      "Henry Stanley" ],
6    name [
7      "Edmund Morel" ],
8    name [
9      "Roger Casement" ]
10 ]
```

Listing 1.14. The result of executing the query program P

```
1  IMPORT file:student.mx AS stud
2
3  GOAL
4    access_allowed []
5  FROM
6    IN stud (
7      students [[ name [
8        "William Sheppard" ] ]]
9    )
10 END
```

Listing 1.15. Failing to query module B

```
1  IMPORT file:student.mx AS stud
2
3  GOAL
4    intrusion_achieved []
5  FROM
6    students [[
7      name [
8        "Roger Casement" ]
9    ]]
10 END
```

Listing 1.16. Failing to query module B

<error>no results</error> (empty results), but for different reasons. The program in Listing 1.15 correctly uses the IN-MODULE construct, but queries data that is not part of the interface of the imported module (cf. module in Listing 1.12). The program in Listing 1.16 queries data that is "visible" wrt. the imported module, but fails to actually query the imported module using the provided IN-MODULE construct.

5.4.2 Role Models as Ontology Components

As we mentioned in Section 5.3.2, many existing ontology modularization approaches focus on enabling distributed reasoning (e.g. [9,12,15]), while other focus on extracting partial ontologies (modules) from already developed ontologies (e.g. [20,50]). However, comparatively few approaches address the issue of how to compose a single monolithic ontology (that can be used by applications) from components. If composed from components, the final monolithic ontology can easily be changed according to new requirements or changes. Thus, different vendors can compose their own ontology that fit their existing infrastructure and data. Such an approach does not have to modify existing reasoners (the most costly artifact to modify), while languages or tools might have to be modified. While this does not allow for distributed reasoning, this scenario covers a niche in ontology modularization that has not received much attention. We have investigated one such approach in [28], based on the notion of *role modeling* and *role models* as the fundamental component type.

In conceptual modeling it has long been known that there is a fundamental distinction between different kinds of concepts: some stand on their own (e.g. *Person*), while others depend on the existence of some other concept (e.g. *Borrower*, who must be related to the borrowed item). Making this distinction explicit is favored in the role modeling community (see e.g. [48,49] and references therein), with successful applications— for example, in object-oriented programming [29]. In role modeling, concepts that can stand on their own are called *natural types*, while dependent concepts are called *role types*. Distinguishing different kinds of concepts is not only important for a better understanding of the modeled domain, but also for ontology reuse. This second application of role types has—to the best of our knowledge—never been investigated by the ontology community. Related role types and their relationships form abstraction units that can be studied and defined on their own. They can intuitively be seen as contexts. Such abstraction units are traditionally called *role models*. As role models often transcend domains, they can be reused in different ontologies.

We here briefly exemplify the main idea through an example given in Manchester OWL syntax [31], which has been extended for the purpose of defining and composing role models; the keywords of the extended constructs are underlined. Listing 1.17 shows an ontology that models a faculty, introducing main concepts (natural types) such as *Professor*, *FacultyMember*, and *PhDStudent*. The faculty is managed by a board which is described in the role model in Listing 1.18. A board consists of board members that elect a chairman.[16] The chairman can appoint one of the members as secretary. The ontology in Listing 1.17 imports the board role model and can so use the concepts it defines. Concepts and properties defined in the role model are marked with ′ to distinguish them from the concepts introduced in the base ontology.

One might ask why the board is described in a role model. The reason is that boards have a recognizable structure with a typical set of relationships that hold between entities in that context, regardless of the particular underlying domain. It therefore makes sense to detach the description of the board from the faculty ontology. The ontology in Listing 1.17 is made up of standard DL constructs, save the *ImportRoles* and *CanPlay* constructs. The meaning of the *ImportRoles* construct is the obvious, making the role model available to the ontology. The *CanPlay* constructs are crucial since they define the relations between the base ontology and the role model. We refer to such connecting statements as *bridge axioms*, and they can be given different semantics (see [28] for details). The role model in Listing 1.18 makes use of two additional constructs, *RoleModel* and *Role* that have the obvious meaning (defining a role model and a role, respectively). The URL of a *RoleModel* can be used to import it using the *ImportRoles* construct. By separating out such contexts—or role models as understood in the role modeling paradigm—they can be reused across ontologies and be composed together to form the resulting ontology that can be deployed in systems and applications. The

---

[16] A 'chairman' is here a person designated to preside over a meeting.

```
 1  Ontology: http://ex.org/Faculty
 2    ImportRoles:
 3        http://ex.org/Board
 4    Class: FacultyMember
 5      CanPlay: BoardMember'
 6    Class: Professor
 7      SubClassOf: FacultyMember
 8      CanPlay: ChairMan'
 9    Class: PhDStudent
10      SubClassOf: FacultyMember
11    Individual: smith
12      Types: Professor, Chairman'
13    Individual: mike
14      Types: PhDStudent,
             BoardMember'
```

```
 1  RoleModel: http://ex.org/Board
 2    Role: BoardMember'
 3    Role: Chairman'
 4      SubClassOf: BoardMember' and
 5        electedBy' some BoardMember'
 6    Role: Secretary'
 7      SubClassOf: BoardMember'
 8    ObjectProperty: electedBy'
 9      Domain: Chairman'
10      Range: BoardMember'
11    ObjectProperty: appointedBy'
12      Domain: Secretary'
13      Range: Chairman'
14
15
```

Listing 1.17. Role-based ontology   **Listing 1.18.** Role model

notion of role models is closely related to collaboration-based design, which has been investigated in object-oriented modeling, but not in ontology design and engineering. For further details we refer the reader to [28].

5.5   Universal Component Models

In this section we present our approach towards creating *universal* invasive component models. We use the term 'universal' to mean applicable to arbitrary formal languages. We do not attempt the creation of a single component model covering every language and situation, which is an untenable idea. Instead, we present a *language-driven* and *generative* approach for creating component models in a very flexible manner that fit a specific language and need. By creating component models, in its extension we define composition systems. We will then explain how we can design composition systems that can address and support the need for components on the Semantic Web (in particular the ones discussed in Section 5.4).

A universal approach is achieved by building on previous composition approaches, most notably grammar-based modularization (GBM) and invasive software composition (ISC) (cf. Section 5.3.3). First we define *universal* GBM (U-GBM) which allows to create explicit interfaces (slots) for arbitrary nonterminals of a grammar. Second we build upon U-GBM to define *universal* ISC (U-ISC). In U-ISC, any formal language defined by a context-free grammar can be adapted to the composition techniques of ISC, which allow developers to scale between explicit and implicit interfaces. Finally, we define *embedded* ISC (E-ISC). In E-ISC it is possible to define intuitive abstractions (or component types) for the benefit of end users. This advance is illustrated in Figure 2. The realization of modules for Xcerpt and role models for OWL (cf. Section 5.4) have been achieved using E-ISC–based techniques.

The approach is, as ISC, static. This means that composition takes place at compile-time. This is in contrast to advanced ASPECTJ features which can also compose at

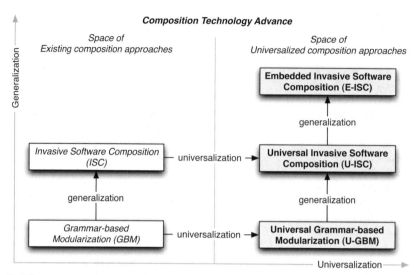

Fig. 2. We present a three-staged advance over previous work (GBM and ISC): *universal grammar-based modularization, universal invasive software composition and embedded invasive software composition*

runtime. In ASPECTJ composition can happen during runtime, for example, based on a particular value of a variable. Not only is our approach static, but we require that the final composition result a valid instance of the base language. This has the benefit that existing compilers, interpreters, or virtual machines, can directly be reused.

5.5.1   Universal Grammar-Based Modularization (U-GBM)

Grammar-based modularization (GBM) [36,37,39] is a composition technique that defines itself by referring to grammar formalisms, in particular *context-free grammars* (CFGs) [35]. A context-free grammar describes the valid programs for some language. Formally, a context-free grammar (CFG) is a 4-tuple [35]:

$$G = (N, \Sigma, P, S)$$

where N a finite set of nonterminal symbols, sometimes called *syntactic categories*, Σ is a finite set of terminal symbols (disjoint from N), P a finite set of production rules $N \times (N \cup \Sigma)^*$ and $S \in N$ the start symbol. Each production rule $N \times (N \cup \Sigma)^*$ can be used to rewrite N by $(N \cup \Sigma)^*$. Any string in $(N \cup \Sigma)^*$ derivable from the start symbol S is called a *sentential form*. A sentential form that does not contain any nonterminal symbols is called a *sentence* (it only contains terminal symbols, that is, it is in Σ^*). All strings that can be derived by a CFG G is called the language $L(G)$ generated by the CFG. Most programming languages can be specified as a context-free grammar.

As we saw in Section 5.3.3, GBM essentially provides a way of programming with sentential forms. The technique is attractive because of its simplicity, which is partly due to only considering explicit interfaces; fragments cannot be modified other than through their declared slots. The only known realization of GBM, the Mjølner System [39, Chapter 17], is only realized for one particular language, namely, BETA [39]. Also, because the Mjølner System supports separate compilation of fragments (into binary code), slots are only supported for a carefully selected set of syntactic categories.

We develop a lightweight, grammar-driven, approach to GBM:[17] Given a context-free grammar G, and a set of nonterminals N_{slot} of G, a function $\psi : CFG \rightarrow CFG$ constructs a "reuse grammar" that describes a language that accepts slots for each of the nonterminals in N_{slot}.

Example 4 (A reuse grammar for Datalog). Consider the following Datalog-like abstract syntax grammar specification with the usual EBNF interpretation of the cardinality constraint + (at least one):

$\langle Datalog \rangle ::= \langle Unit \rangle$

$\langle Unit \rangle \quad ::= \langle Program \rangle$

$\langle Program \rangle ::= \langle Statement \rangle +$

$\langle Statement \rangle ::= \langle Rule \rangle \mid \langle Fact \rangle$

$\langle Rule \rangle \quad ::= \langle Head \rangle \, \langle Body \rangle$

$\langle Fact \rangle \quad ::= \langle Head \rangle$

$\langle Head \rangle \quad ::= \langle Atom \rangle$

$\langle Body \rangle \quad ::= \langle Atom \rangle +$

$\langle Atom \rangle \quad ::= \langle Predname \rangle \, \langle Term \rangle +$

$\langle Term \rangle \quad ::= \langle Variable \rangle \mid$
$\qquad\qquad\quad \langle Constant \rangle \mid \langle Num \rangle$

$\langle Predname \rangle ::= \text{STRING}$

$\langle Variable \rangle ::= \text{CAP\_STRING}$

$\langle Constant \rangle ::= \text{STRING}$

$\langle Num \rangle \quad ::= \text{NUM\_STRING}$

We do not here specify the concrete syntax but it can be assumed to be the standard Datalog syntax (see for example (7)). The last four grammar rules define what predicate names, variables, constant symbols and numbers look like. They are defined by special tokens not further specified here (predicate names and constant symbols are character strings starting with a lower-case letter, variables are capitalized character strings, and numbers are strings of numerals).

The fragment in (8) would be valid wrt. the above Datalog grammar transformed via ψ on the set $N_{slot} = \{Num, Atom\}$.

$$
\begin{array}{r}
\texttt{bonus(X, «SLOT value:}\textit{Num}\texttt{») :- employee(X),} \\
\texttt{«SLOT condition:}\textit{Atom}\texttt{» .}
\end{array} \tag{8}
$$

Not only can we write programs with slots, but we can also define fragments of certain *types*. For example, to be able to use (8) in a composition process we must be able to define fragments of type $\langle Atom \rangle$, so that such a fragment can replace the slot condition, and so on. Which fragment types may be defined is also dictated by the selected set N_{slot}.

To enable the use of slots for a particular base language, we transform the production rules of the base language's grammar appropriately. We can allow (representatives of) non-derived nonterminals $\langle n \rangle$ to appear as slots in fragments by a set of grammar

---

[17] We call it 'lightweight' since we do not consider separate compilation of fragments.

Table 1. Abstract syntax of the SLOT-grammar. The concrete syntax can be chosen appropriately depending on the underlying language grammar into which it will be incorporated.

$\langle Slot' \rangle$	::=	$\langle Ident' \rangle \langle Type' \rangle$
$\langle Ident' \rangle$	::=	STRING
$\langle Type' \rangle$	::=	STRING

transformations via function: $\psi : (CFG, n) \rightarrow CFG$, where n is a nonterminal of the input CFG. For a given input base grammar G, and nonterminal n, ψ is defined by the following transformation steps, resulting in grammar G':

1. Union the SLOT-grammar (cf. Table 1) with G.[18] This means: Union the two disjoint sets of nonterminals, (disjoint) terminal token symbols, (disjoint) production rules, but retain the start symbol of G.
2. For each production rule in G defining nonterminal n ($\langle n \rangle$ on the left-hand side), rename n to (previously non-existing nonterminal) n'. We denote the original n nonterminal n_0 and strings generated by the original nonterminal n for n_0-strings (or $L(n_0)$).
3. Introduce the new unit production rule: $\langle n \rangle ::= \langle n' \rangle$.
4. Introduce the new unit production rule: $\langle n \rangle ::= \langle Slot' \rangle$.

Since the SLOT-grammar and G are disjoint wrt. their nonterminals, the only effect made by steps 1–3 is that the derivation of a string derivable by G from any nonterminal (of G) defined via n is one step longer when derived by G', via the additional unit production rule $\langle n \rangle ::= \langle n' \rangle$.

Being able to transform a given language grammar in the above mentioned way allows to use GBM techniques for arbitrary languages. We do not further detail these techniques but refer the reader to [25,26,27]. We summarize:

– We describe how a base grammar G can be transformed via $\psi(G, \{N_1, \ldots, N_n\}) = G'$, on a set $\{N_1, \ldots, N_n\}$ of nonterminals from G, into a reuse grammar G'. The reuse grammar describes a language where slots may appear. Hence, it is possible to program with the required fragments of GBM: concrete and practical sentential forms.
– Our toolset REUSEWARE[19] is able to generate parsers, handle slot constructs, compose fragments and pretty-print composed programs back such that they can be compiled or interpreted by existing tools.
– We define the reuse grammars such that the fragments' interfaces can be described by referring to the nominal semantics of CFGs, namely the languages they generate.
– We define requirements for *safe compositions*. Safe compositions ensures that each intermediate composition result during the composition process remains valid wrt.

---

[18] If ψ is applied to the same base grammar more than once, this step is only performed once.
[19] http://www.reuseware.org

the underlying reuse grammar. That is, it is not possible to compose syntactically incorrect programs (wrt. the underlying grammar) without getting an error. These safety conditions are derived from the base grammar, and are hence general. In particular we define that final composition results must belong to the language generated by the underlying base grammar. This ensures that existing tools can work with the composition results.

5.5.2   Universal Invasive Software Composition (U-ISC)

Invasive software composition (ISC) is a technique that extends GBM by also considering *implicit* fragment interfaces [5]. In this respect ISC is a more powerful and flexible composition technique, but also arguably harder to use. ISC has previously been applied and tailored for specific languages, mainly Java and XML, in the COMPOST environment [52]. Our goal is to make ISC grammar-driven and universal in the same way as for universal GBM. The main difficulty of achieving a universal approach for ISC is to find a good way of handling the implicit interfaces. Since the components under consideration are (source-code) fragments, the possible implicit interfaces depend on the *component language*, the base language in which the fragments are written. When the component language is fixed, as in [52], this is easier handled. In a grammar-driven approach, not beforehand knowing the component language, the problem is more intricate.

The implicit interfaces of fragments do not only depend on the component language. It is also important that there are restrictions in place that exactly details what those interfaces are. COMPOST addressing Java allows, for example, method entry and exit points to be transformed implicitly [5,52]. This can be used to separate out debugging or logging code from the production code and separately weave such statements into Java methods, in the style of aspect-oriented programming. The particular supported implicit interfaces for Java was a decision made by the developers of COMPOST. For other component languages and systems, these decisions will be different. How can implicit interfaces be supported in a universal approach, without making assumptions on the component language? To better understand the issue, we first look at a simple example of how this is done in COMPOST. Listing 1.19 shows what a *composition program* can look like in COMPOST. The program in Listing 1.19 loads (Line 2) the class box fragment in Listing 1.20, and binds the "methodEntry" implicit point with the value "Debug.println("In Lifecycle");" (Line 3). The result of the composition is shown in Listing 1.21. Naturally more complex composition programs can be written, but from this simple example we notice two things:

1. The *composition language*—the language used to write the composition program—is Java. But we have to be careful about what we mean with the "composition language." There is for example a Java type ClassBox for declaring and using fragments corresponding to Java classes, which is not really Java as much as it is a Java library. In this case we call the Java library for the *core composition language*, and Java itself for the *host composition language* (used as a platform on which the core composition language is realized).

```
1 public void compositionProgram {
2     ClassBox cBox = new ClassBox("RecursiveRobot"); // load a fragment
3     cBox.findHook("lifeCycle.methodEntry").bind(
4         "Debug.println(\"In Lifecycle\");"); // bind method entry point with a
          statement
5     ...
6 }
```

Listing 1.19. Composition program in COMPOST

```
1 public class RecursiveRobot {        1 public class RecursiveRobot {
2 public void lifeCycle() {           2 public void lifeCycle() {
3     work(...);                       3     Debug.println("In Lifecycle");
4     lifeCycle();                     4     work(...);
5   }                                  5     lifeCycle();
6 }                                    6 } }
```

Listing 1.20. A Java fragment   **Listing 1.21.** After composition

⟶ If the Java type ClassBox is provided, users can work with class boxes, etc. If the type does not exist, that fragment type cannot be defined.

2. The implicit interfaces are predefined. For example, the name "methodEntry" on Line 3 in Listing 1.19 is interpreted in a special way by the COMPOST system.

⟶ If the name "methodEntry" carries meaning for COMPOST, such implicit points can be accessed. Implicit points for which no name and interpretation is given cannot be accessed.

Based on these observations we present a *generative* approach to handle grammar-driven ISC and its implicit interfaces. It is generative because a core composition language is generated for each addressed component language, while Java is always used as the host composition language. The approach extends U-GBM from Section 5.5.1 by allowing to specify another nonterminal set N_{impl} from a base grammar G to specify which parts of fragments can be accessed implicitly. We illustrate the idea with an example.

Example 5 (Restricting implicit fragment access in grammar-driven ISC). Imagine we have a grammar Datalog specifying the rule language Datalog (containing nonterminals such as ⟨Rule⟩, ⟨Atom⟩, ⟨Variable⟩, ⟨Num⟩ etc, cf. Example 4). From the specification of a set of nonterminals, for example:

$$N_{impl} = \{Rule, Atom, Num\} \tag{9}$$

Table 2. Main operations on generated Java types, corresponding to grammar nonterminals

Operator	Explanation
Start implicit transformation	
accept(Visitor)	Starts an implicit fragment traversal, defined by the Visitor.
Transform	
bind(Fragment)	Replaces the operand fragment with the value fragment.
bind(String,Fragment)	Replaces a slot (named by String) with a fragment.
extend(Fragment)	Extends the operand fragment with the value fragment.
Boolean context queries	
inContextOf(NonTerminal)	True if the fragment is in context of the NonTerminal.
isFirst() / isLast()	True if the fragment is first / last of a fragment list.
Fragment results	
print(String)	Prints the contents of the fragment to a specified file.

a core composition language is generated that supports dealing with fragments of types *Rule*, *Atom* and *Num*. If some nonterminal is not specified in N_{impl}, fragments of that type cannot be transformed implicitly during composition. Based on (9), variables (defined by nonterminal $\langle Variable \rangle$) can for example not be implicitly accessed in fragments.

Generating core composition languages. For each nonterminal specified in such a set, N_{impl}, two Java types are generated, named: I<N> and I<N>Impl, where <N> is the name of the nonterminal. The former is a Java interface (the "interface"), while the latter is a concrete class (the "class") implementing the interface. The class provides a static method load(String) that can be used to load a fragment (either directly from a string, or from a file). The interface specifies the methods in Table 2, which are implemented by the class.

So, based on the selection in (9), and after having generated the core composition language, we can write the composition program in Listing 1.22. A Datalog fragment lacking any explicit interfaces is defined and declared on Line 2. That fragment is then transformed via its implicit interfaces between Lines 3–17 (by transforming its AST, as in ISC). The resulting fragment is printed on Line 18. The discussed program performs the following composition:

$$\text{bonus(X) :- employee(X).} \xrightarrow{\text{composes into}} \tag{10}$$
$$\text{bonus(X, 100) :- employee(X), efficient(X).}$$

As can be seen from Listing 1.22, certain auxiliary classes are also generated (such as DatalogUtil and DatalogVisitor). The DatalogVisitor class can be sub-classed to override visit(Fragment)-methods that dictates how a fragment should be transformed.[20] One such visit-method exists for each nonterminal specified in N_{impl}. Hence, N_{impl} restricts fragments' implicit interfaces. The accept(Visitor)-method,

---

[20] The default transformation is to do nothing.

```
1  public void compositionProgram() {
2      IRule rule = IRuleImpl.load("bonus(X) :- employee(X).");
3      rule.accept(new DatalogVisitor() { // transform fragment implicitly
4          public boolean visit(IAtom atom) { // transform ⟨Atom⟩s
5              if (atom.inContextOf(DatalogUtil.HEAD)) {
6                  atom.accept(new DatalogVisitor() {
7                      public boolean visit(INum num) { // transform ⟨Num⟩s
8                          if (num.isLast())
9                              num.extend(INumImpl.load("100"));
10                         return true;
11                     }
12                 });
13             } else
14                 if (atom.isLast())
15                     atom.extend(IAtomImpl.load("efficient(X)"));
16             return true;
17  } });
18      rule.print("file:out.datalog"); // "bonus(X,100) :− employee(X), efficient(X)."
19  }
```

Listing 1.22. Composition program transforming a Datalog fragment implicitly

to which a visitor object is passed, traverses the operand fragment implicitly an invokes the appropriate visit(Fragment)-methods depending on the contents of the fragment. For example, assume a fragment corresponding to a Datalog rule is being traversed because the accept(Visitor) method was invoked on that rule fragment (cf. Listing 1.22, Line 3). If a sub-fragment of type ⟨Atom⟩ is encountered during the traversal, visit(IAtom) will be invoked, passing the encountered atom as the argument, hence providing an opportunity to transform it. In Listing 1.22 (Line 3 and 6) we use the *anonymous instance* concept from Java as a simplified way of sub-classing DatalogVisitor.

Since the traversal of a fragment's AST is implicit, we need a way of querying the current fragment context. For example, to tell if an encountered variable occurs in a rule head or body. There are predefined methods for this purpose (cf. Table 2). On Line 5 in Listing 1.22, for example, we check if the encountered atom is in the rule head or body.

Using this approach, we are able to provide a grammar-driven way of transforming fragments implicitly, with restrictions in place for how this may be done (hence, there is a component model). Let us recall what we are able to do with universal ISC, as an extension of universal GBM:

- By referring to a base grammar, we can define which language constructs (represented by a nonterminal) should be "slotable," and which language constructs should be addressable implicitly during composition.
- The approach is *generative*. Based on a comparatively small specification wrt. some base grammar, an appropriate Java library (an API) is generated that can be used to write composition programs for the component language that the base grammar specifies. That is, we generate the *core composition language*. Using the generated

core composition language, and using our composition framework REUSEWARE, we can:

- Define and parse fragments of the component language (containing or not containing slots).
- Bind slots with value fragments.
- Traverse fragments (their ASTs) in an intuitive and semi-declarative way (using visitors). The allowed traversals are restricted by the developer. Hence, there is always a *component model* present.
- Extend fragments implicitly.
- Pretty-print composition results.

- It is important to *control* how compositions may be specified. This is predefined, or specified manually, in COMPOST and ASPECTJ. We automate this process via our generative approach, making it easier to experiment with different restrictions and for different component languages.

5.5.3 Universal Syntactic Abstractions with Embedded ISC

The techniques outlined in Sections 5.5.1 and 5.5.2 can be used to adapt, in a grammar-driven manner, any language to the flexible and powerful composition technique of ISC. This gives the possibility to compose programs in the fashion of Listing 1.22, but for any desirable component language (here we used Datalog as an example). From a developer's point-of-view it does not seem very attractive to construct software in this way, since the composition is described on a very primitive level.[21] But neither GBM, nor ISC, is able to avoid this undesirable primitiveness. One of the reasons is that in both approaches fragments and their interfaces are treated as first-class software entities; they are the basic units of modularization. If this is the level on which components are described, this is also the level on which they must be used.

To address this situation, while still leveraging ISC's generality, we have a twofold objective:

1. Use this generic composition technology to address a well-defined problem. In our case, the problem is to enable component-based development for NE-DSLs (or other languages in need of a component concept).
2. Strive for intuitive "components," or notions of abstractions, rather than having fragments as first-class entities.

Our hypothesis is that software developers want to specify and use components that have an intuitive *raison d'être*. One attractive feature of components is their reusability. Wegner states that "abstraction and reusability are two sides of the same coin" ([54, p. 30] as cited in [38]). Krueger paraphrases Wegner explaining that "every abstraction describes a related collection of reusable entities and that every related collection of reusable entities determines an abstraction" [38, p. 134]. So, a reusable component can be seen as an abstraction; the abstraction describes a set of solutions for which the component can be configured to solve. Krueger goes on to explain that an abstraction

---

[21] Also supported by Aßmann, see [5, p. 278].

```
1 IMPORT file:sales.md AS sales
2 bonus(X, 200) :-
3    IN sales ( employee(X) ).
4 bonus(X, 100) :- employee(X).
5 employee(john).
```

Listing 1.23. Rule program importing a module

```
1 MODULE sales
2 @ employee(X) :-
      sales_employee(X).
3 sales_employee(steve).
4 sales_employee(marco).
```

Listing 1.24. Rule module at file:sales.md

can be described in terms of the abstraction *specification* and its corresponding *realization* [38, p. 134]. Using this terminology, we shall exemplify what we would like to achieve, using Datalog as the example component language.

– *Abstraction specification.* An abstraction specification is arguably more than a plain fragment, notwithstanding that it does have an interface as concerns ISC (it is arguably too primitive). An abstraction specification becomes truly useful when the developer may use intuitive syntax to describe the abstraction (i.e. a component). Consider the Datalog 'module' in Listing 1.24, defining the employees of a fictitious company's sales department (steve and marco), and stating that they are also employees in a more general sense (via the first rule). The module is merely a collection of Datalog rules, with some additional intuitive syntax (MODULE) so as to let developers know what is being defined, namely, a module. There is also an interface construct (@) that defines how the module may be used. The @ in front of the head of the rule means: this rule can be queried by a program or a module importing this module. Thus, it defines the *provided* interface of the module. Notice that this interface definition goes beyond that of GBM and ISC – it's understandable by a developer unaware of either GBM or ISC. It's intuitive, not technical.
– *Abstraction realization.* How the Datalog module in Listing 1.24 can be used is shown in Listing 1.23. The module is imported using the IMPORT-AS construct, and referred to (queried) using the IN construct. The first rule only queries the employees of the sales department module (using IN), while the second rule only queries the "local" employees (the only local employee here is john). Posing the query bonus(X, Y) to the program in Listing 1.23 should give the following results:

$$\{X = steve, Y = 200\}, \{X = marco, Y = 200\}, \{X = john, Y = 100\}$$

The query sales_employee(X) should give no answers (or an error). The reason is that the queried predicate is not directly available, but is encapsulated in the imported module. The query employee(X) would give the single answer:

$$\{X = john\}$$

since only the local employees are directly accessible. All the keywords in red and underlined in Listings 1.23 and 1.24 represent constructs that go beyond what is available in the underlying component language, Datalog. Hence, this program and module are written in an *extension* of Datalog. For example, the abstract grammar

Table 3. Definition of constructs for importing and using modules

---

⟨*ImportAs*⟩ ::= ⟨*File*⟩ ⟨*Predname*⟩

⟨*InModule*⟩ ::= ⟨*Predname*⟩ ⟨*Atom*⟩

⟨*File*⟩ ::= LOCATION

---

snippet in Table 3 could be integrated into the Datalog grammar (cf. Example 4) to define syntax for importing and querying Datalog modules. Our tool REUSEWARE provides means for injecting such syntactic extensions into a base grammar.

The intended semantics of such syntactic extensions can be realized by composition. The extended constructs can hence be given a *compositional semantics* (or *translational* semantics). This can be done by associating the extended constructs with composition programs written in an U-ISC environment (see Section 5.5.2) developed for the component language. For Datalog and its module-related constructs briefly discussed above, these semantics specifications would be similar in style to Listing 1.22. The composition would need to compose extended programs into semantically equivalent programs of the underlying, plain, component language. This way already developed tooling for the component language can be reused (e.g., query engines). What 'semantically equivalent' means can differ depending on the component language and the developed component type (e.g. what properties should the components have?). For our Datalog module example, we would for example want to ensure proper module encapsulation and separation. Because of the connection between extended syntax, for the benefit of programmers, and composition programs providing the semantics of the extended syntax, we refer to this technical approach as *Embedded ISC* (E-ISC).

Relationships between extended syntax and composition programs. The realization of an abstraction—appropriate transformations to handle the component type and ensure certain properties such as encapsulation—can often be a non-trivial task. However, as suggested, such a realization can be specified by a composition program, hence using ISC's basic composition operators *bind()* and *extend()*. We refer to a composition program used for this purpose as a *complex (composition) operator*. A complex operator is atomic and always assumed to be executed in its entirety, or not at all. It should also be noted that a complex operator does not only contain calls to ISC's basic operators, but can also contain *internal fragments* that are needed for the realization of the abstraction the operator is implementing (see fragments $f_1, \ldots, f_n$ in Figure 3). An example of such an internal fragment could be a fragment containing an identifier (a name) which is used in some renaming scheme during composition (for example, for renaming predicate names for our above-discussed Datalog modules).

As discussed, certain constructs are introduced for the purpose of abstraction specification, while others are introduced for the purpose of using the specifications (hence for the abstraction realization). We can categorize these constructs into the following:

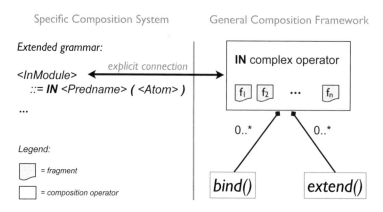

Fig. 3. A complex composition operator is connected to an active syntax construct in an extended grammar

1. *Passive syntax.* This is syntax that is used by the programmer to define components or describe how they should be used. For our rule-based modules the **MODULE** and @ (module interface) constructs are examples of passive syntax.
2. *Active syntax.* This is syntax that is used to deploy components and that takes an active part in the composition of those components. For our rule-based modules the **IMPORT-AS** and **IN** constructs are examples of active syntax.

The notion of active syntax is closely related to complex composition operators. The relation is that an active syntax construct (e.g. **IN**) delegates its work to a complex composition operator. Or seen the other way around, a complex composition operator implements an active syntax construct. The passive syntax constructs are not composition operators, but are used to guide the composition and are hence equally important.

The relation between an active syntax construct and a complex composition operator has to be made explicit by a developer. This is done on the grammar level. For example, since the **IN** construct used above is an active syntax construct, its corresponding nonterminal in the grammar should be connected to some complex composition operator implementing its functionality. This is illustrated in Figure 3, where the nonterminal ⟨InModule⟩ represents the **IN** construct (concrete syntax is shown for clarify).

The only difference between a complex operator and a composition program is that complex operators take some external input, in form of the fragments they are supposed to work on. This external input directly relates to the definition of the active syntax construct that the complex operator is implementing. For example, the **IN** construct (defined by nonterminal ⟨InModule⟩) from Table 3 would need to know the name of the referred module, as well as the atom querying the module, represented by nonterminals ⟨Predname⟩ and ⟨Atom⟩, respectively (cf. Table 3). Since we are using Java as our host composition language, these special composition programs can be realized as Java methods. Considering our module extension to Datalog and an appropriately generated core composition language, the signature of a complex composition operator method implementing the **IN** construct would be (possibly using a different name):

```
1  public IAtom inModule(final IPredname module, final IAtom atom) {
2    atom.accept(new DatalogVisitor() { // traverse atom being operated on
3      public boolean visit(IPredname node) {
4        // transform atom being operated on
5        node.bind(IPrednameImpl.load(node.toString() + "_" +
6              // uses the name of the referred module; stored outside this method
7              names.get(module.toString()) +
8              "_" + interfaceOut.toString())));
9        return true;
10     } });
11   return atom;
12 }
```

Listing 1.25. Simplified complex operator for the **IN** construct

public IAtom inModule(IPredname name, IAtom atom)

The parameter types directly correspond to the definition of the **IN** construct. The fragment returned from a complex composition operator replaces the active syntax that invoked it. This method will be invoked when a construct appears in a composition program that corresponds to the considered nonterminal. That is, if ⟨*InModule*⟩ is connected to the above method, the method will be invoked whenever an **IN** construct is used. A simplified example implementing the **IN** construct is shown in Listing 1.25. The composition operator renames the considered predicate name to ensure separation of the modules in the composition result.

What can we do with this technology? Several things, but in particular we can address the use-cases discussed in Section 5.4. That is, we can combine a grammar-driven composition technique with composition requirements for several different languages, e.g. Semantic Web languages. Syntactic grammar extensions can support component specification, and we can use the above-described composition techniques to support those extensions by implementing suitable complex composition operators. We summarize:

- We essentially provide technology for universal *syntactic abstractions*. Macros (e.g. in Lisp) is an example of an syntactic abstraction concept that can be used to syntactically extend a language. E-ISC differs from traditional macros on several levels, but the comparison is helpful. Our main goal remains to provide component-based development opportunities to NE-DSLs.
- We leverage the generality of ISC, in particular its universalization (U-ISC), but make this technology approachable by a wider audience by providing the opportunity to use the technology "under the hood." Hence, by addressing a particular problem we can make the general, and often hard to understand, techniques of ISC practical. This idea lies at the heart of embedded ISC (E-ISC).

We have implemented several examples that make use of the presented technique. In Section 5.6 we detail the underlying composition idea for one such example, Modular Xcerpt.

Table 4. A store consists of three sections: out, private and in

Section	Data stored in the section
out	*Data part of the provided interface – data to be queried.*
private	*Data for the internal use of the module – not accessible from outside the module.*
in	*Data injected used by the required rules of the module.*

5.6   Example Application: Modular Xcerpt

In this section we present the transformations that are carried out under the hood during composition of Modular Xcerpt programs. Due to lack of space we discuss the composition strategy without detailing how the composition system is concretely realized. In particular, we explain how module encapsulation can be retained in the composition result via the notion of "stores." This enables the reuse of the existing Xcerpt interpreter for Modular Xcerpt.

The purpose of the *store* is to ensure proper module encapsulation. A store can be likened to a virtual (XML) database associated with a unique identifier. Every module is assumed to be associated with a store. A store is divided into three sections: out, private and in (see Table 4 for their explanations). In order for a module to be properly encapsulated, every data term internally constructed by the module (not part of the module's interface) should only be usable by other rules of the same module. If this is not the case then the encapsulation of the module is violated. Or, if rules within the module can query rules outside the module that do not intentionally provide data to the module, then encapsulation is also violated. By directing any data terms constructed into the suitable section of the considered module's store, proper separation, and hence encapsulation, can be ensured.

The extended constructs (IMPORT-AS, IN-MODULE, TO-MODULE) that the module programmers make use of are responsible for directing rules to the correct section of the considered module's store.

– *Module import.* When importing a module using the IMPORT-AS construct, every top-level construct term and query term of each rule is directed to the appropriate section of the module's store. This is in general done by transforming each module rule in the following way:

```
1 CONSTRUCT <head>              1 CONSTRUCT <STORE: <head> >
2 FROM <body>          ⟶       2 FROM <STORE: <body> >
3 END                          3 END
```

where each construct enclosed within < and > is as of yet unspecified. In general the <STORE: <term> > construct can be expanded to the following:

```
store [ id [ <module-id> ], section [ <section> ], <term> ]
```

where <module-id> is a string containing a unique identifier of the module (a URI or the location of the module), <section> a string indicating which section of the store is being referred to (out, private or in), and <term> is the construct term or query term considered. We refer to a *term* when it is irrelevant if we mean a construct or a query term. If the considered term is not part of the module interface, then the section used is private. If the term is a construct term adorned with the public keyword, and is hence part of the *provided* interface, then we use the out section of the store. In the same manner, if we consider an adorned query, then we use the in section. The following is an example of a single rule with an adorned construct term:

```
1 CONSTRUCT
2   public data_out [
3       var X
4   ]
5 FROM
6   data [
7     var X
8   ]
9 END
```

$\longrightarrow$

```
1 CONSTRUCT
2     store [ id [ <module-id> ],
3         section [ "out" ],
4         data_out [ var X ] ]
5 FROM
6     store [ id [ <module-id> ],
7         section [ "private" ],
8         data [ var X ] ]
9 END
```

– *Module querying*. The IN-MODULE construct is provided for querying specific modules. Such queries are meant to query the data terms constructed by a module as part of its *provided* interface, and are hence referred to the out section of the referred module's store. The following transformation is performed:

```
1 CONSTRUCT ...
2 FROM
3   IN mod (
4     <query>
5   )
6 END
```

$\longrightarrow$

```
1 CONSTRUCT ...
2 FROM
3     store [ id [ <module-id> ],
4         section [ "out" ],
5         <query> ]
6 END
```

In the above, mod is the alias given to the imported module (using IMPORT-AS) and <module-id> in the transformed rule is the identifier of the considered module (the exact identifier is an implementation detail, and could e.g. be the module filename).

– *Module provision*. The TO-MODULE construct can be used for providing data to be used by a module. For this data to be made available to the module, it needs to be directed to the in section of the module's store. This is done via the following transformation:

```
1 CONSTRUCT
2   TO mod (
3     <construct term>
4   )
5 FROM ...
6 END
```

$\longrightarrow$

```
1 CONSTRUCT
2     store [ id [ <module-id> ],
3         section [ "in" ],
4         <construct term> ]
5 FROM ...
6 END
```

Example 6 (Simple module composition). The following example makes use of a simple identity module, that is, a module that simply returns the data terms it receives as

```
1  IMPORT file:ident.mx AS ident
2  GOAL result [ var X ]
3  FROM
4      IN ident (
5          ident_out [ var X ] )
6  END
7
8  CONSTRUCT
9      TO ident (
10         ident_in [ "value" ] )
11 END
```

Listing 1.26. A query program using the identity module

```
1  MODULE identity_module
2  CONSTRUCT
3      public ident_out [ var X ]
4  FROM internal [ var X ]
5  END
6
7  CONSTRUCT
8      internal [ var X ]
9  FROM
10     public ident_in [ var X ]
11 END
```

Listing 1.27. An identity module at `file:ident.mx`

```
1  GOAL
2  result [
3      var X
4  ]
5  FROM
6  store [
7      id [ "file:ident.mx" ],
8      section [ "out" ],
9      ident_out [ var X ]
10 ]
11 END
12
13 CONSTRUCT
14 store [
15     id [ "file:ident.mx" ],
16     section [ "in" ],
17     ident_in [ "value" ]
18 ]
19 END
```

Listing 1.28. Program in Listing 1.26 composed to Xcerpt

```
1  CONSTRUCT
2  store [ id [ "file:ident.mx" ],
3          section [ "out" ],
4          ident_out [ var X ] ]
5  FROM
6  store [ id [ "file:ident.mx" ],
7          section [ "private" ],
8          internal [ var X ] ]
9  END
10
11 CONSTRUCT
12 store [ id [ "file:ident.mx" ],
13         section [ "private" ],
14         internal [ var X ] ]
15 FROM
16 store [ id [ "file:ident.mx" ],
17         section [ "in" ],
18         ident_in [ var X ] ]
19 END
```

Listing 1.29. Module in Listing 1.27 composed to Xcerpt

input (Listing 1.27). The identity module could be implemented using a single rule, but we use two rules to show how the internal communication (between the two rules) is directed to the `private` section of the module's store. The program in Listing 1.26 makes use of the identity module by sending it some data, expecting to get the same output as result when querying the module. The result of the query program is, as expected: result ["value"]. The query program and the module, written in the Modular Xcerpt language, are composed to the plain Xcerpt programs show in Listings 1.28 and 1.29, respectively.

All the constructs belonging to the extended language, Modular Xcerpt, have been removed in the composed results. The programs in Listings 1.28 and 1.29 can be merged into a single program and executed by the Xcerpt interpreter to yield the expected result shown above. Thus, the composed and merged program is the realization of the abstractions used in Listings 1.26 and 1.27. The realization ensures that modules are encapsulated using the concept of stores.

Certain special cases are handled while composing modular query programs:

1. *External resources.* If a module queries an external resource, then the query is not transformed, since the query must match the format of that resource.
2. *Complex queries.* We remember that queries are not simply query terms, but sets of query terms joined by logical connectors, such as or or and. When transforming a conjunctive or disjunctive query, the transformations are done on the top-level of each involved query term.
3. *Module nesting.* It is possible for modules to import other modules, so-called module nesting. During the transformation of a module, encountered IN-MODULE and TO-MODULE constructs are transformed wrt. the modules they are referring to.

The transformations described above that ensures that the modules retain their encapsulation in the unified composition result can be implemented using an E-ISC–based technique. A E-ISC–based composition system can be created by a *developer*, while an *end-user* can deploy the benefits of having modules when writing Xcerpt programs (cf. Listings 1.26–1.27). Hence, there are different user roles to consider when enabling component-based development for languages. Further examples of Modular Xcerpt can be found in [7].

5.7   Conclusions

We have in this chapter made a connection between the need for components in non-embedded domain-specific languages (NE-DSLs), in particular those found on the Semantic Web, and invasive component models. It seems that invasive component models have a useful existence together with the declarative and descriptive languages that are abundant on the Semantic Web.

Our composition approach and methodological framework requires that the final composition results are instances of the underlying base languages (reduction semantics). The main reason for this is the desire to reuse existing tools; but also to be able to provide a general solution to a general problem (enable component-based development for NE-DSLs), instead of having to provide language-specific solutions throughout the whole language development process (from NE-DSL requirements, through their design and to their implementation). One immediate consequence of the reduction semantics for E-ISC–based composition systems is that the core expressiveness of the addressed language is never really extended. We do not claim that the expressiveness of NE-DSLs are always at fault, rather their ability to support developers in defining and using reusable entities—components. This is supported by a remark from Clemens

Szyperski, which suggests that there is not always a need to extend the core expressiveness of the base language in order to support components:

> "[...] from a purely formal point of view, there is nothing that could be done with components that could not be done without them."

> – *Clemens Szyperski, Component Software [51, p. 10]*

Nevertheless, components *are* important for *reuse*, that is, for cost reduction and quality improvement in the development process. Could the dream of McIlroy [40] finally become true for Semantic Web languages? This was the goal of this chapter: to introduce a universal approach to modularity for all Semantic Web languages, not only one, and to show a lightweight way to component-based engineering with any DSL on the Web. The technology presented is ready to be applied to more languages. The REUSEWARE toolset is available and will be supported in the following years, and the authors are open to collaborate in experiments with interested readers. Grammar-based tools are like machine tools: they need to be used, employed, and applied to produce a massive amount of products and other tools.

One of the most attractive application areas is the componentization of operational Web languages, that is, workflow and Web service languages, such as BPEL or OWL-S. In the field of programming, gray-box component techniques, such as aspect-orientation or invasive composition, have considerably changed the way we think about software. Since operational Web languages are basically not different from programming languages, the question remains open how gray-box component techniques can be transferred to them. How can BPEL views and aspects be specified? How can OWL-S templates be defined? The technology presented in this chapter should provide answers to these questions and is ready to be applied and tested.

Future work must also treat the open problem of *fragment component contracts*. A fragment component, apart from being composed in a well-formed way, may pose requirements on the components it is composed with. Relations from static semantics could be employed to write fragment component contracts that define composition and substitutability constraints. However, most of these constraints are language-specific. Hence, a universal technology should integrate specification technologies for static semantics, such as attribute grammars or natural semantics.

Of course, on a larger scale, component-based development for Semantic Web languages is still an open issue. In this chapter, we have presented one possible approach for Web-DSL modularization, but other aspects, like application architecture, context-aware adaptation, or application component models [4], have not been solved yet. Also, there are different motivations for *modularization*. Some approaches want to establish connections between different parts (e.g., ontology modules) to enable a shared understanding in a largely distributed environment. Others introduce techniques for heterogeneous or hybrid reasoning to treat heterogeneous ontologies written in diverse languages. Other approaches, such as ours, attempt to create large monolithic entities in a modular and reusable fashion. The aims are different, but each approach strives to improve reuse and modularization in some way. Modularization on the Semantic Web—and in general—will be an exciting field to follow in the coming years.

Acknowledgement

The authors want to thank Sven Karol for valuable comments to the written text. This research has been co-funded by the European Commission and by the Swiss Federal Office for Education and Science within the 7th Framework Programme project MOST number 216691 (cf. http://most-project.eu).

References

1. RIF Working Group - RIF. WWW Page (November 2008), http://www.w3.org/2005/rules/wiki/RIF_Working_Group (accessed November 6, 2008)
2. The AspectJ Project (October 2008), http://www.eclipse.org/aspectj/
3. Apel, S., Batory, D.: How AspectJ is Used: An Analysis of Eleven AspectJ Programs. Technical Report MIP-0801, Department of Informatics and Mathematics, University of Passau, Germany (2008)
4. Aßmann, U.: Composing Frameworks and Components for Families of Semantic Web Applications. In: Bry, F., Henze, N., Małuszyński, J. (eds.) PPSWR 2003. LNCS, vol. 2901, pp. 1–15. Springer, Heidelberg (2003)
5. Aßmann, U.: Invasive Software Composition. Springer, New York (2003)
6. Aßmann, U.: Reuse in Semantic Applications. In: Eisinger, N., Małuszyński, J. (eds.) Reasoning Web 2005. LNCS, vol. 3564, pp. 290–304. Springer, Heidelberg (2005)
7. Aßmann, U., Bartho, A., Drabent, W., Henriksson, J., Wilk, A.: Composition Framework and Typing Technology Tutorial. Technical Report IST506779/Dresden/I3-D14/D/PU/b1, Technical University of Dresden (2008)
8. Aßmann, U., Berger, S., Bry, F., Furche, T., Henriksson, J., Johannes, J.: Modular Web Queries—From Rules to Stores. In: Meersman, R., Tari, Z., Herrero, P. (eds.) OTM-WS 2007, Part II. LNCS, vol. 4806, pp. 1165–1175. Springer, Heidelberg (2007)
9. Bao, J., Caragea, D., Honavar, V.: Towards Collaborative Environments for Ontology Construction and Sharing. In: International Symposium on Collaborative Technologies and Systems (CTS 2006), pp. 99–108 (2006)
10. Batory, D., Lofaso, B., Smaragdakis, Y.: JTS: Tools for Implementing Domain-Specific Languages. In: Proceedings of the 5th International Conference on Software Reuse, pp. 143–153. IEEE, Los Alamitos (1998)
11. Bergel, A., Ducasse, S., Nierstrasz, O.: Classbox/J: Controlling the Scope of Change in Java. In: OOPSLA 2005: Proceedings of the 20th annual ACM SIGPLAN conference on Object oriented programming, systems, languages, and applications, pp. 177–189. ACM, New York (2005)
12. Grau, B.C., Parsia, B., Sirin, E.: Combining OWL ontologies using $\mathcal{E}$-connections. Journal of Web Semantics 4(1), 40–59 (2006)
13. Berners-Lee, T.: The semantic web (May 2001), http://www.sciam.com/article.cfm?articleID=00048144-10D2-1C70-84A9809EC588EF21
14. Boag, S., Chamberlin, D., et al.: XQuery 1.0: An XML Query Language. W3C Recommendation (January 23, 2007), http://www.w3.org/TR/xquery/
15. Borgida, A., Serafini, L.: Distributed Description Logics: Assimilating Information from Peer Sources. Journal of Data Semantics 1, 153–184 (2003)
16. Bravenboer, M., Visser, E.: Concrete Syntax for Objects: Domain-Specific Language Embedding and Assimilation without Restrictions. In: OOPSLA 2004: Proceedings of the 19th annual ACM SIGPLAN conference on Object-oriented programming, systems, languages, and applications, pp. 365–383. ACM Press, New York (2004)

17. Bry, F., Schaffert, S.: A Gentle Introduction into Xcerpt, a Rule-based Query and Transformation Language for XML. In: Proceedings of International Workshop on Rule Markup Languages for Business Rules on the Semantic Web, Sardinia, Italy, June 14 (2002)
18. Bry, F., Schaffert, S.: The XML Query Language Xcerpt: Design Principles, Examples, and Semantics. In: Chaudhri, A.B., Jeckle, M., Rahm, E., Unland, R. (eds.) NODe-WS 2002. LNCS, vol. 2593, pp. 295–310. Springer, Heidelberg (2003)
19. Clark, J.: XSL transformations (XSLT) (November 1999), http://www.w3.org/TR/xslt
20. Cuenca Grau, B., Horrocks, I., Kazakov, Y., Sattler, U.: Just the Right Amount: Extracting Modules from Ontologies. In: Proceedings of the Sixteenth International World Wide Web Conference (WWW 2007) (2007)
21. Dennis, J.B.: Modularity. In: Advanced Course on Software Engineering. Lecture Notes in Economics and Mathematical Systems, vol. 81, pp. 128–182. Springer, Heidelberg (1973)
22. Deursen, A., Klint, P., Visser, J.: Domain-specific Languages: An Annotated Bibliography. ACM SIGPLAN Notices 35(6), 26–36 (2000)
23. Filman, R.E., Friedman, D.P.: Aspect-Oriented Programming is Quantification and Obliviousness. Technical Report 01.12 (2000)
24. Fowler, M.: MF Bliki: DomainSpecificLanguage (July 2008), http://www.martinfowler.com/bliki/DomainSpecificLanguage.html   (accessed July 2, 2008)
25. Henriksson, J.: A Lightweight Framework for Universal Fragment Composition – with an Application in the Semantic Web (to appear). PhD thesis, Technical University of Dresden (December 2008)
26. Henriksson, J., Heidenreich, F., Johannes, J., Zschaler, S., Aßmann, U.: Extending Grammars and Metamodels for Reuse – The Reuseware Approach. IET Software, Special Issue on Language Engineering (2007)
27. Henriksson, J., Johannes, J., Zschaler, S., Aßmann, U.: Reuseware – Adding Modularity to Your Language of Choice. Proceedings of TOOLS EUROPE 2007: Special Issue of the Journal of Object Technology (2007)
28. Henriksson, J., Pradel, M., Zschaler, S., Pan, J.Z.: Ontology Design and Reuse with Conceptual Roles. In: Calvanese, D., Lausen, G. (eds.) RR 2008. LNCS, vol. 5341, pp. 104–118. Springer, Heidelberg (2008)
29. Herrmann, S.: A Precise Model for Contextual Roles: The Programming Language ObjectTeams/Java. Applied Ontology 2(2), 181–207 (2007)
30. Hofer, C., Ostermann, K., Rendel, T., Moors, A.: Polymorphic Embedding of DSLs. To appear in Proceedings of the 7th International Conference on Generative Programming and Component Engineering (GPCE 2008). ACM Press, New York (2008)
31. Horridge, M., Patel-Schneider, P.F.: Manchester Syntax for OWL 1.1. In: International Workshop OWL: Experiences and Directions (OWLED 2008) (2008)
32. Kiczales, G., Lamping, J., Menhdhekar, A., Maeda, C., Lopes, C., Loingtier, J.-M., Irwin, J.: Aspect-Oriented Programming. In: Akşit, M., Matsuoka, S. (eds.) ECOOP 1997. LNCS, vol. 1241, pp. 220–242. Springer, Heidelberg (1997)
33. Kiselev, I.: Aspect-Oriented Programming with AspectJ. Sams, Indianapolis (2002)
34. Klyne, G., Carroll, J.J.: Resource Description Framework (RDF): Concepts and Abstract Syntax. W3C Recommendation, February 10 (2004), http://www.w3.org/TR/rdf-concepts/
35. Kozen, D.C.: Automata and Computability. Springer, New York (1997)
36. Kristensen, B.B., Madsen, O.L., Møller-Pedersen, B., Nygaard, K.: Abstraction mechanisms in the BETA programming language. In: POPL 1983: Proceedings of the 10th ACM SIGACT-SIGPLAN symposium on Principles of programming languages, pp. 285–298. ACM, New York (1983)

37. Kristensen, B.B., Madsen, O.L., Møller-Pedersen, B., Nygaard, K.: Syntax Directed Program Modularization. In: Degano, P., Sandewall, E. (eds.) Interactive Computing Systems (1983)

38. Krueger, C.W.: Software reuse. ACM Computing Surveys 24(2), 131–183 (1992)

39. Madsen, O.L., Møller-Pedersen, B., Nygaard, K.: Object-Oriented Programming in the BETA Programming Language. Addison-Wesley, Reading (1993)

40. McIlroy, D.M.: Mass-Produced Software Components. In: Buxton, J.M., Naur, P., Randell, B. (eds.) Software Engineering Concepts and Techniques (1968 NATO Conference of Software Engineering), pp. 88–98. NATO Science Committee, Brussels (1969)

41. Nilsson, U., Małuszyński, J.: Logic, Programming, and PROLOG. John Wiley & Sons, Inc., New York (1995)

42. Pan, J.Z., Serafini, L., Zhao, Y.: Semantic Import: An Approach for Partial Ontology Reuse. In: Proc. of the ISWC 2006 Workshop on Modular Ontologies (WoMO) (2006)

43. Parnas, D.L.: On the Criteria to Be Used in Decomposing Systems into Modules. Communications of the ACM 15(12), 1053–1058 (1972)

44. Patel-Schneider, P.F., Hayes, P., Horrocks, I.: OWL Web Ontology Language Semantics and Abstract Syntax. W3C Recommendation, February 10 (2004), http://www.w3.org/TR/owl-semantics/

45. Schaffert, S.: Xcerpt: A Rule-Based Query and Transformation Language for the Web. Dissertation/Ph.D. thesis, Institute of Computer Science, LMU, Munich (2004)

46. Schaffert, S., Bry, F., Fuche, T.: Simulation Unification. Technical Report IST506779/Munich/I4-D5/D/PU/a1, Institute for Informatics, University of Munich (2005)

47. Smaragdakis, Y., Batory, D.S.: Mixin Layers: an Object-Oriented Implementation Technique for Refinements and Collaboration-Based Designs. Software Engineering and Methodology 11(2), 215–255 (2002)

48. Steimann, F.: On the Representation of Roles in Object-Oriented and Conceptual Modelling. Data and Knowledge Engineering 35(1), 83–106 (2000)

49. Steimann, F.: The role data model revisited. Applied Ontology 2(2), 89–103 (2007)

50. Stuckenschmidt, H., Klein, M.: Structure-Based Partitioning of Large Concept Hierarchies. In: McIlraith, S.A., Plexousakis, D., van Harmelen, F. (eds.) ISWC 2004. LNCS, vol. 3298, pp. 289–303. Springer, Heidelberg (2004)

51. Szyperski, C.: Component Software: Beyond Object-Oriented Programming. Addison-Wesley, New York (1998)

52. The COMPOST Consortium. The COMPOST system, http://www.the-compost-system.org

53. Van Wyk, E., Krishnan, L., Schwerdfeger, A., Bodin, D.: Attribute Grammar-based Language Extensions for Java. In: Ernst, E. (ed.) ECOOP 2007. LNCS, vol. 4609, pp. 575–599. Springer, Heidelberg (2007)

54. Wegner, P.: Varieties of Reusability. In: Proceedings of Workshop on Reusability in Programming, September 1983, pp. 30–44 (1983)

Chapter 6
Controlled English for Reasoning on the Semantic Web

Juri Luca De Coi[1], Norbert E. Fuchs[2], Kaarel Kaljurand[2], and Tobias Kuhn[2]

[1] L3S, University of Hanover
decoi@l3s.de
http://www.l3s.de
[2] Department of Informatics and Institute of Computational Linguistics,
University of Zurich
{fuchs,kalju,tkuhn}@ifi.uzh.ch
http://attempto.ifi.uzh.ch

Abstract. The existing Semantic Web languages have a very technical focus and fail to provide good usability for users with no background in formal methods. We argue that controlled natural languages like Attempto Controlled English (ACE) can solve this problem. ACE is a subset of English that can be translated into various logic based languages, among them the Semantic Web standards OWL and SWRL. ACE is accompanied by a set of tools, namely the parser APE, the Attempto Reasoner RACE, the ACE View ontology and rule editor, the semantic wiki AceWiki, and the Protune policy framework. The applications cover a wide range of Semantic Web scenarios, which shows how broadly ACE can be applied. We conclude that controlled natural languages can make the Semantic Web better understandable and more usable.

6.1 Why Use Controlled Natural Languages for the Semantic Web?

The Semantic Web proves to be quite challenging for its developers: there is the problem of adequately representing domain knowledge, there is the question of the interoperability of heterogeneous knowledge bases, there is the need for reliable and efficient reasoning, and last but not least the Semantic Web requires generally acceptable user interfaces.

Languages like RDF, OWL, SWRL, RuleML, R2ML, SPARQL etc. have been developed to meet the challenges of the Semantic Web. The developers of these languages are predominantly researchers with a strong background in logic. This is reflected in the languages, all of which have syntaxes that conspicuously show their logic descent. Domain experts and end-users, however, often do not have a background in logic. They shy away from logic notations, and prefer to use notations familiar to them — which is usually natural language possibly complemented by diagrams, tables, and formulas.

The developers of Semantic Web languages have tried to overcome the usability problem by suggesting alternative syntaxes, specifically for OWL. However,

F. Bry and J. Maluszynski (Eds.): Semantic Techniques for the Web, LNCS 5500, pp. 276–308, 2009.

even the Manchester OWL Syntax [21], which is advertised by its authors as "easy to read and write", lacks the features that would bring OWL closer to domain experts. The authors of [33,23] list the problems that users encounter when working with OWL, and as a result of their investigations express the need for a "pedantic but explicit" paraphrase language. Add to this that many knowledge bases require a rule component, often expressed in SWRL. The proposed SWRL syntax, however, is completely different from any of the OWL syntaxes. Query languages for OWL ontologies introduce yet other syntaxes.

The syntactic complexity of Semantic Web languages can be hidden to some extent by front-end tools such as Protégé[1] that provides various graphical means to view and edit knowledge bases. While the subclass hierarchy of named classes can be concisely presented graphically, for more complex expressions users still have to rely on one of the standard syntaxes.

Thus the languages developed for the Semantic Web do not seem to meet all of its challenges. Though by and large they fulfill the requirements of knowledge representation and reasoning, they seem to fail the requirement of providing general and generally acceptable user interfaces.

Concerning user interfaces, natural language excels as the fundamental means of human communication. Natural language is easy to use and to understand by everybody, and — other than formal languages — does not need an extra learning effort. Though for particular domains there are more concise notations, like diagrams and formulas, natural language can be and is used to express any problem: only listen to scientists paraphrasing complex formulas, or to somebody explaining the way to the station. For this reason, we will in the following focus only on natural language, and not discuss complementary notations. Since natural language is highly expressive, and is used in any application domain, some researchers even consider natural language "the ultimate knowledge representation language" [37]. This claim should be taken with reservations since we must not forget that natural language is highly ambiguous and can be very vague.

Thus there seems to be a conflict: on the one side the Semantic Web needs logic-based languages for adequate knowledge representation and reasoning, and on the other side the Semantic Web requires natural language for generally acceptable user interfaces.

This conflict was already encountered before the advent of the Semantic Web, for instance in the fields of requirements engineering and software specification. Their researchers proposed to use controlled natural languages[2] to solve the conflict — where a controlled natural language is a subset of the respective natural language specifically designed to be translated into first-order logic. This translatability turns controlled natural languages into logic languages and enables them to serve as knowledge representation and reasoning languages, while preserving readability. As existing controlled natural languages show, the ambiguity and vagueness of full natural language can be avoided.

---

[1] http://protege.stanford.edu/

[2] http://www.ics.mq.edu.au/~rolfs/controlled-natural-languages/

Therefore it is only natural that researchers have proposed to use controlled natural language also for the Semantic Web [35]. In fact, several studies have shown that controlled natural languages offer domain experts improved usability over working with OWL [27,14,18].

Controlled natural languages, for instance Attempto Controlled English that we present in the following, can be translated into various Semantic Web languages, thus providing the features of these languages in one and the same user-friendly syntax. In our view, this demonstrates that ACE and similar controlled natural languages have the potential to optimally meet the challenges of the Semantic Web.

This chapter is structured as follows. Section 2 gives an overview of controlled natural languages. In section 3 we present Attempto Controlled English (ACE), and describe how ACE texts can be translated into first-order logic. Section 4 shows how ACE fits into the Semantic Web, concretely how ACE can be translated into OWL and SWRL, how ACE can be used to express rules and policies, and briefly how ACE can be translated into the languages RuleML, R2ML and PQL. Section 5 is dedicated to tools developed for the ACE language, namely the Attempto Reasoner RACE, the ACE View ontology and rule editor, the semantic wiki AceWiki, and the front-end for the Protune policy language. In section 6 we summarize our experiences, and assess the impact of controlled natural languages on the Semantic Web.

6.2   Controlled Natural Languages: State of the Art

Besides Attempto Controlled English (ACE) that we will describe in detail in the next section, there are several other modern controlled natural languages:

PENG [36] is a language that is similar to ACE but follows a more light-weight approach in the sense that it covers a smaller subset of natural English. Its incremental parsing approach makes it possible to parse partial sentences and to look-ahead to find out how the sentence can be continued.

Common Logic Controlled English (CLCE) [38] is another ACE-like language that has been designed as a human interface language for the ISO standard Common Logic[3].

Computer Processable Language (CPL) [7] is a controlled English developed at Boeing. Instead of applying a small set of strict interpretation rules, the CPL interpreter resolves various types of ambiguities in a "smart" way that should lead to acceptable results in most cases.

E2V [32] is a fragment of English that corresponds to a decidable two-variable fragment of first-order logic. In contrast to the other languages, E2V has been developed to study the computational properties of certain linguistic structures and not to create a real-world knowledge representation language.

While the languages presented above have no particular focus on the Semantic Web, there are several controlled natural languages that are designed specifically for OWL:

---

[3] http://cl.tamu.edu/

Sydney OWL Syntax (SOS) [10] builds upon PENG and provides a syntactically bidirectional mapping to OWL. The syntactic sugar of OWL is carried over one-to-one to SOS. Thus, semantically equivalent OWL statement that use different syntactical constructs are always mapped to different SOS statements.

Rabbit [18] is a controlled English developed and used by Ordnance Survey (Great Britain's national mapping agency). Rabbit is designed for a scenario where a domain expert and an ontology engineer work together to produce ontologies. Using Rabbit is supported by the ROO (Rabbit to OWL Ontology construction) editor [11]. ROO allows entering Rabbit sentences, helps to resolve possible syntax errors, and translates them into OWL.

Lite Natural Language [2] is a controlled natural language that maps to DL-Lite which is one of the tractable fragments of OWL. Lite Natural Language can be seen as a subset ACE.

CLOnE [13] is a very simple language defined by only eleven sentence patterns which roughly correspond to eleven OWL axiom patterns. For that reason, only a very small subset of OWL is covered.

ACE is unique in the sense that it covers both aspects: It is designed as a general-purpose controlled English providing a high degree of expressivity. At the same time, ACE is fully interoperable with the Semantic Web standards, since a defined subset of ACE can bidirectionally be mapped to OWL.

6.3 Attempto Controlled English (ACE)

6.3.1 Overview of Attempto Controlled English

This section contains a brief survey of the syntax of the language Attempto Controlled English (ACE). Furthermore, we summarize ACE's handling of ambiguity, and show how sentences can be interrelated by anaphoric references.

Syntax of ACE. The vocabulary of ACE comprises predefined function words (e.g. determiners, conjunctions), predefined fixed phrases (e.g. 'it is false that', 'for all'), and content words (nouns, proper names, verbs, adjectives, adverbs).

The grammar of ACE — expressed as a set of construction rules and a set of interpretation rules — defines and constrains the form and the meaning of ACE sentences and texts.

An ACE text is a sequence of declarative sentences that can be anaphorically interrelated. Furthermore, ACE supports questions and commands. Declarative sentences can be simple or composite.

Simple ACE sentences can have the following structure:

subject + verb + complements + adjuncts

A customer inserts two cards manually in the morning.

Every sentence of this structure has a subject and a verb. Complements (direct and indirect objects) are necessary for transitive verbs ('insert something') and ditransitive verbs ('give something to somebody'), whereas adjuncts (adverbs, prepositional phrases) that modify the verb are optional.

Alternatively, simple sentences can be built according to the structure:

'there is'/'there are' + noun phrase

There is a customer.

Every sentence of this structure introduces only the object described by the noun phrase.

Elements of a simple sentence can be elaborated upon to describe the situation in more detail. To further specify the nouns, we can add adjectives, possessive nouns and *of*-prepositional phrases, or variables as appositions.

A bank's trusted customer X inserts two valid cards of himself.

Other modifications of nouns are possible through relative clauses

A customer who is trusted inserts two cards that he owns.

Composite sentences are recursively built from simpler sentences through coordination, subordination, quantification, and negation.

Coordination by 'and' is possible between sentences and between phrases of the same syntactic type.

A customer inserts a card and an automated teller checks the code.
A customer inserts a card and enters a code.

Coordination by 'or' is possible between sentences, verb phrases, and relative clauses.

A customer inserts a card or an automated teller checks the code.
A customer inserts a card or enters a code.
A customer owns a card that is invalid or that is damaged.

Coordination by 'and' and 'or' is governed by the standard binding order of logic, i.e. 'and' binds stronger than 'or'. Commas can be used to override the standard binding order.

There are three constructs of subordination: *if-then*-sentences, modality, and sentence subordination. With the help of *if-then*-sentences we can specify conditional situations, e.g.

If a card is valid then a customer inserts it.

Modality allows us to express possibility and necessity.

A trusted customer can insert a card.
A trusted customer must insert a card.
It is possible that a trusted customer inserts a card.
It is necessary that a trusted customer inserts a card.

Sentence subordination means that a complete sentence is used as an object, e.g.

> It is false that a customer inserts a card.
> A clerk believes that a customer inserts a card.

Sentences can be existentially or universally quantified. Existential quantification is typically expressed by indefinite determiners ('a man', 'some water', '3 cards'), while universal quantification is typically expressed by the occurrence of 'every' — but see below for the quantification within *if-then*-sentences. In the example

> Every customer inserts a card.

the noun phrase 'every customer' is universally quantified, while the noun phrase 'a card' is existentially quantified, i.e. every customer inserts a card that may, or may not, be the same card that another customer inserts. Note that this sentence is logically equivalent to the sentence

> If there is a customer then the customer inserts a card.

which shows that noun phrases occurring in the *if*-part of an *if-then*-sentence are universally quantified.

Negation allows us to express that something is not the case, e.g.

> A customer does not insert a card.

To negate something for all objects of a certain class one uses 'no'.

> No customer inserts more than 2 cards.

To negate a complete statement one uses sentence negation.

> It is false that a customer inserts a card.

ACE supports two forms of queries: *yes/no*-queries and *wh*-queries. *Yes/no*-queries ask for the existence or non-existence of a specified situation.

> Does a customer insert a card?

With the help of *wh*-queries, i.e. queries with query words, we can interrogate a text for details of the specified situation. If we specified

> A trusted customer inserts a valid card manually.

we can ask for each element of the sentence with the exception of the verb, e.g.

> Who inserts a card?
> Which customer inserts a card?
> What does a customer insert?
> How does a customer insert a card?

Finally, ACE also supports commands. Some examples:

> John, go to the bank!
> John and Mary, wait!
> Every dog, bark!
> A brother of John, give a book to Mary!

Constraining Ambiguity. To constrain the ambiguity of full English ACE employs three simple means

- some ambiguous constructs are not part of the language; unambiguous alternatives are available in their place
- all remaining ambiguous constructs are interpreted deterministically on the basis of a small number of interpretation rules
- users can either accept the assigned interpretation, or they must rephrase the input to obtain another one

Here is an example how ACE replaces ambiguous constructs by unambiguous constructs. In full English relative clauses combined with coordinations can introduce ambiguity, e.g.

> A customer inserts a card that is valid and opens an account.

In ACE the sentence has the unequivocal meaning that the customer opens an account. To express the alternative meaning that the card opens an account the relative pronoun 'that' must be repeated, thus yielding a coordination of relative clauses.

> A customer inserts a card that is valid and that opens an account.

However, not all ambiguities can be safely removed from ACE without rendering it artificial. To deterministically interpret otherwise syntactically correct ACE sentences we use a small set of interpretation rules.

Here is an example of an interpretation rule at work. In

> A customer inserts a card with a code.

'with a code' attaches to the verb 'inserts', but not to 'a card'. To express that the code is associated with the card we can employ the complementary interpretation rule that a relative clause always modifies the immediately preceding noun phrase, and rephrase the input as

> A customer inserts a card that carries a code.

Anaphoric References. Usually an ACE text consists of more than one sentence, e.g.

> A customer enters a card and a code. If a code is valid then an automated teller accepts a card.

To express that all occurrences of 'card' and 'code' should mean the same card and the same code, ACE provides anaphoric references via the definite article, i.e.

> A customer enters a card and a code. If the code is valid then an automated teller accepts the card.

During the processing of the ACE text, all anaphoric references are replaced by the most recent and most specific accessible noun phrase that agrees in gender and number.

What does "most recent and most specific" mean? Given the sentence

A customer enters a red card and a blue card.

then

The card is correct.

refers to the second card, which is the textually closest noun phrase that matches the anaphor 'the card', while

The red card is correct.

refers to the first card that is the textually closest noun phrase that matches the anaphor 'the red card'.

What does "accessible" mean? Like in full English, noun phrases introduced in *if-then*-sentences, universally quantified sentences, negations, modality, and subordinated sentences cannot be referenced anaphorically in subsequent sentences. Thus for each of the sentences

If a customer owns a card then he enters it.
A customer does not enter a card.

we cannot refer to 'a card' with

The card is correct.

Anaphoric references are also possible via personal pronouns

A customer enters his own card and its code. If it is valid then an automated teller accepts the card.

or via variables

A customer X enters X's card Y and Y's code Z. If Z is valid then an automated teller accepts Y.

Note that proper names always denote the same object.

6.3.2 From Attempto Controlled English to First-Order Logic

ACE texts can be mapped to Discourse Representation Structures (DRS) [24,5]. DRSs use a syntactic variant of the language of standard first-order logic which we extended by some non-standard structures for modality, sentence subordination, and negation as failure. This section gives a brief overview of the DRS representation of ACE texts. Consult [12] for a comprehensive description. DRSs consist of a domain and of a list of conditions, and are usually displayed in a box notation:

The domain is a set of discourse referents (i.e. logical variables) and the conditions are a set of first-order logic predicates or nested DRSs. The discourse referents are existentially quantified with the exception of boxes on the left-hand side of an implication where they are universally quantified. We are using a reified (or "flat") notation for the predicates. For example, the noun 'a card' that normally would be represented in first-order logic as

```
card(A)
```

is represented as

```
object(A,card,countable,na,eq,1)
```

relegating the predicate 'card' to the constant 'card' used as an argument in the predefined predicate 'object'. In that way, we can reduce the potentially large number of predicates to a small number of predefined predicates. This makes the processing of the DRS easier and allows us to include some linguistic information, e.g. whether a unary relation comes from a noun, from an adjective, or from an intransitive verb. Furthermore, reification allows us to quantify over predicates and thus to express general axioms needed for reasoning over ACE text in the Attempto Reasoner RACE that is presented in Section 6.5.1.

Proper names, countable nouns, and mass nouns are represented by the object-predicate:

John drives **a car** and buys **2 kg of rice**.

```
A B C D E
object(A,'John',named,na,eq,1)
object(B,car,countable,na,eq,1)
predicate(C,drive,A,B)
object(D,rice,mass,kg,eq,2)
predicate(E,buy,A,D)
```

Adjectives introduce `property`-predicates:

A **young** man is **richer than** Bill.

```
A B C D
object(A,'Bill',named,na,eq,1)
object(B,man,countable,na,eq,1)
property(B,young,pos)
property(C,rich,comp_than,A)
predicate(D,be,B,C)
```

As shown in the examples above, verbs are represented by `predicate`-predicates. Each verb occurrence gets its own discourse referent which is used to attach modifiers like adverbs (using `modifier_adv`) or prepositional phrases (using `modifier_pp`):

John **carefully** works **in** an office.

```
A B C
object(A,'John',named,na,eq,1)
object(B,office,countable,na,eq,1)
predicate(C,work,A)
modifier_adv(C,carefully,pos)
modifier_pp(C,in,B)
```

The `relation`-predicate is used for *of*-constructs, Saxon genitive, and possessive pronouns:

A brother **of** Mary**'s** mother feeds **his own** dog.

```
A B C D E
object(A,'Mary',named,na,eq,1)
object(B,brother,countable,na,eq,1)
relation(C,of,A)
object(C,mother,countable,na,eq,1)
relation(B,of,C)
relation(D,of,B)
object(D,dog,countable,na,eq,1)
predicate(E,feed,B,D)
```

There are some more predicates which are not discussed here, but are described in [12]. The examples so far have been simple in the sense that they contained no universally quantified variables and there was no negation, disjunction, or implication. For such more complicated statements, nested DRSs become necessary. In the case of negation, a nested DRS is introduced that is prefixed by a negation sign:

A man **does not** buy a car.

```
A
object(A,man,countable,na,eq,1)

        B C
  ¬     object(B,car,countable,na,eq,1)
        predicate(C,buy,A,B)
```

Note that 'a man' is not in the scope of the negation. In ACE, scopes are determined on the basis of the textual order of the sentence elements. In the following example, 'a man' is also under negation:

It is false that a man buys a car.

```
        A B C
  ¬     object(A,man,countable,na,eq,1)
        object(B,car,countable,na,eq,1)
        predicate(C,buy,A,B)
```

The ACE structures 'every', 'no', and 'if ... then' introduce implications that are denoted by arrows between two nested DRSs.

Every woman owns a house.

```
  A                                        B C
  object(A,woman,countable,na,eq,1)   ⇒    object(B,house,countable,na,eq,1)
                                           predicate(C,own,A,B)
```

As stated before already, discourse referents that are introduced in a DRS box that is on the left-hand side of an implication are universally quantified. In all other cases, they are existentially quantified. Disjunctions — which are represented in ACE by the coordination 'or' — are represented in the DRS by the logical sign for disjunction:

John works **or** travels.

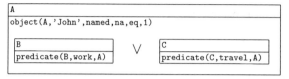

The modal constructs of possibility ('can') and necessity ('must') are represented by the standard modal operators (see [6] for details):

Sue **can** drive a car.

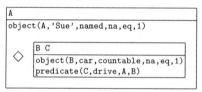

Bill **must** work.

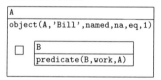

Finally, *that*-subordination can lead to the situation where a discourse referent stands for a whole sub-DRS:

John knows **that** his brother works.

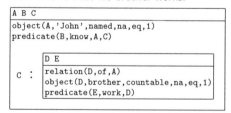

Every ACE sentence can be mapped to exactly one DRS using the introduced DRS elements. DRSs are a convenient and flexible way to represent logical statements.

6.3.3   Attempto Parsing Engine (APE)

The Attempto Parsing Engine (APE) is a tool that translates an input ACE text into a DRS, provides various technical feedback (tokenization and sentence splitting of the input text, tree-representation of the syntactic structure of the input), and various logical forms and representations derived from the DRS: standard first-order logic form, DRS in XML, OWL/SWRL. An ACE paraphrase of the input text is also offered, by translating (verbalizing) the obtained DRS into a subset of ACE.

If the input text contains syntax errors or unresolvable anaphoric references then the translation into a DRS fails and a message is output that pinpoints the location and the cause of the error. Furthermore, APE tries to suggest how to resolve the problem.

APE implements the ACE syntax in the form of approximately 200 definite clause grammar rules using feature structures. APE comes with a large lexicon containing almost 100'000 English words. User defined lexica can be used in addition or in place of this large lexicon.

APE has been implemented in SWI-Prolog and released under the LGPL open source license. The distribution also includes the DRS verbalizer, translator from ACE to OWL/SWRL, and more[4]. APE has a command-line client and can be also used from Java, or over HTTP as a REST web service[5] or from its demo client[6]. Figure 1 shows a screenshot of the APE web client.

6.4   Fitting ACE into the Semantic Web

6.4.1   OWL and SWRL

In order to make ACE interoperable with some of the existing Semantic Web languages, mappings have been developed to relate ACE to OWL and SWRL (see a detailed description in [22]). For example, the mapping of ACE to OWL/SWRL translates the ACE text

> Every employee that does not own a car owns a bike.
> Every man that owns a car likes the car.
> Which car does John own?

into a combination of OWL axiom, SWRL rule and DL-Query (an OWL class expression).

$$employee \sqcap \neg (\exists\, own\, car) \sqsubseteq \exists\, own\, bike$$
$$man(?x) \wedge own(?x, ?y) \wedge car(?y) \rightarrow like(?x, ?y)$$
$$car \sqcap \exists\, own^- \{John\}$$

Note that the mapping is performed on the DRS level, meaning that all ACE sentences that share their corresponding DRS are mapped into the same OWL/SWRL form. ACE provides a lot of linguistically motivated syntactic sugar, e.g. the following sentences have the same meaning (because they have the same DRS).

---

[4] http://attempto.ifi.uzh.ch/site/downloads/
[5] http://attempto.ifi.uzh.ch/site/docs/ape_webservice.html
[6] http://attempto.ifi.uzh.ch/ape/

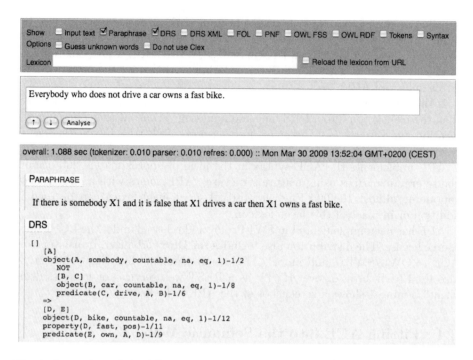

Fig. 1. Screenshot of the APE web client, showing the DRS and the paraphrase of the sentence 'everybody who does not drive a car owns a fast bike'

> John knows every student.
> Every student is known by John.
> If there is a student then John knows the student.
> For every student John knows him/her.

In order to keep the mappings simple and immediately reversible, they currently support only a fragment of ACE. Notably, there is no support for modifiers such as adjectives, adverbs, and prepositional phrases. The covered ACE fragment, however, is so large and syntactically and semantically expressive, that it covers almost all of OWL 2 (some aspects of data properties are not handled) and SWRL (again, data properties are not completely covered). ACE questions that contain exactly one query word ('what', 'which', 'whose', 'who') are mapped to DL-Queries.

The OWL→ACE mapping allows us to verbalize existing OWL ontologies as ACE texts. This mapping is not just the reverse of the ACE→OWL as it also covers OWL axiom and expression types that the ACE→OWL mapping does not generate. For example

PropertyDomain(write author)

is verbalized as

> Everything that writes something is an author.

Table 1. Examples of verbalizing OWL property and class expressions as ACE verbs and noun phrases (including common nouns and proper names), where R is a named property, C, C_1, ..., C_n are (possibly complex) class expressions, a is an individual, n is a natural number larger than 0. In the actual verbalizations, the word 'something' is often replaced by a noun representing a conjoined named class, e.g. *IntersectionOf(cat ExistsSelf(like))* would be verbalized as 'cat that likes itself'.

OWL properties and classes	Examples of ACE verbs and noun phrases
Named property	Transitive verb, e.g. 'like'
InverseProperty(R)	Passive verb, e.g. 'is liked by'
Named class	Common noun, e.g. 'man'
owl:Thing	'something', 'thing'
ComplementOf(C)	'something that is not a car', 'something that does not like a cat'
IntersectionOf($C_1 \ldots C_n$)	something that is a person and that owns a car and that ...
UnionOf($C_1 \ldots C_n$)	something that is a wild-animal or that is a zoo-animal or that ...
OneOf(a)	Proper name, e.g. 'John', 'something that is John'
SomeValuesFrom(R C)	something that loves a person
ExistsSelf(R)	something that likes itself
MaxCardinality(n R C)	something that has at most 2 spouses

The resulting ACE sentence can be handled by the ACE→OWL mapping by converting it into a general class inclusion axiom with the same semantics as the property domain axiom.

The subset of ACE used in these mappings provides a corresponding ACE content word (proper name, common noun, transitive verb) for each OWL entity, whereas complex OWL class and property expressions map to ACE phrases, and OWL axioms map to ACE sentences. At the entity level, OWL individuals are denoted by ACE proper names, named classes by common nouns, and (object) properties by transitive verbs and relational nouns (e.g. 'part of'). In OWL, it is possible to build complex class expressions from simpler ones by intersection, union, complementation and property restriction. Similarly, ACE allows building complex noun phrases via relative clauses which can be conjoined (by 'and that'), disjoined (by 'or that'), negated (by 'that is/are/does/do not') and embedded (by 'that'). OWL anonymous inverse properties map to ACE passive verbs. This proves that in principle, each OWL structure can be mapped to a corresponding ACE structure.[7] Table 1 shows some examples of mapping OWL classes and properties.

---

[7] Only very complex structures that would require parentheses to denote the scope of their constructors cannot be directly mapped to ACE as ACE does not offer a similar parentheses mechanism for grouping. In order to enable the verbalization in such cases, one can replace part of the complex structure by a named class to simplify the structure.

Table 2. Examples of verbalizing OWL axioms as ACE sentences, where $R_1, \ldots, R_n$, and S are object property expressions; C and D are class expressions; and a, a_1 and a_2 are individuals.

OWL axiom types	Examples of ACE sentences
SubClassOf(C D)	Every man is a human.
SubPropertyOf(PropertyChain($R_1 \ldots R_n$) S)	If X knows Y and Y is an editor of Z then X submits-to Z.
DisjointProperties(R_1 R_2)	If X is a child of Y then it is false that X is a spouse of Y.
SameIndividual(a_1 a_2)	Bill is William.
DifferentIndividuals(a_1 a_2)	Bill is not John.
ClassAssertion(C a)	Bill is a man that owns at least 2 cars.

OWL axioms are mapped to ACE sentences (see table 2 for some examples). Apart from sentences that are derived from the axioms about individuals (*SameIndividual*, *DifferentIndividuals*, *ClassAssertion*, *PropertyAssertion*), all sentences are *every*-sentences or *if-then*-sentences, meaning that they have a pattern '*NounPhrase VerbPhrase*' or 'If X ... then X ... Y' where *NounPhrase* starts with 'every' or 'no'. Of course, in the ACE to OWL/SWRL direction one can use *if-then*-sentences instead of *every*-sentences and has also otherwise more flexibility.

In a nutshell, the mappings between ACE and OWL/SWRL provide an alternative syntax for OWL and SWRL. This syntax is readable as standard English, it makes the difference between OWL and SWRL invisible, and provides linguistically motivated syntactic sugar. This syntax is mainly intended for structurally and semantically complex knowledge bases for which visual methods and the official OWL/SWRL syntaxes fail to provide a user-friendly front-end.

6.4.2   AceRules: Rules in ACE

AceRules is a multi-semantics rule engine using ACE as input and output language. AceRules has been introduced in [26] and is designed for forward-chaining interpreters that calculate the complete answer set. The following is a simple exemplary program (we use the term "program" for a set of rules and facts):

> John is a customer.
> John is a friend of Mary.
> Mary is an important customer.
> Every customer is a person.
> Every customer who is a friend of Bill gets a discount.
> If a person is important then he/she gets a discount.
> Every friend of Mary is a friend of Bill.

Submitting this program to AceRules, we get the following answer (we use the term "answer" for the set of facts that can be derived from a program):

Mary is important.
Mary is a customer.
John is a customer.
Mary is a person.
John is a person.
John is a friend of Mary.
John is a friend of Bill.
Mary gets a discount.
John gets a discount.

The program and the answer are both represented in ACE and no other formal notation is needed for the user interaction.

AceRules is designed to support various semantics. Depending on the application domain, the characteristics of the available information, and on the reasoning tasks to be performed, different rule semantics are needed. At the moment, AceRules incorporates three different semantics: courteous logic programs [17], stable models [15], and stable models with strong negation [16]. Only little integration effort would be necessary to incorporate other semantics into AceRules.

Negation is a complicated issue in rule systems. In many cases, two kinds of negation [39] are required. Strong negation (also called "classical negation" or "true negation") indicates that something can be proven to be false. Negation as failure (also called "weak negation" or "default negation"), in contrast, states only that the truth of something cannot be proven.

However, there is no such general distinction in natural language. It depends on the context, what kind of negation is meant. This can be seen with the following two examples in natural English:

1. *If there is no train approaching then the school bus can cross the railway tracks.*
2. *If there is no public transport connection to a customer then John takes the company car.*

In the first example (which is taken from [16]), the negation corresponds to strong negation. The school bus is allowed to cross the railway tracks only if the available information (e.g. the sight of the bus driver) leads to the conclusion that no train is approaching. If there is no evidence whether a train is approaching or not (e.g. because of dense fog) then the bus driver is not allowed to cross the railway tracks.

The negation in the second sentence is most probably to be interpreted as negation as failure. If one cannot conclude that there is a public transport connection to the customer on the basis of the available information (e.g. public transport schedules) then John takes the company car, even if there is a special connection that is just not listed.

As long as only one kind of negation is available, there is no problem to express this in controlled natural language. As soon as two kinds of negation are supported, however, we need to distinguish them somehow. We found a natural way to represent the two kinds of negation in ACE. Strong negation is represented with the common negation constructs of natural English:

- 'does not', 'is not' (e.g. 'John is not a customer')
- 'no' (e.g. 'no customer is a clerk')
- 'nothing', 'nobody' (e.g. 'nobody knows John')
- 'it is false that' (e.g. 'it is false that John waits')

To express negation as failure, we use the following constructs:

- 'does not provably', 'is not provably' (e.g. 'a customer is not provably trustworthy')
- 'it is not provable that' (e.g. 'it is not provable that John has a card')

This allows us to use both kinds of negation side by side in a natural looking way. The following example shows a rule using strong negation and negation as failure at the same time.

> If a customer does not have a credit-card and is not provably a criminal then the customer gets a discount.

This representation is compact and we believe that it is well understandable. Even persons who have never heard of strong negation and negation as failure can understand it to some degree.

The original stable model semantics supports only negation as failure, but it has been extended to support also strong negation. Courteous logic programs are based on stable models with strong negation and support both forms of negation.

None of the two forms of stable models guarantee a unique answer set. Thus, some programs can have more than one answer. In contrast, courteous logic programs generate always exactly one answer. In order to achieve this, priorities are introduced and the programs have to be acyclic. The AceRules system demonstrates how these different rule semantics can be expressed in ACE in a natural way.

6.4.3   The Protune Policy Language

The term "policy" can be generally defined as a "statement specifying the behavior of a system", i.e., a statement which describes which decision the system should take or which actions it should perform according to specific circumstances.

Some of the application areas where policies have been lately used are security and privacy. A security policy defines security restrictions for a system, organization or any other entity. A privacy policy is a declaration made by an organization regarding its use of customers' personal information (e.g., whether third parties may have access to customer data and how that data will be used).

The ability of expressing policies in a formal way can be regarded as desirable: the authority defining policies would have to express them in a machine-understandable way whereas all further processing of the policies could take place in an automatic fashion.

For this reason a number of policy languages have been defined in the last years (cf. [9] for an extensive comparison among them). Nevertheless a major hindrance to widespread adoption of policy languages are their shortcomings in terms of usability: in order to be machine-understandable all of them rely on a formal syntax, which common users find unintuitive and hard to grasp.

We think that the use of controlled natural languages can dramatically improve usability of policy languages. This section describes how we exploited (a subset of) ACE in order to express policies and how we developed a mapping between ACE policies and the ones written in the Protune policy language. The Protune policy language is extensively described in Chapter 4. This section only provides a general overview of the Protune policy language and especially focuses on its relevant features w.r.t. the ACE → Protune mapping.

Protune is a Logic Programming-based policy language and as such a Protune policy has much in common with a Logic Program. For instance the Protune policy

$$A \leftarrow B_{11}, \ldots, B_{1n}.$$
$$\ldots$$
$$A \leftarrow B_{m1}, \ldots, B_{mn}.$$

can be read as follows: A holds if one of

- $(B_{11}$ and $\ldots$ and $B_{1n})$
- $\ldots$
- $(B_{m1}$ and $\ldots$ and $B_{mn})$

holds. In this overview we only introduce two additional features of Protune policies w.r.t. usual logic programs, namely actions and complex terms.

A policy may require that under some circumstances some actions are performed: a typical scenario in an authorization context requires that access to some resource is allowed only if the requester provides a valid credential. For this reason the Protune language allows to define actions like in the following example.

$$allow(action_1) \leftarrow action_2.$$

The rule above can be read as follows: $action_1$ *can be* executed if $action_2$ *has been* executed. Notice the different semantics of the actions according to the side of the rule they appear in: in order to stress this semantic difference we force the policy author to write actions appearing in the left side of a rule into the $allow/1$ predicate.

The evaluation of a policy may require to deal with entities which can be modeled as sets of $(attribute, value)$ pairs. This is the case with the credentials mentioned in the example above. The Protune language allows to refer to $(attribute, value)$ pairs of such an entity by means of the following notation.

$$identifier.attribute : value$$

Only a subset of the ACE language needs to be exploited when defining policies: data (i.e., integers, reals and strings), nouns, adjectives (in positive, comparative and superlative form), (intransitive, transitive and ditransitive) verbs and prepositional phrases (in particular *of*-constructs) can be used with some restrictions, the most remarkable of which is that plural noun phrases are not allowed. This means that neither expressions like 'some cards' or 'at least two cards' nor sentences like 'John and Mary are nice' are supported. However notice that some of such sentences (although not all) can be rewritten as sets of sentences (e.g., the previous example can be split into 'John is nice' and 'Mary is nice'). The complete set of restrictions can be found in [8].

ACE provides a number of complex structures to combine simple sentences into larger ones, whereas only few of them (namely negation as failure, possibility and implication) can be exploited in order to express policies. Moreover whilst ACE complex structures can be arbitrarily nested, in ACE policies nesting is allowed only according to given patterns. Roughly speaking (more on this in [8]) ACE policies must have one of the following formats

- If B_1 and ... and B_n then H.
- H.

where B_i ($1 \leq i \leq n$) may contain a negation-as-failure or possibility construct and H may contain a possibility construct. For example, only the first one of the following sentences is a correct ACE policy: 'if it is not provable that John has a forged credit-card then John can access "myFile"' and 'it is not provable that John has a forged credit-card'.

The restrictions listed above allow to straightforwardly map ACE sentences into Protune rules: it should be easy to figure out that the ACE implication (resp. negation as failure) construct maps to Protune rules (resp. negated literals). On the other hand the ACE possibility construct is meant to convey the semantics of the *allow*/1 Protune predicate. Other remarkable mapping rules are accounted for in the following list.

- A programmer asked to formalize the sentence 'John gives Mary the book' as a logic program would most likely come up with a rule like *give(john, mary, book)*. Indeed in many cases translating verbs into predicate names can be considered the most linear approach, and we pursued this approach as well. However the arity of a Protune predicate can be arbitrary, whereas intransitive (resp. transitive, ditransitive) verbs can be naturally modeled as predicates with arity one (resp. two, three). For this reason we decided to exploit ACE prepositional phrases (except *of*-constructs) for providing further parameters to a Protune predicate. For instance, sentence 'John gets "A" in physics.' translates into '*get#in*'('*John*', '*A*', *physics*).
- A statement like 'John is Mary's brother' can be seen as asserting some information about the entity "Mary", namely that the value of her property "brother" is "John". It should be then intuitive exploiting Protune complex terms to map such ACE sentence to '*Mary*'.*brother* : '*John*'.
- When translating noun phrases like 'a user' it must be decided if it really matters whether we are speaking about a "user" (in which case the noun phrase could be translated as *user(X)*) or not (in which case the noun phrase

could be translated as a variable X). According to our experience, policy authors tend to use specific concepts even if they actually mean generic entities. For this reason we followed the second approach, according to which the sentence 'if a user owns a file then the user can access the file' is translated into the Protune rule: $allow(access(User, File)) \leftarrow own(User, File)$. If it is needed to point out that the one owning a file is a user, the sentence can be rewritten e.g., as follows: 'if X is a user and X owns a file then X can access the file' which gives the translation: $allow(access(X, File)) \leftarrow user(X)$, $own(X, File)$.

6.4.4   Other Web Languages

ACE has also been translated into other Semantic Web languages. A translator has been implemented that converts ACE texts into the Rule Markup Language (RuleML) [19]. Another translator has been developed that translates a subset of ACE into the REWERSE Rule Markup Language (R2ML) [30]. R2ML integrates among others the Semantic Web Rule Language (SWRL) and the Rule Markup Language (RuleML) [40]. Furthermore, ACE has been used as a front-end for the Process Query Language (PQL) that allows users to query MIT's Process Handbook. It has been shown that queries expressed in ACE and automatically translated into PQL provide a more user-friendly interface to the Process Handbook [4,3].

6.5   ACE Tools for the Semantic Web

6.5.1   Attempto Reasoner RACE

The Attempto Reasoner RACE supports automatic reasoning in the first-order subset of ACE that consists of all of ACE with the exception of negation as failure, modality, and sentence subordination. For simplicity, the first-order subset of ACE is simply called ACE in this section.

RACE proves that theorems expressed in ACE are the logical consequence of axioms expressed in ACE, and gives a justification for the proof in ACE. If there is more than one proof, then RACE will find all of them. If a proof fails, then RACE will indicate which parts of the theorems could not be proved. Variations of the basic proof procedure permit query answering and consistency checking.

The current implementation of RACE is based on the model generator Satchmo [31]. The Prolog source code of Satchmo is available — which allows us to easily add modifications and extensions. The two most important extensions are an exhaustive search for proofs and a tracking mechanism.

- exhaustive search: while Satchmo stops once it finds the first inconsistency, RACE will find all inconsistencies
- tracking mechanism: RACE will report for each successful proof which minimal subset of the axioms is needed to prove the theorems.

Currently, we employ RACE only for theorem proving. To better answer *wh*-questions we plan to utilize RACE also as model generator.

RACE works with the clausal form of first-order logic. ACE axioms A and ACE theorems T are translated — via DRSs generated by APE — into their first-order representations FA, respectively FT. Then the conjunction $(FA \wedge \neg FT)$ is translated into clauses, submitted to RACE and checked for consistency. RACE will find all minimal inconsistent subsets of the clauses and present these subsets using the original ACE axioms A and theorems T. If there is no inconsistency, RACE will generate a minimal finite model — if there is one.

RACE is supported by auxiliary axioms expressed in Prolog. Auxiliary axioms implement domain-independent linguistic knowledge that cannot be expressed in ACE since this knowledge depends on the DRS representations of ACE texts. A typical example is the relation between the plural and the singular forms of nouns. Auxiliary axioms can also act as meaning postulates for ACE constructs that are under-represented in the DRS, for example generalized quantifiers and indefinite pronouns. Finally, auxiliary axioms could also be used instead of ACE to represent domain-specific knowledge.

ACE is undecidable. Technically this means that RACE occasionally would not terminate. To prevent this situation, RACE uses a time-out with a time limit that is calculated on the size of the input.

In the spirit of the Attempto project, running RACE should not require any knowledge of theorem proving in general, and of the working of RACE in particular. Nevertheless, RACE offers a number of parameters that let users control the deductions from collective plurals, enable or disable the output of the auxiliary axioms that were used during a proof, and limit the search for proofs. These parameters have default settings that allow the majority of the users to ignore the parameters.

RACE processes clauses by forward-chaining whose worst-case time complexity is $O(n^2)$, where n is the number of clauses. To reduce the run-time of RACE we need to reduce primarily the number of clauses that participate:

- we profit from simplifications introduced in the DRS representation that lead to fewer clauses
- we use clause compaction
- we eliminate after the first round of forward reasoning the clauses with the body *true* that cannot be fired again
- we apply intelligent search for clauses that could be fired in the next round of forward reasoning
- we use complement splitting — given a disjunction $(A \vee B)$, one investigates $(A \wedge \neg B)$, respectively $(\neg A \wedge B)$ — though complement splitting is not guaranteed to increase the efficiency in each case

RACE can be called via a SOAP web service[8] and can conveniently be accessed via a web client[9]. Figure 2 is a typical screenshot of the RACE web client.

---

[8] http://attempto.ifi.uzh.ch/site/docs/race_webservice.html
[9] see http://attempto.ifi.uzh.ch/race/ and
  http://attempto.ifi.uzh.ch/site/docs/race_webclient_help.html

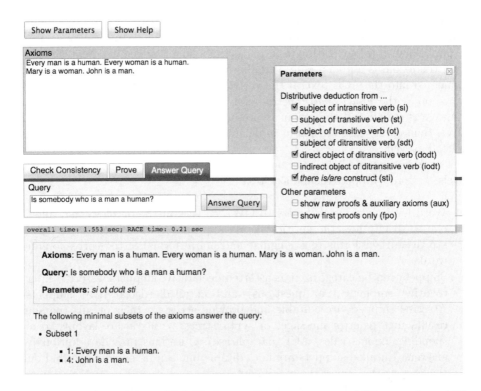

Fig. 2. The web interface of RACE showing how to answer an ACE query from ACE axioms with the default setting of the parameters

6.5.2   ACE View Ontology and Rule Editor

The ACE View ontology and rule editor[10] allows users to develop OWL ontologies and SWRL rulesets in ACE. The ACE View editor lets the user create, browse, and edit an ACE text, and query both its asserted and automatically entailed content.

In the context of ACE View, an ACE text is a set of ACE snippets where each snippet is a sequence of one or more (possibly anaphorically linked) ACE sentences. In general, each snippet corresponds to an OWL or SWRL axiom, but complex ACE sentences that involve sentence conjunction can map to several axioms. When a snippet is added to the text, it is automatically parsed and converted into OWL/SWRL. If the translation fails then the snippet is still accepted, but as it does not have any logical axioms attached, it cannot be considered as part of the text during reasoning. In case the translation succeeds, the snippet is mapped to one or more OWL axioms and SWRL rules which are in turn merged with the underlying knowledge base representation. In case a snippet is deleted from the text, its corresponding axioms (if present) are removed from the knowledge base.

---

[10] http://attempto.ifi.uzh.ch/aceview/

ACE View is implemented as an extension to the Protégé editor[11]. Therefore, the ACE View user can alternatively switch to one of the default Protégé views to perform an ontology editing task. In case an axiom is added in a Protégé view, then the axiom is automatically verbalized into an ACE snippet which in turn is merged into the ACE text. If the verbalization fails (e.g. the verbalizer does not support the *FunctionalProperty*-axiom with data properties) then an error message is stored and the axiom is preserved in the ACE text in Manchester OWL Syntax. In case an axiom is deleted, then its corresponding snippet is deleted as well.

The ACE text (and thus the ontology) can be viewed and edited at several levels — word, snippet, vocabulary, text.

- Word level provides an access to OWL entities in the ontology and allows one to specify how the entities should appear in ACE sentences, i.e. what are the surface forms (e.g. singular and plural forms) of the corresponding words.
- Snippets can be categorized as asserted declarative snippets, asserted interrogative snippets (i.e. questions) and entailed (declarative) snippets. Asserted snippets are editable and provide access to their details (parsing results such as error messages or syntax trees/syntax aware layout, corresponding axioms/rules, ACE paraphrase). Questions provide additionally answers. Entailed snippets are not editable but can be explored to find out the reasons that cause the entailment.
- Vocabulary is a set of ACE content words. It can be sorted alphabetically or by frequency of usage. As content words correspond to OWL entities, standard Protégé views offer even more presentation options, e.g. the "back-bone hierarchy" of subclass and "part of" relations; separation of the vocabulary into classes, properties, individuals. The vocabulary level provides a quick access to the word level, each selected/searched word (entity) can be automatically shown in the word level, or its corresponding snippets in the text level.
- An ACE text is a set of ACE snippets. This set can be filtered, sorted, and searched. Reasoning can be performed on the whole text to find out about its (in)consistency. A new text can be generated by filling it with snippets that the asserted text entails.

The ACE View user interface comprises several "views" that allow for browsing and editing of the ACE text at all the described levels (see figures 3 and 4). In the "Lexicon view" and "Words view", the complete content word vocabulary of the ACE text is presented, sorted either alphabetically or by frequency of usage. The "Lexicon view" allows the user to edit the surface forms (singular, plural, past participle) of words and make sure that they all correspond to the same OWL entity. When a new entity is generated in the standard Protégé views, the surface forms of its corresponding content word are automatically

---

[11] http://protege.stanford.edu/

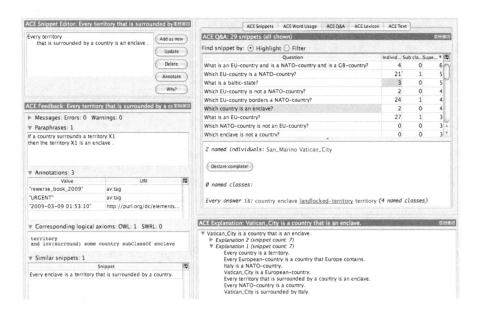

Fig. 3. One possible layout of the ACE View editor. Several views are shown: ACE Snippet Editor shows the currently selected snippet; ACE Feedback shows its paraphrase, annotations, the corresponding OWL axiom, and a list of syntactically similar snippets; Q&A view shows all the entered questions, and the answers to the question 'Which country is a an enclave?'; ACE Explanation shows the justification for the answer 'Vatican_City is a country that is an enclave'. The justification contains two sets of snippets (i.e. different explanations), one of which is expanded.

generated based on rules of English morphology. The user can override these forms if needed.

The "Snippets view" organizes all the asserted snippets in a table. With each snippet a set of its features are presented: snippet length (in content words), creation time, number of annotations, etc. The table rows can be highlighted and filtered based on the selected word, presenting only the snippets that contain the word. The "Snippet Editor" lets the user to edit an existing snippet, or create a new one. The "Feedback view" shows the logical and linguistic properties of the selected snippet, and meta information such as annotations for the snippet. For sentences that fail to map to OWL/SWRL, error messages are provided. Error messages point to the location of the error and explain how to deal with the problem.

The "Q&A view" lists ACE questions and answers to them. These questions correspond to DL-Queries which are essentially (possibly complex) class expressions. The answers to a DL-Query are named individuals (members of the queried class) or named classes (named super and subclasses of the queried class). In ACE terms, the answers are ACE content words — proper names and common nouns. While the answers to DL-Queries are representation-wise

Asserted class hierarchy: encl...				
▼ Thing				
▼ EU-country				
baltic-state				
G8-country				
▶ NATO-country				
baltic-state				
body-of-water				
enclave				
doubly-landlocked-territory				
island-country				
landlocked-territory				
▶ neutral-country				
▼ territory				
▼ country				
▶ European-country				

ACE Snippets: 231 snippets (7 shown)

Find snippet by: ◯ Highlight ⦿ Filter

Snippet	Words ▼	Timestamp
Every territory that is surrounded by a country is an enclave.	4	2009-03-09 22:54:40
Every enclave is a territory that is surrounded by a country.	4	2009-03-09 22:54:40
No territory that is bordered by at least 2 countries is an encl...	4	2009-03-09 22:54:40
Which enclave is bordered by at least 2 countries?	3	2009-03-09 22:54:53
Which enclave is not a country?	2	2009-03-09 22:54:53
Which country is an enclave?	2	2009-03-09 23:00:45
What is an enclave?	1	2009-03-09 22:54:53

ACE Metrics:

Snippets	231
Sentences	231
Questions	29
SWRL snippets	0
Non OWL/SWRL snippets	0
Unverbalized axioms	0
Snippets that contain nothing but	2
Content words (CN + TV + PN)	78
Common nouns (CN)	14
Transitive verbs (TV)	5
Proper names (PN)	59
Unused content words	0
Wordforms	102

Fig. 4. Another possible layout of the ACE View editor. Several views are shown: the standard Protégé tree view shows the subclass hierarchy of named classes; ACE Snippets view shows the snippets that reference the selected entity 'enclave', the number of content words and the creation time is shown for each snippet; Metrics view shows various (mostly linguistic) metrics of the ACE text.

identical in the ACE view and in the standard Protégé view, the construction of the query is potentially much simpler in the ACE view, as one has to construct a natural language question.

The "Entailments view" provides a list of ACE sentences that follow logically from the ACE text, i.e. these sentences correspond to the entailed axioms of the ontology. Such axioms are generated by the integrated reasoner on the event of classification. The "Explanation view" provides an "explanation" for a selected entailed snippet. Such an explanation is a (minimal) sequence of asserted snippets that justify the entailment. Presenting a tiny fragment of the ontology which at the same time is sufficient to cause the entailment greatly improves the understanding of the reason behind the entailment.

ACE View is implemented as a plug-in for Protégé 4 and relies heavily on the OWL API [20] that provides a connection to reasoners, entailment explanation support, storage of OWL axioms and SWRL rules in the same knowledge base, etc. The main task of the ACE View plug-in, translating to and from OWL/SWRL, is performed by two external translators — APE web service (see section 6.3.3) and OWL verbalizer[12]. The entity surface forms are automatically generated using SimpleNLG[13].

6.5.3   AceWiki: ACE in a Semantic Wiki

AceWiki[14] is a semantic wiki that uses ACE to represent its content. Figure 5 shows a screenshot of the AceWiki interface. Semantic wikis combine the philosophy of wikis (i.e. quick and easy editing of textual content in a collaborative way over the Web) with the concepts and techniques of the Semantic Web (i.e. giving information well-defined meaning in order to enable computers and people to work in cooperation). The general goal of semantic wikis is to manage formal representations within a wiki environment.

---

[12] http://attempto.ifi.uzh.ch/site/docs/owl_to_ace.html

[13] http://www.csd.abdn.ac.uk/~ereiter/simplenlg/

[14] See [27], [28], and http://attempto.ifi.uzh.ch/acewiki

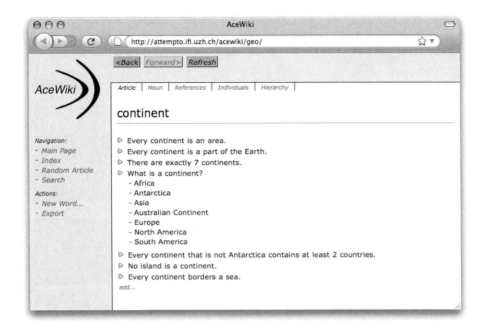

Fig. 5. This screenshot shows an AceWiki article about the concept 'continent'. The content of the article is written in ACE.

There exist many different semantic wiki systems. Semantic MediaWiki [25], IkeWiki [34], and OntoWiki [1] belong to the most mature existing semantic wiki engines. Unfortunately, none of the existing semantic wikis supports expressive ontology languages in a general way. For example, none of them allows the users to define general concept inclusion axioms like 'every country that borders no sea is a landlocked country'. Furthermore, most of the existing semantic wikis fail to hide the technical aspects and are hard to understand for people who are not familiar with the technical terms.

AceWiki tries to solve these problems by using controlled natural language. Ordinary people who have no background in logic should be able to understand, modify, and extend the formal content of a wiki. Instead of enriching informal content with semantic annotations (as many other semantic wikis do), AceWiki treats the formal statements as the primary content of the wiki articles. The use of controlled natural language allows us to express also complex axioms in a natural way.

The goal of AceWiki is to show that semantic wikis can be more natural and at the same time more expressive than existing semantic wikis. Naturalness is achieved by representing the formal statements in ACE. Since ACE is a subset of natural English, every English speaker can immediately read and understand the content of the wiki. In order to enable easy creation of ACE sentences, the users are supported by an intelligent predictive text editor [29] that is able to

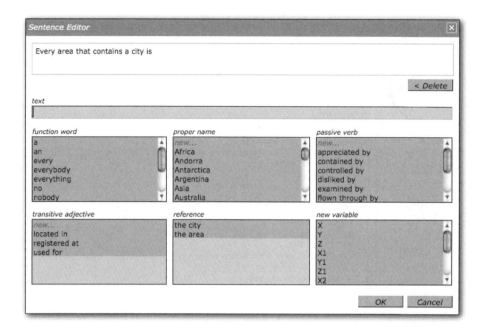

Fig. 6. A screenshot of the predictive editor of AceWiki. The partial sentence 'Every area that contains a city is ...' has already been entered and now the editor shows all possibilities to continue the sentence. The possible words are arranged by their type in different menu boxes.

look ahead and to show the possible words to continue the sentence. Figure 6 shows a screenshot of this editor.

In AceWiki, words have to be defined before they can be used. At the moment, five types of words are supported: proper names, nouns, transitive verbs, *of*-constructs (i.e. nouns that have to be used with *of*-phrases), and transitive adjectives (i.e. adjectives that require an object). Figure 7 shows the lexical editor of AceWiki that helps the users in creating and modifying word forms in an appropriate way.

Most sentence that can be expressed in AceWiki can be translated into OWL. Some examples are shown here:

  ▷ Every country that borders no sea is a landlocked-country.
  ▷ Switzerland borders exactly 5 countries.
  ▷ No city that is located in Europe is controlled by the USA.
  ▷ If X borders Y then Y borders X.
  ▷ Every moon orbits a planet or orbits a dwarf-planet.

AceWiki relies on the ACE→OWL translation that has been introduced in Section 6.4.1. The OWL reasoner Pellet[15] is seamlessly integrated into AceWiki,

---

[15] http://clarkparsia.com/pellet/

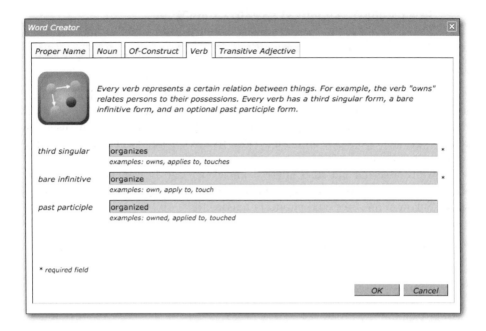

Fig. 7. The lexical editor of AceWiki helps the users to define the word forms. The example shows how a new transitive verb — "organize" in this case — is created.

so that reasoning can be done within the wiki environment. Since only OWL compliant sentences can be considered for reasoning, the sentences that are outside of OWL are marked with a red triangle:

> ▷ No ocean borders every continent.
> ▷ Every person that has a car owns the car or leases the car.
> ▷ If Berlin is a capital then Germany is a stable country.
> ▷ Every trip that starts at X and that ends at X is a round trip.

In this way, it is easy to explain to the users that only the statements that are marked by a blue triangle are considered when the reasoner is used. We plan to provide an interface that allows skilled users to export the formal content of the wiki and to use it within an external reasoner or rule-engine. Thus, even though the statements that are marked by a red triangle cannot be interpreted by the built-in reasoner they can still be useful.

Consistency checking plays a crucial role because any other reasoning task requires a consistent ontology in order to return useful results. In order to ensure that the ontology is always consistent, AceWiki checks every new sentence — immediately after its creation — whether it is consistent with the current ontology. Otherwise, the sentence is not included in the ontology:

▷ Every country is a part of exactly 1 continent.
▷ Every country that borders Switzerland is a part of Europe.
▷ Germany borders Switzerland.
▷ Germany is a part of Asia.

After the user created the last sentence of this example, AceWiki detected that it contradicts the current ontology. The sentence is included in the wiki article but the red font indicates that it is not included in the ontology. The user can remove this sentence, or keep it and try to reassert it later when the rest of the ontology has changed.

Not only asserted but also inferred knowledge can be represented in ACE. At the moment, AceWiki can show inferred class hierarchies and class membership. Furthermore, AceWiki supports queries that are formulated in ACE and evaluated by the reasoner:

▷ Which cities are located in a country that borders Switzerland?
- Berlin
- Milano
- Paris
- Rome
- Vienna

Thus, ACE is used not only as an ontology- and rule-language, but also as a query-language.

A usability experiment [27] showed that people with no background in formal methods are able to work with AceWiki and its predictive editor. The participants — without receiving instruction on how to use the interface — were asked to add general and verifiable knowledge to AceWiki. About 80% of the resulting sentences were semantically correct and sensible statements (in respect of the real world). More than 60% of those correct sentences were complex in the sense that they contained an implication or a negation.

6.5.4 Protune

Chapter 4 describes the Protune framework in detail. Here we simply provide a general overview of the Protune framework, and especially focus on the role of ACE in this framework by building on the concepts introduced in Section 6.4.3.

The Protune framework aims at providing a complete solution for all aspects of policy definition and policy enforcement. Special attention has been given to the interaction with users, be they policy authors or end-users whose requests have been accepted or rejected. For policy authors a set of tools is available to ease the task of defining policies. For end-users a number of facilities are provided to explain why a request was accepted or rejected.

In the following we describe the tools provided by the Protune framework for policy authors, namely

Protune editor: It allows advanced users to exploit the full power of the Protune language by directly providing Protune code

ACE front-end for Protune: It enables users familiar with ACE but not with
Protune to define Protune policies

Predictive editor: It provides a user interface which guides non-expert users
toward the definition of syntactically correct policies (under development)

Advanced users can exploit the Protune editor for policy authoring. The editor helps them to avoid annoying syntactical errors, and provides facilities like
syntax highlighting, visualization of error/warning/todo messages, automatic
completion, outlining, as well as other facilities that come for free with a rich
client platform. A demo of the Protune editor can be found online[16].

The Protune editor is intended for users who already have some knowledge of
the Protune policy language. For others users an ACE front-end for Protune has
been developed that allows them to define policies by means of the subset of ACE
described in Section 6.4.3. Such natural language policies can then be automatically translated into semantically equivalent Protune policies, and enforced by
the Protune framework. The ACE→Protune compiler provides a command-line
interface that translates an input ACE policy into the corresponding Protune
policy, or if an error occurs, shows error messages like

Within the scope of negation-as-failure only one single predicate is
allowed.

or

Only "be" can be used as relation in the "then"-part of an implication.

Messages like these are shown if a syntactically correct ACE sentence cannot be
translated into a valid policy. For incorrect ACE sentences the error messages
provided by APE (cf. 6.3.3) are returned to the user.

The command-line interface that we just described assumes that the user is
already familiar with ACE. For unexperienced users a predictive editor like the
one described in Section 6.5.3 would be more advisable. A predictive editor for
the subset of ACE defined in Section 6.4.3 is in development.

Although the facilities described above have been designed in order to target
different categories of users, they can benefit any user. Expert users might want
to exploit the ACE front-end for Protune in order to define policies in a more
intuitive way and maybe fine-tune the automatically generated Protune policies
later. On the other hand, novice users might want to switch from the predictive
editor to the command-line interface as soon as they get sufficiently familiar with
the ACE language.

6.6   Conclusions

We showed how controlled natural languages in general and ACE in particular
can bridge the usability gap between the complicated Semantic Web machinery

---

[16] http://policy.l3s.uni-hannover.de:9080/policyFramework/protune/

and potential end users with no experience in formal methods. Many tools have been developed around ACE in order to use it as a knowledge representation and reasoning language for the Semantic Web, and for other applications.

The ACE parser is the most important tool. It translates ACE texts into different forms of logic, including the Semantic Web standards OWL and SWRL. AceRules shows how ACE can be used as a practical rule language. We presented RACE that is a reasoner specifically designed for reasoning in ACE. AceWiki demonstrates how controlled natural language can make semantic wikis at the same time expressive and very easy to use. We showed how ACE can help in defining policies by providing a front-end for the Protune policy language. Last but not least, ACE View is an ontology editor that shows how ontologies can be managed conveniently in ACE. The large number of existing tools exhibits the maturity of our language.

Evaluation of the AceWiki system showed that ACE is understandable and usable even for completely untrained people. More user studies are planned for the future.

If the vision of the Semantic Web should become a reality then we have to provide user-friendly interfaces. The formal nature of controlled natural languages enables to use them as knowledge representation languages, while preserving readability. Our results show how controlled natural language can bring the Semantic Web closer to its potential end users.

References

1. Auer, S., Dietzold, S., Riechert, T.: OntoWiki — A Tool for Social, Semantic Collaboration. In: Cruz, I., Decker, S., Allemang, D., Preist, C., Schwabe, D., Mika, P., Uschold, M., Aroyo, L.M. (eds.) ISWC 2006. LNCS, vol. 4273, pp. 736–749. Springer, Heidelberg (2006)
2. Bernardi, R., Calvanese, D., Thorne, C.: Lite Natural Language. In: IWCS-7 (2007)
3. Bernstein, A., Kaufmann, E., Fuchs, N.E.: Talking to the Semantic Web — A Controlled English Query Interface for Ontologies. AIS SIGSEMIS Bulletin 2(1), 42–47 (2005)
4. Bernstein, A., Kaufmann, E., Fuchs, N.E., von Bonin, J.: Talking to the Semantic Web — A Controlled English Query Interface for Ontologies. In: 14th Workshop on Information Technology and Systems, December 2004, pp. 212–217 (2004)
5. Blackburn, P., Bos, J.: Working with Discourse Representation Structures. In: Representation and Inference for Natural Language: A First Course in Computational Linguistics, vol. 2 (September 1999)
6. Bos, J.: Computational Semantics in Discourse: Underspecification, Resolution, and Inference. Journal of Logic, Language and Information 13(2), 139–157 (2004)
7. Clark, P., Harrison, P., Jenkins, T., Thompson, J., Wojcik, R.H.: Acquiring and Using World Knowledge Using a Restricted Subset of English. In: FLAIRS 2005, pp. 506–511 (2005)
8. De Coi, J.L.: Notes for a possible ACE → Protune mapping. Technical report, Forschungszentrum L3S, Appelstr. 9a, 30167 Hannover (D) (July 2008)
9. De Coi, J.L., Olmedilla, D.: A Review of Trust Management, Security and Privacy Policy Languages. In: Proceedings of the 3rd International Conference on Security and Cryptography (SECRYPT 2008). Springer, Heidelberg (2008)

10. Cregan, A., Schwitter, R., Meyer, T.: Sydney OWL Syntax — towards a Controlled Natural Language Syntax for OWL 1.1. In: Golbreich, C., Kalyanpur, A., Parsia, B. (eds.) 3rd OWL Experiences and Directions Workshop (OWLED 2007). CEUR Proceedings, vol. 258 (2007)

11. Dimitrova, V., Denaux, R., Hart, G., Dolbear, C., Holt, I., Cohn, A.: Involving Domain Experts in Authoring OWL Ontologies. In: Sheth, A.P., Staab, S., Dean, M., Paolucci, M., Maynard, D., Finin, T., Thirunarayan, K. (eds.) ISWC 2008. LNCS, vol. 5318, pp. 1–16. Springer, Heidelberg (2008)

12. Fuchs, N.E., Kaljurand, K., Kuhn, T.: Discourse Representation Structures for ACE 6.0. Technical Report ifi-2008.02, Department of Informatics, University of Zurich, Zurich, Switzerland (2008)

13. Funk, A., Davis, B., Tablan, V., Bontcheva, K., Cunningham, H.: D2.2.2 Report: Controlled Language IE Components version 2. Technical report, University of Sheffield (2007)

14. Funk, A., Tablan, V., Bontcheva, K., Cunningham, H., Davis, B., Handschuh, S.: CLOnE: Controlled Language for Ontology Editing. In: Aberer, K., Choi, K.-S., Noy, N., Allemang, D., Lee, K.-I., Nixon, L.J.B., Golbeck, J., Mika, P., Maynard, D., Mizoguchi, R., Schreiber, G., Cudré-Mauroux, P. (eds.) ISWC 2007. LNCS, vol. 4825, pp. 142–155. Springer, Heidelberg (2007)

15. Gelfond, M., Lifschitz, V.: The stable model semantics for logic programming. In: Proceedings of the 5th International Conference on Logic Programming, pp. 1070–1080. MIT Press, Cambridge (1988)

16. Gelfond, M., Lifschitz, V.: Classical negation in logic programs and disjunctive databases. New Generation Computing 9, 365–385 (1990)

17. Grosof, B.N.: Courteous logic programs: Prioritized conflict handling for rules. Technical Report RC 20836, IBM Research, IBM T.J. Watson Research Center (December 1997)

18. Hart, G., Johnson, M., Dolbear, C.: Rabbit: Developing a Controlled Natural Language for Authoring Ontologies. In: Bechhofer, S., Hauswirth, M., Hoffmann, J., Koubarakis, M. (eds.) ESWC 2008. LNCS, vol. 5021, pp. 348–360. Springer, Heidelberg (2008)

19. Hirtle, D.: TRANSLATOR: A TRANSlator from LAnguage TO Rules. In: Canadian Symposium on Text Analysis (CaSTA), Fredericton, Canada (October 2006)

20. Horridge, M., Bechhofer, S., Noppens, O.: Igniting the OWL 1.1 Touch Paper: The OWL API. In: Golbreich, C., Kalyanpur, A., Parsia, B. (eds.) 3rd OWL Experiences and Directions Workshop (OWLED 2007). CEUR Proceedings, vol. 258 (2007)

21. Horridge, M., Drummond, N., Goodwin, J., Rector, A., Stevens, R., Wang, H.H.: The Manchester OWL Syntax. In: 2nd OWL Experiences and Directions Workshop (OWLED 2006) (2006)

22. Kaljurand, K.: Attempto Controlled English as a Semantic Web Language. PhD thesis, Faculty of Mathematics and Computer Science, University of Tartu (2007)

23. Kalyanpur, A., Parsia, B., Sirin, E., Grau, B.C.: Repairing unsatisfiable concepts in OWL ontologies. In: Sure, Y., Domingue, J. (eds.) ESWC 2006. LNCS, vol. 4011, pp. 170–184. Springer, Heidelberg (2006)

24. Kamp, H., Reyle, U.: From Discourse to Logic. Introduction to Modeltheoretic Semantics of Natural Language, Formal Logic and Discourse Representation Theory. Kluwer Academic Publishers, Dordrecht (1993)

25. Krötzsch, M., Vrandečić, D., Völkel, M., Haller, H., Studer, R.: Semantic Wikipedia. Web Semantics: Science, Services and Agents on the World Wide Web 5(4), 251–261 (2007)

26. Kuhn, T.: AceRules: Executing Rules in Controlled Natural Language. In: Marchiori, M., Pan, J.Z., de Sainte Marie, C. (eds.) RR 2007. LNCS, vol. 4524, pp. 299–308. Springer, Heidelberg (2007)
27. Kuhn, T.: AceWiki: A Natural and Expressive Semantic Wiki. In: Semantic Web User Interaction at CHI 2008: Exploring HCI Challenges (2008)
28. Kuhn, T.: AceWiki: Collaborative Ontology Management in Controlled Natural Language. In: Proceedings of the 3rd Semantic Wiki Workshop. CEUR Proceedings, vol. 360 (2008)
29. Kuhn, T., Schwitter, R.: Writing Support for Controlled Natural Languages. In: Proceedings of the Australasian Language Technology Workshop (ALTA 2008) (2008)
30. Lukichev, S., Wagner, G., Fuchs, N.E.: Deliverable I1-D11. Tool Improvements and Extensions 2. Technical report, REWERSE (2007), http://rewerse.net/deliverables.html
31. Manthey, R., Bry, F.: SATCHMO: A Theorem Prover Implemented in Prolog. In: Lusk, E.'., Overbeek, R. (eds.) CADE 1988. LNCS, vol. 310, pp. 415–434. Springer, Heidelberg (1988)
32. Pratt-Hartmann, I.: A two-variable fragment of English. Journal of Logic, Language and Information 12(1), 13–45 (2003)
33. Rector, A.L., Drummond, N., Horridge, M., Rogers, J., Knublauch, H., Stevens, R., Wang, H., Wroe, C.: OWL Pizzas: Practical Experience of Teaching OWL-DL: Common Errors & Common Patterns. In: Motta, E., Shadbolt, N.R., Stutt, A., Gibbins, N. (eds.) EKAW 2004. LNCS, vol. 3257, pp. 63–81. Springer, Heidelberg (2004)
34. Schaffert, S.: IkeWiki: A Semantic Wiki for Collaborative Knowledge Management. In: Proceedings of the First International Workshop on Semantic Technologies in Collaborative Applications (STICA 2006), pp. 388–396 (2006)
35. Schwitter, R., Kaljurand, K., Cregan, A., Dolbear, C., Hart, G.: A Comparison of three Controlled Natural Languages for OWL 1.1. In: 4th OWL Experiences and Directions Workshop (OWLED 2008), DC, Washington, April 1-2 (2008)
36. Schwitter, R., Tilbrook, M.: Let's Talk in Description Logic via Controlled Natural Language. In: Logic and Engineering of Natural Language Semantics 2006 (LENLS 2006), Tokyo, Japan, June 5-6 (2006)
37. Sowa, J.F.: Knowledge Representation: Logical, Philosophical, and Computational Foundations. Brooks Cole Publishing Co., Pacific Grove (2000)
38. Sowa, J.F.: Common Logic Controlled English. Technical report (2007), http://www.jfsowa.com/clce/clce07.htm (Draft, March 15, 2007)
39. Wagner, G.: Web Rules Need Two Kinds of Negation. In: Bry, F., Henze, N., Maɫuszyński, J. (eds.) PPSWR 2003. LNCS, vol. 2901, pp. 33–50. Springer, Heidelberg (2003)
40. Wagner, G., Giurca, A., Lukichev, S.: A Usable Interchange Format for Rich Syntax Rules Integrating OCL, RuleML and SWRL. In: Hitzler, P., Wache, H., Eiter, T. (eds.) RoW 2006 Reasoning on the Web Workshop at WWW 2006 (2006)

Chapter 7
Semantic Search with GoPubMed

Andreas Doms and Michael Schroeder

Technical University of Dresden, Germany
adoms@biotec.tu-dresden.de
http://biotec.tu-dresden.de

Abstract. Searching relevant information on the web is a main occupation of researchers nowadays. Classical keyword-based search engines have limits. Inconsistent vocabulary used by authors is not handled. Relevant information spread over multiple documents can not be found. An overview over an entire document collection can not be given by the means of ranked lists. Question answering requiring semantic disambiguation of occurring terminology is not possible. Trends in the literature can not be followed if vocabulary is evolving over time.

GoPubMed is a semantic search engine using the background knowledge of ontologies to index the biomedical literature. In this chapter we discuss how semantic search can contribute to overcome the limits of classical search paradigms.

Keywords: Search Engines, Ontologies, Life Sciences, Question Answering, Expert Knowledge, Literature Trends.

7.1 Biomedical Literature Search

A major goal in the post-genome era is the exploration of the order and logic of genetic programs (11). Advances in sequencing technology made genomes of many organisms available. High-throughput experiments create masses of data which can be mined for new insights into biological programs.

Despite the fact that ever more biomedical knowledge is stored in structured databases (58) including genome sequences, molecular structures, biological pathways, protein interactions and gene expression arrays most of the biomedical knowledge available nowadays is still only accessible through unstructured text in scientific publications.

The biomedical literature grows at a tremendous rate and PubMed comprises already over 18.000.000 abstracts. Finding relevant literature is an important and difficult problem with up to 5,000 new citations in PubMed every day. Biomedical textming aims to manage this information blast (7).

Recently much research has been devoted to the analysis of the biomedical literature. This interest has been sparked by the growth in literature, but also by the availability of abstracts, full papers, and bibliometric data. Researchers have been specifically interested in automatically extracting information from free text such as protein names (22; 60; 31), ontology terms (55; 8; 16), and

F. Bry and J. Maluszynski (Eds.): Semantic Techniques for the Web, LNCS 5500, pp. 309–342, 2009.

protein interactions (57; 21; 32). These techniques are then applied to aid or even automate the annotation of the representations of proteins in databases.

Ontologies are increasingly used to capture biological knowledge. (26) defines an ontology as an explicit specification of a conceptualization. In (38) the authors give examples of biological ontologies and ontology-based knowledge. A prominent example of an ontology is the Gene Ontology (GO) (24), which provides a hierarchical vocabulary for function, processes and cellular locations. GO is used to annotate proteins in biological databases such as the sequence database UniProt (3) and the protein structure databank PDB (6).

7.1.1 Limits of Classical Search

Finding relevant literature is an important and difficult problem. The amount of literature available online today is enormous. Ingenta (`www.ingenta.com`), an online index of 17,000 periodicals, has 7 million articles going back to 1988. Infotrieve (`www.infotrieve.com`) indexes over 20,000 journals with 15 million citations. CiteSeer (`www.citeseer.com`), a digital library, covers over one million articles and 22 million citations. Other important examples of literature search engines are Google Scholar, Scopus, Scirus and Forschungsportal.Net. Databases for scientific literature are growing at an astonishing rate. PubMed, a biomedical literature database, has grown by 754,003 cited documents in the last year and covers now more than 18 million abstracts of scientific literature, although about half of them are retractions and corrections (55). With 871 million searches in PubMed in 2007 it is clear that researchers spend a considerable amount of time searching the scientific literature.

Great quantities of knowledge and information are available to researchers through millions of documents. But without effective ways of access a lot of it will remain unnoticed by the readers due to the overwhelming amounts of text. Classical search engines have the following limitations:

- Classical search engines do not provide search results spread over multiple documents. The answers are a ranked list of single documents represented as a text excerpt and a URL.
- The search result is a ranked hitlist. The costs for the user, the time spent to find relevant documents, quickly add up if the keywords are relevant for multiple topics. To increase the precision[1] the user has to refine his query iteratively, which typically reduces recall[2] at the same time.
- Results comprising many relevant documents are poorly represented by a few documents in the beginning of a hitlist.
- Classical search engines are unaware of synonyms and relational information of common terminology. Some engines aim at producing a higher recall by expanding the query with synonyms, this technique typically reduces the precision of the systems.
- Meta information characterizing the entire collection of relevant documents is not part of the response of a classical search engine.

---

[1] The precision measures of the ability of an algorithm to retrieve only relevant entities.
[2] The recall measures the ability of an algorithm to extract all relevant entities.

Table 1. Categorization of Biomedical Search Engines

Information Retrieval		Knowledge Retrieval	
Improved Querying	Results Processing	Tools Integration	Semantic Processing
askMedline	BioIE	HubMed	EBIMed
PubMedInteract	ReleMed	PubFocus	GOAnnotator
PICO Linguist	PubMed PubReMiner	Harvester	Info-PubMed
BabelMeSH	ClusterMed		XploreMed
PubFinder	BioMetaCluster		iHop
CiteXplore			ExpertMapper
			Textpresso
			AliBaba
			Chilibot
			PubGene
			MedStory
			GoPubMed

The following section introduces online available biomedical search engines which provide functionality beyond classical search engines.

7.1.2   Biomedical Search Engines

With the fast growth of the biomedical literature the number of specialized search engines tailored to the needs of medicals and biologists has increased. In 1997, the US government decided to make MEDLINE, the citation catalog of the National Library of Medicine, publicly accessible via the World Wide Web. In 2001, a new URL was introduced www.pubmed.gov. Since then it has become the most popular literature database online. With the introduction of the Entrez Programming Utilities and the availability of citations in XML format since 2000 the number of alternative PubMed interfaces has increased quickly. Biomedical search engines can be classified according to their focus on Information Retrieval support and Knowledge retrieval support (36). However, it is not always possible to separate clearly. Table 1 shows how the biomedical search engines discussed in this section are categorized.

Search engines focusing on Information Retrieval. Information Retrieval is the process of searching for documents or information in documents executed by a human user or automated agent. A system supporting a human user querying is aimed at increasing the ratio between relevant and non-relevant documents upon a query, e.g web search engines. An automatic system's task is the aggregation of filtered information to reduce the number of documents requiring further processing, e.g. customizable RSS feed services.

Improved Querying. PubMed expands user queries using MeSH[3] headings and additional vocabularies such as drugs or chemicals. If a query contains such a

---

[3] The Medical Subject Headings (MeSH) thesaurus is a controlled vocabulary comprising biomedical and health-related topics.

term the query is expanded with the option to include also articles which were manually annotated with this term. In the PubMed interface this expanded query can be reviewed by the user. Also the E-Utilities can be called to compute this expansion. This query expansion helps retrieving relevant articles which otherwise would be missed.

Some tools aim to improve the querying of PubMed by supporting the user during query formulation. Features reach from language translation over graphical aims to pre-processing full English questions:

The text-based website askMEDLINE (19) takes a natural language question as input. The system removes irrelevant words and the remaining words are tested to relate to MeSH headings by querying PubMed. Terms classified as "other eligible entries" are eliminated as well if the remaining search results are few. The result is always a list of citation titles and links to the abstract and full text.

PubMedInteract (42) is a web interface to PubMed and presents slider bars to set PubMed search limits and parameters. A "Preview Count" option computes the number of articles to be expected with the current settings.

PICO Linguist (18) offers non English medicals the option to build a structured clinical query with medical terms that may be difficult to express in English by using the PICO framework. The user may specify the patient's problem, the therapy and alternative therapies and the outcome in his/her own language. Primary sources of vocabularies for translation are UMLS, MeSH, WHO EMRO and UMLF. UMLS is a project which combines several terminologies such as ICD-10 (International Classification of Diseases), MeSH (40), SNOMED (45), LOINC (20), Gene Ontology (4) and OMIM (29) into one resource. The Metathesaurus contains concepts, concept names, and other attributes from more than 100 terminologies, classifications, and thesauri; some in multiple editions.

The BabelMeSH (18) website maps search terms to a multilingual MeSH in 12 different languages. Only terms listed in the multilingual vocabulary can be used for the query.

The PubFinder service (25) aims to automatically extract Pubmed abstracts that deal with a specific scientific subject. The user enters a representative set of PubMed ids. Based on the abstracts, a list of discriminating words is calculated which is used for ranking Pubmed abstracts for their probability of belonging to the user defined topic. The first 100 words exhibiting the highest difference in occurrence between both the global PubMed frequency of a word in a reference dictionary and the frequency of a word in the selected abstracts make the list of discriminating words. A set of abstracts dealing with literature mining contains, for example, these words: abstracts, medline, information, articles, names, precision, database, recall, protein, literature, databases, references, system, automatically, interactions, set, mining, scientific, automated, motivation and others.

CiteXplore indexes documents from sources like Medline, European Patent Office, Chinese Biological Abstracts and Citeseer using the Lucene full text index. Advanced searches such as wildcard search on selected attributes is offered. Another option is the expansion with synonyms. Information gathered from other applications such as InterPro, SwissProt/Trembl and Alternative Splicing is cross referenced. The external WhatIsIt textmining service is used to highlight proteins, genes and protein-protein interactions. The references can be exported to EndNote, RIS and Bibtex format.

Results Processing. Some systems process search results further to facilitate browsing of a large number of documents or link to further related citations based on the content of the search result. Examples are evidence highlighting, document re-ranking and information organization. Evidence highlighting visually emphasizes text passages in source documents. For example, the word in a sentence stating a relation of two entities is underlined. Readers are supported when scanning through relevant text passages. Documents can be sorted according to selected criteria such as date, type of citation, usage of vocabulary or reputation. Hyperlinks to documents not in the original search result, for example referenced papers or papers with similar content are linked, support researchers in finding all relevant material. Information organization is the process of organizing information such that it becomes useful. For example tables or network graphs support understanding.

BioIE is a rule-based system that extracts informative sentences from MEDLINE document or uploaded texts. Informative sentences refer to structures, functions, diseases and therapeutic compounds, localisations or familial relationships of biological entities, particularly proteins. The selected text base can be visualized in tabular form as word, MeSH term and word phrase frequency tables. Textual templates are used to identify informative sentences of a selected type, e.g. functional descriptions. The sentences can be further filtered for cooccurrence with additional keywords.

ReleMed (54) expands a user's query automatically using UMLS and MeSH. Names of proteins and genes are expanded as well. Also lexical variants of words are generated. The user has the option to undo these expansions selectively. Matches in separate sentences are highlighted. ReleMed uses the relational MySql database to implement a full text index over single sentences. MeSH headings associated with the abstracts are concatenated and treated as an additional sentence. The relevance of an article is defined in eight levels depending on the cooccurrence of all keywords in one or more sentences.

PubMed PubReMiner (37) shows the user journals in which his/her keywords are mentioned the most. It displays authors publishing the most articles mentioning the keywords. It shows words that have been used most in the title and abstract of the articles. Queries can be refined based on document attributes such as address, substances, MeSH headers, publication year, author and others.

Vivisimo applies clustering methods in ClusterMed and BioMetaCluster (56). In ClusterMed PubMed results are clustered in various ways. Document distances are computed based on strings in (1) title, abstract, and Medical Subject Headings, (2) title, abstract only, (3) MeSH only, (4) author's name only, (5) affiliation only and (6) date of publication only. Vivisimo uses words found in this document's attributes to label clusters. The clusters are ordered by the number of documents contained in them. The cluster hierarchy is computed using statistical language processing. For the query *rab5* ClusterMed returned several clusters such as Vacuoles, Phagosomes, Rabaptin-5, Rab5a and others. The labels are computed on the basis of word occurrence statistics in the retrieved article abstracts. The cluster "Rabaptin-5" contains sub-clusters such as "Ubiquitin", "GAT domain", "Vesicular transport", "Nucleotide exchange", "Dimerization Of Rabaptin-5", "Endocytic membrane fusion", "Correlated, Tissue", "FRET microscopy", "Cleaved in apoptotic" and other labels.

ClusterMed gives the option to compute the clusters only on the MeSH headings. The same string based clustering techniques are applied but using only words from MeSH. A clustering result for the query *rab5* displays clusters labeled with MeSH headings such as "Guanine Nucleotide Exchange Factors", "Virology" and "Pathology" but also concatenated labels such as "Analysis, Liver", "Chromatography, Affinity, Cattle", "Phagosomes, Microbiology" which do not correspond to a single MeSH heading or sub-heading but to a combination of them. A cluster does not necessarily comprise sub-clusters reflected by a relation in the UMLS. In the examples the cluster "Guanine Nucleotide Exchange Factors" comprises labels of cellular components, diseases, peptides, proteins and algorithms. The clustering algorithm grouped them on the basis of statistical co-occurrence in the result set. No information about relations between headings is used.

Another feature of ClusterMed is the clustering by authors. Here, the strings of the last name plus the initials are clustered. Sub-clusters contain co-authors. The clusters may contain PubMed citations of different authors with same last name and initials.

BioMetaCluster is a meta search engine based on the Vivisimo clustering architecture. It queries 22 web resources relevant for the biomedical domain using string based clustering of the search results.

Search Engines Focusing on Knowledge Retrieval. In (59) the authors compare knowledge retrieval systems and define their task as finding knowledge from information and organizing it into structures that humans can use. Following (1), the content of the human mind can be classified into data, information, knowledge, understanding and wisdom. (1) Data simply exists and has no significance beyond its existence. Symbols such as raw numbers are data. (2) Information is data that has been given meaning by way of relational connection. It adds context. This meaning does not need be useful. A relational database holds structured relational data. (3) Knowledge is the appropriate collection of information, such that its intention is to be useful. Computer programs modeling or simulating some process apply knowledge. (4) Understanding is an interpolative and probabilistic process. With understanding one can synthesize new knowledge or at least information. (5) argue that some artificial intelligence systems can generate new knowledge and are therefore "understanding". (5) Wisdom is an extrapolative and non-deterministic, non-probabilistic process. Read the essay from (53) for an interesting philosophical discussion. (5) suggest an interpretation of the concepts as shown in figure 1. The transitions from data to information, to knowledge, and finally to wisdom is achieved by understanding. Some systems focus on the integration of external programs to achieve these transitions.

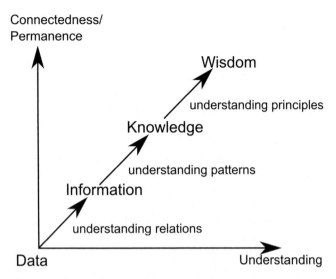

Fig. 1. The diagram shows the transitions from data to information, to knowledge, and finally to wisdom. Understanding is not a separate level of its own, it is necessary for the transition from each stage to the next. Figure adopted from (5). (1) points out that information ages quickly, knowledge has a longer time span, only wisdom is permanent.

Tools Integration. Biomedical search engines that process search results externally call API methods or web services to analyze single citations or batches of citations. These features are influenced by the ideas of the Semantic Web. Some services allow for various data exchange formats such as RDF, XML, BibTex, RIS, Endnote and plain text.

HubMed (17) offers a battery of external tools to process PubMed results. The search behaves exactly like the original in PubMed if one chooses to use the option sort by date. An alternative is the option sort by relevance. Here, an Apache Lucene index is employed but no MeSH headings are expanded as is done via the PubMed search. The option sort by relevance seems to favor citations containing all query keywords. Some internal tools help managing references. A clipboard stores a set of citations for later reuse. The history function enables the recovery of previous searches. Citations can be tagged with arbitrary keywords for later filtering for them. Moreover, tags of other users may be used.

The search can be narrowed or widened with the most closely related words. The relatedness is computed using a tf-idf ranking (35) of the words of the first 500 citations. Another option to manipulate the original query is the clustering feature of HubMed. The Lingo algorithm (39) is used to cluster the first 200 citations of the original query. The clusters are linked back to the first 20 citations of each cluster.

An external utility called by HubMed is the Entrez' ELink utility. For some articles, generally if an article refers to the discovery or sequencing of a gene or protein, the inter-database links are presented by HubMed. Another tool employed is the Whatizit web service (51), which recognizes terms such as protein names and biological processes, linking them to services such as UniProt and Gene Ontology. Citations in the clipboard can be visualized with the Touch-Graph Java applet[4]. HubMed focuses on improved querying but also on tool integration and thereby supporting knowledge retrieval.

PubFocus (48) is a bibliometric statistics tool integrating external data and web services to process PubMed searches and provide ranked lists of prominent authors, cities and journals. The ranking of citations is based on journal impact factors, volume of forward references, referencing dynamics and authors' contribution level. PubFocus uses the non-free Journal Citation Reports Impact Factors published by Thomson Scientific. Forward citation information is based on PubMed Central and Google Scholar. The data retrieval is executed online by parsing the external website's HTML output. The authors define several indexes such as citations-over-age index, the Combined Impact Factor, the Cumulative Impact Factor and the Author's Rank. Terms of the NCI thesaurus and the MGD mammalian gene ontology database occurring within titles and/or abstracts of citations are extracted using a MySQL full text search. Statistics for such terms are also integrated in the web interface and serve for further refinement of the initial query.

Harvester crawls and cross-links the following web resources: BLAST, CDART, CDD, ensEMBL, Entrez, GenomeBrowser, gfp-cDNA, Google-Scholar, GoPubMed, H-Inv, HomoloGene, Hwiki-Forum, iHOP, IPI, MapView, Mitocheck, OMIM, PolyMeta, PSORT II, SMART, SOSUI, SOURCE, STRING, Unigene, UniprotKB, Wikipedia. Harvester cross-links public bioinformatic databases and prediction servers to provide fast access to protein specific bioinformatic information.

Semantic Processing. Various techniques to overcome the semantic gap between text and its meaning are employed by biomedical search engines. A strong assumption made by some tools is that co-occurrence of biological entities in a sentence potentially indicates an observation or hypothesis of an interaction in vivo. Biological entities have a highly ambiguous terminology. Disambiguation techniques aim at solving this problem. Some tools make use of relations of concepts in taxonomies or ontologies and employ reasoning techniques. If findings can be confirmed to be significant such information is aggregated and can be seen as knowledge and used for question answering. A form of dialog can guide knowledge retrieval by directing what kind of knowledge is requested. Finally, some tools experiment with hypothesis generation, used knowledge and some statistical signals to present potentially new knowledge which needs to be confirmed.

With EBIMed (52) identify associations between protein/gene cellular components, biological processes, molecular functions, drugs and species. Results are presented in tabular form. Sentences supporting the associations are cited. The tabular form of presenting many relations between biological entities

---

[4] sourceforge.net/projects/touchgraph

ordered by their frequency is a way of providing the user with a quick overview. Such an accumulated form of information can support literature search for associations because the user reads mainly relevant sentences. The authors claim that EBIMed is complementary to PubMed in the sense that large result sets in PubMed are tedious to read through while EBIMed's tables are expected to be of more help for larger sets of citations.

GOAnnotator aims to support database curators. Their task is to confirm automatic database curations by linking annotations to experimental results described in peer-review publications. The tool provides textual evidences for gene products which have already uncurated automatically generated annotations and links the uncurated annotations to texts extracted from literature thus supporting GO curators in choosing the correct terms. GOAnnotator is utilising the hierarchical structurte of GO and can also suggest alternative, i.e. more precise annotations. The precision of the system is high due to the focused search with the previously annotated similar concept.

Info-PubMed provides information from Medline on protein-protein interactions. Given the name of a gene or protein, it shows a list of the names of other genes/proteins which co-occur in sentences from Medline, along with the frequency of co-occurrence. Information Extraction techniques are used to identify a set of sentences which clearly indicate interactions. The user interface allows to collect statements in a drag&drop manner and to visualize them using an external tool.

XplorMed (46) allows the user to explore a set of Medline abstracts. Three entry points are offered: Medline search, a set of document PMIDs or via a database entry with associations to Medline. The system then classifies the abstracts based on the top hierarchy of the MeSH terms associated to each MEDLINE entry. One or more top level categories of MeSH can be selected. The associated articles are now analyzed for words that are significantly related to others in this subset of articles. A main contribution of XplorMed is the functionality to explore the vocabulary used in a set of articles. The context of prominent words from the abstracts can be visualized. This is a list of other words in the texts which appear frequently together with such words. Based on this analysis, chains of words, an ordered set of words where the 2nd depends on the 1st, the 3rd depends on the 2nd, an so on, can be selected. The selected chain of words may be used to re-rank the initial Medline citations. The more chain words appear in an abstract the higher it is ranked. The authors claim that XplorMed can be used for sets of articles for which the user does not initially know what to expect. Prominent vocabulary is revealed, which later is used for re-ranking.

Information Hyperlinked over Proteins (33) is a website offering hyperlinked navigation of PubMed abstracts via gene/protein mentions in sentences. Upon a search the user is presented a list of sentences containing concurrent gene/protein mentions. An interaction of the concurring entities is assumed and the predicted type of the interaction is highlighted. In case of existence of large scale experimental evidence of an interaction this is indicated as well by a link to the experimental results. Gene/protein name disambiguation is a difficult task (60).

iHop enables users to verify its findings by highlighting the entity names in the original sentences. Thus, research will be able to confirm the findings. Three levels of confidence of the algorithm are indicated. The user may create gene models by appending interactions of genes/proteins to a graphical representation. While iHop is an Information Retrieval tool as it displays PubMed sentences for gene/protein mentions it additionally disambiguates such entities. This enables a semantic connection between sentences also via synonymous labels of the same entity. Furthermore, this can be used to create interaction networks manually or automatically. This is a potential source of new insights into previously published data potentially supporting new hypotheses. Therefore, iHop is recognized as a Knowledge Retrieval tool as well.

For a selection of 105 topics Expert Mapper computed prominent authors from Medline citations of the years 1997 to 2006 grouped into geographical regions. The main contribution of Expert Mapper is the accumulation of affiliation information for an author so that it becomes possible to make a reasonable manual prediction about the identity of an individual.

Textpresso (41) is an ontology-based search engine built of scientific literature on C. elegans and selected others domains. The texts are indexed with biological concepts and relations. The labels fall into 33 categories that comprise the Textpresso ontology. On a second level Textpresso maintains ca. 14.500 regular expressions representing known formulations of relations of a parent category with other entities. A selection of full text articles of selected species is indexed. The user can retrieve sentences mentioning keywords and concepts. Currently, the list comprises 101 concepts, some of which are known from the Gene Ontology. For each category Textpresso maintains a list of regular expressions used to index the texts. Textpresso can retrieve abstracts mentioning a life stage in C. elegans and a cell part. The indexation of an article with a descendant concept implies the indexation also with the ancestor concepts. Each concept in the ontology has its own identification algorithm. Textpresso provides ten-thousands of indexed articles containing more than 222.000 facts.

AliBaba (47) visualizes PubMed as a graph. It parses PubMed abstracts for biological objects and their relations as mentioned in the texts. Ali Baba visualizes the resulting network in graphical form, thus presenting a quick overview over all information contained in the abstracts. A variety of relations between proteins, (sub)cellular locations, genes, drugs, tissues, diseases and others are detected. The extracted relations have a confidence value which can be used for filtering less likely correct associations. The interactive graphical representation allows for human interpretation of high dimensional data.

Chilibot (13) searches PubMed abstracts for specific relationships between proteins, genes, or keywords. The user enters a list of two or more genes or other keywords. PubMed searches for citations mentioning those entities or a synonym in one sentence. The resulting sentences are later categorized into six types: (1) Interactive relationship (stimulative), (2) Interactive relationship (inhibitory), (3) Interactive relationship (both stimulative and inhibitory), (4) Interactive relationship (neutral), (5) Non-interactive (i.e. parallel) relationship and

(6) Abstract co-occurrence only. The relations are then visualized in a 2D graph with colored nodes and edges. The nodes denote the biological entity or keywords, and the edges denote the observed type of relation and its count. The graph is hyperlinked with the original set of sentences the relations were derived from. The user can confirm each relation and remove false edges from the graph.

Another feature of Chilibot is the generation of new hypotheses. Two nodes which are not directly connected to each other can be searched for the missing link. A graph of potential indirect interactions is drawn. The hypothesis is made on the basis of common connections to other entities, built by association, a principle previously used in gene function studies (50). PubGene (34) is similar to Chilibot relying on non-directional interactions. PubGene Webtools allow users to analyze gene expression data with literature network information, browse literature neighbors of a given gene, search literature articles for a set of genes, search ontology terms related to a given gene, search MeSH terms found with a set of genes, and search for official nomenclature.

Medstory[5] groups result items into categories. The categories show users how the results distribute. Each category suggests further topics. Selecting of the sub-topics gives the choice of starting a new search with a narrowed query. The main researchers in the area are listed. Medstory is focused on the non-medical expert.

Summary of Comparision. Table 2 summarizes the features of all compared search engines. The features were selected in order to highlight the different approaches of the tools: (1) PubMed query expansion/refinement: expands MeSH headings and additional vocabularies such as drugs or chemicals, citation metadata, (2) expands gene/protein names with synonyms, (3) offers narrowing/expanding with ontology concepts, (4) language translation of terms, (5) full natural language questions handled, (6) querying with other documents/database cross-references, (7) alternative full text index (Lucene/MySQL), (8) refinement based on metadata derived from initial resultset, (9) meta search in separate databases, (10) refinement based on keywords derived from initial resultset, (11) bypassed normal PubMed query expansion/special PubMed queries, (12) entity specific (genes/proteins), (14) Search in UMLS, (15) Search in MeSH, (16) Search in Gene Ontology, (17) Browse within Taxonomy/Ontology hierarchy, (18) Browse within identified text occurences, (19) Query history, (20) Permanent user account, (21) Session clipboard, (23) links to title, (24) links to abstract, (25) provides external links, (26) shows PMID, (27) shows evidence sentence, (28) shows text snippets, (29) calls external web services, (31) highlighted keywords from query, (32) highlighted biomedical entities/relations, (33) highlighted ontology concepts detected, (34) highlighted vocabular (cluster labels/significant words), (36) Re-ranking based on concurrence of keywords, (37) Re-ranking based on concurrence of identified entities, (38) Re-ranking based on external database references or precomputed statistics, (39) Language structure (e.g. conclusive sentences), (41) Cosin similarity based, (42) based on co-authorship, (43)

---

[5] medstory.com

Table 2. Comparision of Biomedical Search Engines. The main features of the tools are marked in bold.

Biomedical Search Engines	Query transformation/Query refinement	Explore controlled vocabulary	Personalization	Link to original resultset	Evidence highlighting	Re-ranking	Connection to other related documents	Information organization	Special features	Single citation processing	Batch processing	Batch export	Entity recognition	Concurrence/Relations	Disambiguation	Subsumption/Reasoning	Summarization	Question answering	User Dialog	Hypotheses generation
PubMed	1	14, 15, 16	**20, 21**	23, 24, 25			41													
askMedline	5	15		23, 25			41													
PubMed Interact	1			23, 24, 25			41		50											
PICO Linguist	4			23, 24, 25			41													
BabelMesh	4			23, 24, 25			41		51											
Pubfinder	6		20	23, 24, 25				41												
ReleMed	7			23, 25	31	36														
PubMed PubReMiner	**8**			26				47												
ClusterMed	1			23, 24, 25	34	36		44												
BioMetaCluster	9			28	34			44												
PubFocus	1			23, 24		38	42	47			60									
HubMed	7, 10		21	23, 24, 25, **29**	32, 33	36	**41**	48	52	57		59	63, 64, 65	67, 68, 69						
BioIE	1			26, 27	31			47						72						
CiteXplore	2, 7		19	23, 24, 25, 26	32, 33		43			57			64, 65	67, 68	72	74				
iHOP	2, 7	**18**	20	25, 26, 27	**32**	**37, 38**	46				60		67	72			83, 84			
Info-PubMed	12		20	26, 27	32			46, 47, 48	55				67	72					90	
EBIMed				24, 25, 26, 27			47			57			67, 68	72			83			
GoPubMed	1, 2, 3, 8	15, 16, **17**	19, 21	23, 24, 25	31, 33	36, 37	41	45					62, 63, 64, 65	67, 68, 69	71, 72	74, 76	81, 82	89		
AliBaba	1	18		25, 26, 27	32		46, 49					58	67	72	75	78			90	
XplorMed	6			23, 25	34	36		44				58		72						
GOAnnotator	12			23, 26		38	47			58, 60			**68**	72		78				
Textpresso		17		23, 24, 25	34	36	41		53				62, 65	67, 68	72					
Chilibot	11		20	27	31, 32	39		46, 49	54				67	72			84		90	92

via author name, (44) hierarchical classification based on distance metrics, (45) hierarchical classification using taxonomies/ontologies, (46) 2D concept graph, (47) tabular statistics, (48) Call external service, (50) graphical sliders, (51) email communication, (52) social tagging, (53) special query language, (54) batch processing, (55) drag&drop GUI, (57) external markup tool, (58) import literature references from external databases curations, (59) visualization using an external tool, (60) external large scale experimental metadata used,

(62) XML, (63) RDF, (64) BibTex, (65) Endnote (RIS), (67) biomedical entities (e.g. gene/proteins), (68) Taxonomy/Ontology terminology, (69) Wikipedia
terminology, (71) within abstracts, (72) within sentences, (74) disambiguation
for bio-entities, (75) disambiguation for taxonomy/ontology terminology, (76)
disambiguation for authors, (78) is-a generalization, (80) significant strings, (81)
significant taxonomy/ontology concepts, (82) expert profiles, (83) significant bio-
entities, (84) textual synopsis, (87) explicit question answering, (89) question
categories, (90) graphical interaction, (92) explicit hypothesis generation.

Table 2 shows that most biomedical search engines already provide extended
functionality for Information Retrieval while about only half of them provide
some support for Knowledge Retrieval. The Life Sciences are early adopters for
semantic technologies. A number of biomedical search engines offer functionality to process results semantically in order to provide more condensed or more
relevant results to the user. Enitity Recognition and Co-Occurrence analysis
are most widely supported. The background knowlegde of ontologies is not yet
widely used.

7.1.3   The Ontology-Based Search Paradigm

The basis of a semantic search engine is the underlying specific expert knowledge.
This knowledge is captured within one or more ontologies. In contrast to classical
keyword search the results are not presented as a long ordered list of documents
but as an hierarchical index. The concepts defined in the ontologies are identified
in the text of each document. This mapping phase is clearly most crucial for the
quality of the results. Efficient concept recognition algorithms are necessary to
cope with varying morphology and syntax of concept labels. Another important
issue is Word Sense Disambiguation as concept labels may have different meaning
in different contexts (2).

After mapping the documents to the mentioned concepts in the text a hi-
erachical index, further named induced ontology, is computed. The idea is to
compute a graph which connects all concepts found in the annotations of the
retrieved documents to the most general concept defined, the root of the ontology. All concepts of the background knowledge which are part of a path from
an annotated concept all the way up to the root are included in the induced
ontology. The graph contains multiple instances of concepts having more than
one parent concept. The set of concepts Θ represented by the induced ontology
can be defined as follows:

$$\Theta = \bigcup_{1..d}^{i} \bigcup_{1..a(i)}^{t} ancestors(O, concept(i, t)) \tag{1}$$

where d is the number of documents of a result set and $a(i)$ is the number of
annotations in a document. The function *concept* returns the ontology concept
associated with a document annotation and *ancestors* returns the set of concepts
transitively related to the given concept in the ontology O.

The resulting acyclic graph functions as a "table of contents" for the search results. The graph can be explored by the user. At any level a concept can be selected. The documents associated with this node are only those which mention the concept or a descendent concept or a synoynm. Figure 2 shows an example of an Induced Ontology for a query to PubMed using the MeSH and Gene Ontology as background knowledge.

Beside the classical keyword based search paradigm and the ontology-based search paradigm there exists the natural language processing based paradigm. One recently created state-of-the-art NLP search engine is Powerset[6]. It uses the background knowledge of Wikipedia to answer full sentence questions. Powerset analyzes the document corpus using Natural Language Processing techniques. The user enters a full question and the system responds with a list of possible answers in the form of text passages from Wikipedia. The background knowledge used is stored in natural language text not in structured ontologies. The user needs to be able to formulate a question containing enough information for the system to find relevant answers and the system must be able to correctly parse the sentence. It is often difficult for the user to formulate a precise question and often it is even more difficult for parser to understand a complex question.

Figure 3 shows keyword-based and natural language based paradigms in relation to the ontology-based paradigm. The answers of keyword and NLP-based searches are typically single documents, while ontology-based searches offer an outline of all relevant documents. NLP and ontology-based searches use semantic technologies to structure text, while keyword-based search is based on character n-grams. Keyword and ontology-based searches take word phrases as input, while NLP search takes full questions as input.

7.2   Answering Biomedical Questions

When users search they have questions in mind. Answering questions in a domain requires the knowledge of the terminology of that domain. Classical approaches to search do not make use of background knowledge during search. This section describes GoPubMed's approach to uses ontological background knowledge when mining a literature corpus to answer biomedical questions. It is shown that the background knowledge can be used to find more relevant documents and to organize the results in order to focus on important aspects. An aspect is a set of similar statements, e.g. a set of sentences describing the relation between a disease and a drug. The goal is to answer biomedical questions with a minimum of user interactions.

The goal of Information Retrieval systems is to maximize the precision of the results, by minimizing the number of irrelevant documents presented to the user, and to maximize the recall, by minimizing the number of missed relevant documents.

After retrieving relevant documents the user is interested to collect specific information from the documents. Ultimately the user wants to answer a question

---

[6] powerset.com

Induced Ontology for PubMed query 'rab5'

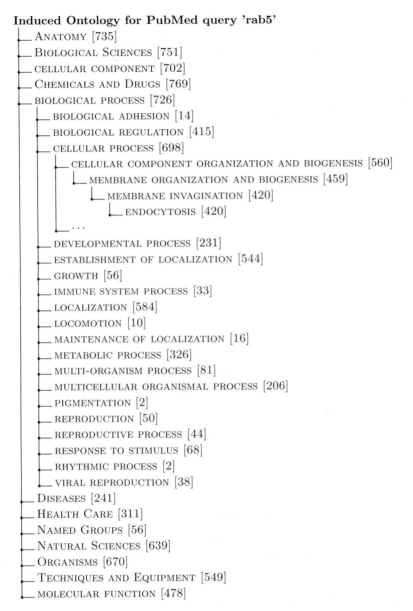

Anatomy [735]

Biological Sciences [751]

cellular component [702]

Chemicals and Drugs [769]

biological process [726]

    biological adhesion [14]

    biological regulation [415]

    cellular process [698]

        cellular component organization and biogenesis [560]

            membrane organization and biogenesis [459]

                membrane invagination [420]

                    endocytosis [420]

        . . .

    developmental process [231]

    establishment of localization [544]

    growth [56]

    immune system process [33]

    localization [584]

    locomotion [10]

    maintenance of localization [16]

    metabolic process [326]

    multi-organism process [81]

    multicellular organismal process [206]

    pigmentation [2]

    reproduction [50]

    reproductive process [44]

    response to stimulus [68]

    rhythmic process [2]

    viral reproduction [38]

Diseases [241]

Health Care [311]

Named Groups [56]

Natural Sciences [639]

Organisms [670]

Techniques and Equipment [549]

molecular function [478]

Fig. 2. This figure shows the induced ontology for the PubMed result to the query "rab5". The top level MeSH and GO categories are alphabetically ordered. The GO branch cellular process is expanded to the concept endocytosis, the topic process most intensively for the protein. The induced ontology shows 16 other biological processes related to the protein rab5. The numbers denote the amount of the linked documents.

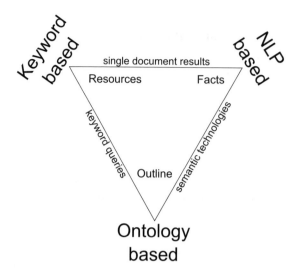

Fig. 3. Three search paradigms. Keyword- and ontology-based search take boolean keyword queries as input while NLP-based search takes full question sentences as input. NLP- and keyword-based search return single documents as results while ontology-based search returns a graph which can be used as an outline of the search results. Ontology- and NLP-bases search use semantic technologies to map documents or parts of documents to formally defined background knowledge. Keyword-based search returns lists of resources, e.g. lists of URLs. NLP-based search returns facts as answers to questions, e.g. the distance between two cities. Ontology-based search returns an outline into a potentially very large result set. The outline organizes the resulting documents hierarchically according to identified topics.

and thereby acquire knowledge. The idea is to organize search results using the structured background knowledge of ontologies, such that its intention is to be useful.

7.2.1 Characterization of Question Types

The user can only expect answers to questions which actually are covered by text passages from the documents or by the defined concepts in the background knowledge. It is assumed that the background knowledge and the documents corpus deal with the same domain. Answers to questions which are short text passages of literature abstracts are named citation answers, answers which can be given in the form of a collection of ontology concepts and their definitions are named glossary answers. The idea in both cases is to reason over the relations captured in the background knowledge to answer questions.

Citation Answers. For example the user wants to know how aspirin works. It is possible to answer this question by providing one or more statements if the corpus contains documents mentioning the drug and an adequate concept of the background knowledge in proximity. The user does not need to know this concept nor its synonyms. The entry point to answer the question will be the categories of the background knowledge. Every level of the hierarchically structured knowledge comprises potential answers. The more familiar a user is with the domain the deeper he or she can select an entry point.

One possible scenario to answer the question is: the user sends the keyword "aspirin" to the system. The system shows a table of content comprising identified concepts in the retrieved documents. The user selects an appropriate concept among the displayed categories in the table of content. An appropriate concept to answer the above question is BIOLOGICAL PROCESS, a rather general concept in the biomedical field. Users familiar with the biomedical domain may select a more specific concepts such as REGULATION OF CELLULAR METABOLIC PROCESS. The system now displays highlighted text passages mentioning the keyword and the selected concepts and its descendants. Possible outcomes for the above question are:

> *Aspirin exerts its unique pharmacological effects by irreversibly acetylating a serine residue in the cyclooxygenase site of prostaglandin-H(2)-synthases (PGHSs). (PMID: 18242581)*

> *NO-donating aspirin inhibits the activation of NF-kappaB in human cancer cell lines and Min mice. (PMID: 18174252)*

In the first passage CYCLOOXYGENASE (PATHWAY), a biological process, is mentioned. In the second passage the ACTIVATION OF NF-KAPPAB, a regulating process in the cell, is mentioned. Aspirin has various effects on biological processes in an organism. Both answers represent correct and useful answers to different audiences depending on their background.

Glossary Answers. A user is interested in learning which diseases are associated to a well known virus. In this scenario documents with mentions of any disease in proximity of the virus are relevant. The answer to the question would be a list of diseases linked to the text passages stating the relation to the virus. The answer might be a ranked list of diseases sorted by the frequency of cooccurrence in the literature.

For the query "HIV" a system might identify the following passage in a publication:

> *Hepatitis B, C seroprevalence and delta viruses in HIV-1 Senegalese patients at HAART initiation (retrospective study). (PMID: 18551596)*

HEPATITIS is a disease defined in the MeSH terminology. MeSH defines 6527 descendants of the concept DISEASE. The citation shows a close relation between HIV infection and this disease.

The following section describes algorithms developed to find answers to biomedical questions based on an semantically annotated literature corpus using ontological background knowledge.

Bibliometric Answers. Another type of questions regards the publishing behaviour in a selected research field. Questions whether a topic is an active research field and which authors are most prominent in a field can only be answered with the entire literature corpus. The problem is that there is no common vocabulary used by researchers. The terminology used varies over time and geographic locations. Section 7.3 shows how ontology-based search can answer bibliometric questions.

7.2.2   Using Background Knowledge to Answer Questions

The TREC Genomics Track 2006 (30) is an annual activity of the information retrieval community aiming to evaluate systems and users. For the evaluation of biomedical search engines a new single task was developed that focused on retrieval of passages (from part to sentence to paragraph in length) with linkage to the source document. Topics are expressed as questions and the systems were measured on how well they retrieve relevant information at the passage, aspect, and document level. The participating systems returned passages linked to source documents. Judges rated the returned passages and grouped them by aspect.

The following questions of the TREC Genomics Track 2006 were answered using GoPubMed. There were three types of queries: (1) questions for the role of a gene or protein in a disease, (2) the inter-relation of biological entities and (3) the biological function of an entity.

For the first type of questions the query keywords were the protein name or synonyms if provided by the algorithm described in (28) which is implemented in GoPubMed. Then clicking in the induced ontology on the respective disease the linked documents ranked according to the aspect are scanned. Only correct evidences on the first page of answers were considered.

For the second type of questions the two entities and their synonyms were entered as keywords and answers from the top categories were considered. The top categories were computed with a "tf-idf" ranking.

For the third type of questions the biological entity was entered as the query string and the respective biological function was selected in the ontology. Again only answers provided on the first page were considered. The TREC questions of all years are numbered. The questions of the year 2006 begin with 160 and end with 187.

Roles of Genes and Proteins in a Diseases: What is the role of PrnP in mad cow disease? (TREC #160). *"Since 2004, significant associations between bovine spongiform encephalopathy (BSE) susceptibility in cattle and frequencies of insertion/deletion (ins/del; indel) polymorphisms within the bovine prion protein gene (PRNP) have been reported. PMID: 18399944"*

GoPubMed presents this answer after searching for *"PrnP"* followed by two intuitive clicks: (1) PrnP is a protein, so we use the protein name expansion by clicking the option: *"Expand your query with synonyms for PrnP"*. (2) The disease branch of MeSH lists the official name among the top concepts, ENCEPHALOPA-THY, BOVINE SPONGIFORM. Figure 4 shows the induced ontology for the query

Fig. 4. This screenshot shows the induced ontology for the query *"PrnP"* in GoP-ubMed. ENCEPHALOPATHY, BOVINE SPONGIFORM, the official name for "Mad Cow Disease" is selected. It links to 77 publications in PubMed. Additionally SCRAPIE, CREUTZFELDT-JAKOB SYNDROME and other diseases are listed in the disease branch of MeSH.

"PrnP". The disease ENCEPHALOPATHY, BOVINE SPONGIFORM, alias Mad Cow Disease, links to 77 publications mentioning the disease and PrnP. Figure 5 shows how GoPubMed visualizes documents in the snippet mode by showing a sentence mentioning the keyword and the selected term or a descendant.

The TREC benchmark lists 32 answers mentioning a mutation in PrnP playing a role in mad cow disease. This shows that the important aspect, mutated variants of PrnP are related to mad cow disease, can be found with a simple search in GoPubMed. Three intuitive user interactions. The induced ontology holds answers to many more aspects related to the topic. For example it links 91 publications to a related disease in humans CREUTZFELDT-JAKOB SYNDROME and SCRAPIE a fatal disease of the nervous system in sheep and goats, 185 links.

71: **Central nervous system extracellular matrix** changes in a **transgenic mouse** model of bovine spongiform encephalopathy.

PMID: 18789733 Related Articles

Costa C et al., Vet J: , 2008

Bovine spongiform encephalopathy (BSE) is a transmissible spongiform encephalopathy characterised by accumulation of resistant prion protein (PrP (BSE)), neuronal loss, spongiosus and glial cell proliferation.

Fig. 5. This screenshot shows the first text snippet the user sees when clicking on ENCEPHALOPATHY, BOVINE SPONGIFORM. The keyword "PrP" is a synonym of "PrnP" and is highlighted in yellow. The disease's synonym, BOVINE SPONGIFORM ENCEPHALOPATHY, is highlighted in green. The snippet is ranked highest because it is a recent publication plus keyword and concept is mentioned in the same sentence.

This shows a major advantage of the ontology-based search: serendipity and overview search. The background knowledge provides an outline of the whole content of the search results.

Inter-relation of Biological Entities: How does BARD1 regulate BRCA1 activity? (TREC #168). " BARD1 regulates BRCA1-mediated transactivation of the p21(WAF1/CIP1) and Gadd45 promoters. (PMID: 18243530)"

GoPubMed presents the title of this publication as an answer after searching for "BRCA1 AND BARD1" followed by one click on the option "Expand your query with synonyms for BRCA1, BARD1". The third text snippet clearly mentions the inter-relation of the two proteins. GoPubMed shows by default not the whole abstract. One sentence is selected as a text snippet which mentions keywords and concepts. In this example no concept was selected but the two keywords are mentioned in the text snippets.

GoPubMed classifies the 132 articles using MeSH and the Gene Ontology. The GO branch BIOLOGICAL PROCESS lists important concepts such as REGULATION OF PROGRESSION THROUGH CELL CYCLE with 67 evidences like: "The BRCA1 tumor suppressor exists as a heterodimeric complex with BARD1, and this complex is thought to mediate many of the functions ascribed to BRCA1, including its role in tumor suppression.". Interestingly the snippet was ranked high because TUMOR SUPPRESSION is a Gene Ontology synonym of REGULATION OF PROGRESSION THROUGH CELL CYCLE.

Another important aspect is represented by the second biological process DNA REPAIR. Clicking on it shows the evidence: "Cells deficient in the Werner syndrome protein (WRN) or BRCA1 are hypersensitive to DNA interstrand cross-links (ICLs), whose repair requires nucleotide excision repair (NER) and homologous recombination (HR). PMID: 16714450". The abstract, shown in figure 6, was ranked high because it mentions the keywords and the concept INTERSTRAND CROSSLINK REPAIR which was identified by the Concept Recognition algorithm. The concept is a synonym of NUCLEOTIDE-EXCISION REPAIR which is a descendant of DNA REPAIR. For the aspect "DNA repair" there are 52 more evidences listed in the induced ontology.

39: **Collaboration of Werner syndrome protein** and BRCA1 in **cellular** responses to **DNA** interstrand cross-links.

PMID: 16714450   Related Articles

Cheng WH, Kusumoto R, Opresko PL, Sui X, Huang S, Nicolette ML, Paull TT, Campisi J, Seidman M, Bohr VA

Journal: Nucleic Acids Res, 34 (9): 2751-60, 2006

Cells deficient in the **Werner syndrome protein** (WRN) or BRCA1 are **hypersensitive** to DNA interstrand cross-links (ICLs), whose repair requires **nucleotide excision repair** (NER) and homologous recombination (HR). However, the roles of WRN and BRCA1 in the repair of DNA ICLs are not understood and the molecular mechanisms of ICL repair at the processing stage have not yet been established. This study demonstrates that WRN **helicase activity**, but not exonuclease **activity**, is required to process DNA ICLs in cells and that WRN cooperates with BRCA1 in the **cellular** response to DNA ICLs. BRCA1 interacts directly with WRN and **stimulates** WRN **helicase** and **exonuclease activities** in vitro. The interaction between WRN and BRCA1 increases in cells treated with **DNA cross-linking** agents. WRN binding to BRCA1 was mapped to BRCA1 452-1079 **amino acids**. The BRCA1/BARD1 complex also associates with WRN in vivo and **stimulates** WRN **helicase activity** on forked and **Holliday junction** substrates. These findings suggest that WRN and BRCA1 act in a coordinated manner to facilitate repair of DNA ICLs.

MeSH: Cell Line, Tumor, **Immunoprecipitation**, **cell proliferation**, **Cell Proliferation**, **DNA Damage**,

Humans, **RNA Interference**, **RNA interference**, **Hela Cells**, **BRCA1 Protein**, **DNA Helicases**,

Cross-Linking Reagents, **RecQ Helicases**, **Ficusin**

Fig. 6. A full abstract shown in the document view of GoPubMed for the query "*BRCA1 AND BARD1*". The abstract is shown when selecting concept DNA RE-PAIR. Note that DNA repair is not mentioned literally in the text. Instead the Concept Recognition algorithm detected INTERSTRAND CROSSLINK REPAIR as a synonym of NUCLEOTIDE-EXCISION REPAIR which is a descendant of DNA REPAIR.

Biological Function of Entities: How do Bop-Pes interactions affect cell growth? (TREC #177). *"The nucleolar PeBoW-complex, consisting of Pes1, Bop1 and WDR12, is essential for cell proliferation and processing of ribosomal RNA in mammalian cells. (PMID: 16738141)"*.

GoPubMed presents this snippet when searching for "*Bop AND Pes*" followed by a click on the option "*Expand your query with synonyms for Bop, Pes*" and filtering for the MeSH heading CELL GROWTH PROCESSES. The advanced search feature in GoPubMed allows for filtering with concept branches. The advanced search *Bop AND Pes +*mesh# "Cell Growth Processes"* retains only documents mentioning cell growth. The above snippet is shown at position 1 out of six documents when using this advanced search query and expanding the protein names.

The TREC gold-standard lists snippets of the full text of PubMed Central documents as valid answers to the questions. For question #177 the benchmark lists 7 snippets which were accepted by the curators as an answer to the question. For example the passage: "*Interestingly, a potential homologous complex of Pes1-Bop1-WDR12 in yeast (Nop7p-Erb1p-Ytm1p) is involved in the control of ribosome biogenesis and S phase entry. In conclusion, the integrity of the PeBoW complex is required for ribosome biogenesis and cell proliferation in mammalian cells. (PMID: 16043514)*" was accepted to answer question number 177.

All questions of the TREC 2006 Genomics Track could be answered with GoPubMed. It is important to note that the answers of GoPubMed are based on the abstracts and not on the full texts as used by the TREC participants. Table 3 shows the summary of results of this evaluation. The column Advanced Query shows the query syntax submited to the system for each query. It contains the keywords of the question plus the concept branch relevant to answer the question. The column "Ex" denotes whether the Protein/Gene name expansion was used or not. The Aspect column contains the concept uder which the answer was found. IS gives the number of documents used to induce the ontology. RS is the number of documents linked to the aspect. Pos is the position in the snippet list the answer was found. UI is the number of user interactions required to find the question.

Despite the fact that this evaluation was carried out with the abstract texts only, in contrast to the full texts used during the TREC evaluation, GoPubMed is able to answer all questions with a minimum of user interactions required. The advanced queries used in this experiment can easily be replaced by a simple keyword search plus a click on the concept in the ontology. In most cases the concepts selected here are top categories of the queries. In the other cases a search in the background knowledge with the option "Find related categories ..." quickly locates the appropriate concepts.

The number of official TREC answers per question varies between 0 and 593 snippets. The answers mostly cover not only one aspect of the topic. For example one TREC answer to the question number 160 is related to mutations: "Nineteen mutations of the PrP gene are associated with inherited human prion disease... (PMID: Pmid: 7642588 Span: 19641-86)" another aspect, the pathogenesis in cattle and sheep, is covered by the correct answer: "bovine spongiform encephalopathy in cattle, and scrapie in sheep are members of a family of infectious neurodegenerative mammalian diseases known as the transmissible spongiform encephalopathies. During disease pathogenesis, a protease-resistant form of prion protein (PrP) accumulates in the brain and other tissues of infected animals... Pmid: 7852415 Span: 4349-483".

In GoPubMed the spectrum of relevant answers to the questions is reflected in the induced ontology. The user can explore the aspects by navigating through the induced hierarchy. The top categories help to identify important aspects without the need to dive into the hierarchy.

7.3   Revealing Trends in the Literature

Recently much research has been devoted to the analysis of the biomedical literature. This interest has been sparked by the growth in literature, but also by the availability of abstracts, full papers, and bibliometric data. Researchers have been specifically interested in automatically extracting information from free text such as protein names (22; 60; 31), ontology terms (55; 8; 16), and protein interactions (57; 21; 32).

Underlying all of the above textmining applications is the literature, which grows overall. But at a closer glance it turns out that some research areas shrink, while others take off. Bibliometric analyses aim to shed light on such developments and to identify emerging trends. Such analyses date back to the 60s

Table 3. QN: Question Number, Ex: Expand Protein/Gene names, IS: Sample Size, RS: Result Size, Pos: Position, UI: User interactions

QN	Advanced Query	Ex	Aspect	IS	RS	Pos	UI
160	PrnP	yes	mesh#"Encephalopathy, Bovine Spongiform"	1000	77	1	1
161	IDE	no	mesh#"Alzheimer Disease"	1000	57	1	1
162	MMS2	no	go#"DNA repair"	1000	1	5	1
163	APC +*mesh#"Colonic Neoplasms"	no	Colonic Neoplasms [210]	1000	208	7	1
164	Nur77 +*mesh#"Parkinson Disease"	no	mesh#"Parkinson Disease"	1000	6	1	1
165	CTSD AND ApoE +*mesh+"Alzheimer Disease"	yes	mesh#"Alzheimer Disease"	5000	18	3	2
166	TGF-beta1 +*mesh#"Cerebral Amyloid Angiopathy"	no	mesh#"Cerebral Amyloid Angiopathy"	1000	3	1	1
167	nucleoside diphosphate kinase +*mesh#"Neoplasms"	yes	"Neoplasm Metastasis"	1000	148	1	3
168	BARD1 AND BRCA1 +*go#"biological process"	yes	go#"regulation of progression through cell cycle"	1000	67	1	3
169	APC +*mesh#Actins	no	mesh#Actins	1000	19	2	5
170	COP2 AND CFTR +*go#"ER-associated protein catabolic process"	yes	go#ER-associated protein catabolic process	1000	3	2	3
171	Nurr-77 +*mesh#"T-Lymphocytes"	yes	mesh#"T-Lymphocytes"	1000	100	2	2
172	p53 +*go#apoptosis	no	go#"induction of apoptosis"	1000	409	1	1
173	alpha7 nicotinic receptor	no	go#"ethanol metabolic process"	1000	1	1	2
174	BRCA1 +*go#"ubiquitin binding"	no	mesh#Neoplasms	1000	2	1	3
175	L1 AND L2 AND HPV11 +*go#"viral capsid"	no	go#"viral capsid"	1000	2	2	2
176	Sec61 +*mesh#"Cystic Fibrosis Transmembrane Conductance Regulator"	no	mesh#"Cystic Fibrosis"	1000	6	1	2
177	Bop AND Pes +*mesh#"Cell Growth Processes"	yes	mesh#"Cell Growth Processes"	1000	6	1	3
178	IGF +*mesh#Skin +*mesh#"Receptors, Insulin"	no	go#"regulation of keratinocyte migration"	1000	2	1	1
179	HNF4 +*mesh#"COUP Transcription Factors"	yes	mesh#"Carcinoma, Hepatocellular"	1000	9	1	2
180	Ret AND GDNF	yes	mesh#"Hepatocytes"	1000	3	1	3
181	huntingtin +*mesh#Mutation	no	go#autosome	1000	48	1	2
182	sonic hedgehog +*mesh#Mutation	no	mesh#"Holoprosencephaly"	1000	17	1	1
183	NM23 +*mesh#Mutation	no	mesh#"Genes, Tumor Suppressor"	1000	78	1	1
184	Pes +*go#"cell growth" +*mesh#Mutation	yes	mesh#"Cell Growth Processes"	1000	4	1	1
185	hypocretin receptor 2 +*mesh#Mutation	yes	mesh#Narcolepsy	1000	15	3	3
186	Presenilin-1 +*mesh#Mutation	yes	mesh#"Alzheimer Disease"	1000	466	1	3
187	FHM1 +*mesh#Mutation	no	mesh#Hippocampus	1000	2	1	1

(49) and typically focus on research topics (23), specific journals (9), or the researchers themselves (43; 44; 49; 27; 12; 10). (23) investigate e.g. the research on programmed cell death. Despite programmed cell death being described 25 years ago, it took some 15 years until journals such as "Cell Death and Differentiation" emerged and the number of publications in the field in general took off. As an example for an analysis of a specific journal Boyack analysed the emergence and development of topics covered by PNAS (9). A very active area of research aims to understand the social process of publishing by investigating co-author and co-citation networks (43; 44; 49; 27; 12; 10). Such analyses allow one to identify authors in an organisation, who work interdisciplinary and connect otherwise unconnected co-author networks (44), to animate citations of key publications over time (12), evolution of author and publication networks (10), and to understand how groups form (27). All of these analyses are useful to take a birds-eye view onto research. This section links such analyses to the ontology-based literature search engine GoPubMed to support the discovery of trends on topics interesting for molecular biologists.

GoPubMed extracts GO terms from all 18.000.000 PubMed abstracts and allows users to explore their search results with the Gene Ontology. GoPubMed's association of GO concepts with abstracts is a valuable resource to understand how a research topic - represented by a GO concept - develops. It shows how many articles were published over time, which authors are most prolific for the topic, which journals cover the topic best, and which countries publish most on the topic. The use of an ontology for these analyses is very important as it includes synonyms and subconcepts. As an example, (23) point out that during the 60s and 70s researchers in the US used "programmed cell death" while their European colleagues used "apoptosis". In this analysis, these are treated as equivalent with the help of the underlying ontology. Also it is important to consider subconcepts as some papers may mention GTPases in general, while others refer to specific GTPases such as Ran, Rac, Rho, etc. Again, the use of the ontology ensures that an analysis of GTPases will include all specific GTPases.

Besides research topics, authors and places were analysed. The results show their publishing activity over time and the topics covered. Finally, the whole biomedical literature was analysed to identify the journals, which mention most GO concepts. Assuming that GO captures the background knowledge of a molecular biologist, these are the most important journals for the molecular biologist.

The experiment was structured as follows: a bibliometric analysis focusing on "important topics" investigating apoptosis and endosome was carried out. Second, an investigation of "important places" shows how organisations and places can be classified, and the analysis of "important journals" summarizes the main topics covered by a journal. Finally, the whole Gene Ontology was analysed for the 20 most important journals for a molecular biologist.

It is important to mention that any attempt of a quantitative analysis of the literature, however sophisticated, must be interpreted by informed judgment. Absolute citation frequencies may be misinterpreted. It is not the intention to judge the significance of the contributions of individual scientists or institutions.

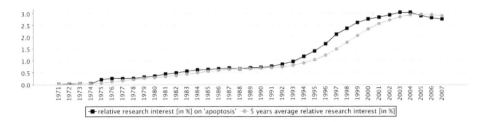

Fig. 7. PubMed abstracts per year mentioning APOPTOSIS (including synonyms and speacialisations of the term)

7.3.1 Important Topic: Apotosis

In 1997, Garfield and Melino analysed the scientific literature on programmed cell death (apoptosis) (23) using the ISI's Science Citation Index. Using the date of publication, frequency, citation and co-citation of papers, they analysed the development of the field, the countries most active, the main journals, and key authors. They found that there was a significant increased impact of articles on programmed cell death after 1990 and that it was one of the hottest topics in 1997. Some of the countries among the most active are the US, UK, Germany, Australia and France. Journals most actively publishing on apoptosis are Immunology, Blood, FASEB Journal Cancer Research, Biological Chemistry, PNAS and Oncogene. The most cited authors are AH Wyllie, SJ Korsmeyer and GT Williams as well as later on position 17 and 22 JC Reed and PH Krammer.

Such information is valuable to quickly get an overview over a new field. Although the author stated in a later addendum that they made a mistake so Strasser and Vaux were not mentioned although they had to be on rank 2 and 3. Also other research fields like nitric oxide and p53 are mentioned as similarly active research fields at this time.

Bibliometric analyses as above are valuable but difficult to produce. Especially at the beginning emerging trends may be known under different names. As Garfield and Melino point out, initially the term apoptosis was used more frequently in Europe while the US coined the same topic programmed cell death. Besides synonyms such analyses should consider papers, which do not mention apoptosis explicitly but implicitly.

As example consider papers, which mention RELEASE OF CYTOCHROME C FROM MITOCHONDRIA or CASPASE ACTIVATION. Since the release of cytochrome c from the mitochondrial intermembrane space into the cytosol leads to caspase activation and is an early step of apoptosis. Hence papers, which mention these terms should also be considered as covering apoptosis.

These two problems - the use of synonyms and the inclusion of specialisations of terms - can be addressed with an ontology. The Gene Ontology defines e.g. the synonyms APOPTOSIS and TYPE I PROGRAMMED CELL DEATH. Furthermore, it defines that CASPASE ACTIVATION is part of APOPTOTIC PROGRAM, which is part of APOPTOSIS.

Since GoPubMed indexes all PubMed abstracts with the GeneOntology our analyses include the consideration of synonyms and specialisations. For apoptosis we find very similar results to Garfield and Melino. The exponential increase of publications after 1990 can be seen in figure 7. The figure also shows that this trend continued after 1997, when Garfield and Melino's analysis was published. While Garfield and Melino claim that APOPTOSIS is a hot topic, we can quantify this claim to some extent.

Considering the number of papers between 1991 and 1997 apoptosis ranks at position 16 in comparison to other GO concepts at the same level as apoptosis or deeper in the GO hierarchy. An analysis of the countries also confirms Garfield and Melino finding though Australia's high rank in 1997 has diminished. All of the relevant journals identified by Garfield and Melino are confirmed by our analysis. Their six journals rank in the top 5 and at position 8 (Blood). The analysis reveals additionally Biochemical and Biophysical Research Communications and Nucleic Acids Research at position 6 and 7 as highly relevant for apoptosis.

The ranking for most actively publishing authors is different in this anaylsis than the results of Garfield and Melino. This is due to the fact that in contrast to them only the number of papers mentioning apoptosis could be used to rank the authors and not the citations and impact of the papers. However the two highly ranked authors Reed and Krammer also rank very high in the results. This shows that considering citations of articles gives in general a significantly different ranking than simply using term mentioning frequency. To improve this information about citations for each abstract would be needed.

One can conclude that this approach of indexing the usage of GO concepts in literature abstracts leads to very similar results when compared with the approach of Garfield and Melino. The main differences are: (1) that this approach is fully automated and always up to date, while their results date back to 1997, (2) synonyms and specialisations are considered, (3) and most of all this analysis is available online for any of the 24,000 GO concepts.

7.3.2 Important Topic: Endosome

Let us consider another example besides apoptosis, namely GO's cellular compo-nent ENDOSOME, which includes subterms such as the early and late endosome. As shown in the bottom of Figure 8, one can see that research in this areas has steadily increased in superlinear fashion. Clearly research related to the en-dosome is a hot topic at the moment. As may be expected the literature is dominated by countries from North America, Europe and Japan. However, a small part is attributed to Singapore, which is significant due to its small size. Table 8 shows the main journals and authors.

7.3.3 Important Place: Dresden

Bibliometric analyses can also be applied to get an overview over organisa-tions and places. Evaluating the topics covered by publications whose affiliation mentions Dresden reveals e.g. that biomedical research in Dresden is focused

Table 4. Statistics for ENDOSOME (Art. = number of articles containing ENDOSOME).
Top left: Most prolific journals. Top right: Most prolific authors.

Journal	Art.	Author	Art.
Biological chemistry	777	P D Stahl	59
Cell biology	420	H J Geuze	52
Cell science	261	T Berg	44
Molecular bio. of the cell	250	J Gruenberg	44
Virology	180	I Mellman	44
PNAS	178	B I Posner	41
Immunology	152	J J Bergeron	40
Bioch. et biophy. acta	144	K Sandvig	36
European j. of cell bio.	134	M Zerial	36
Biochemical journal	131	G Griffiths	35
Traffic	114	B van Deurs	35
EMBO journal	113	R G Parton	35
American j. of physiology	104	A S Verkman	31
Bioch. & biophy.res.comm.	90	H Stenmark	29
Cell and tissue research	90	S R Pfeffer	29
Others journals	6389		

on the following topics: *antiporter activity, pregnancy, apoptosis, cell proliferation, viral nucleocapsid, cytosol, exogen, microtubule, spindle, fever, gastrulation,
lactation, cytokinesis, endosome, autosome, vasodilation, enucleation, phosphorylation, wound healing, dendrite, lipid raft, RNA interference, cytoskeleton,
angiogenesis, cell migration, inflammatory response, mismatch repair, vacuole,
collagen type I, fibrinolysis, insulin secretion, vascular endothelial growth factor
receptor binding, phagocytosis, cellular respiration, pore complex, chromatin.*

This gives an immediate impression and individual topics can be traced back
to researchers and groups. E.g. RNA interference is a hot topic in Dresden with
a high-throughput RNAi screening facility in place, which has lead to numerous
publications including high-profile papers in Nature.

7.3.4 Important Journal: Which Are the 10 Most Frequently Used GO Terms in Nature, Cell and Science?

Similar to the analysis   (9) carried out for PNAS, we can analyse other
journals. As an example, we looked at the 10 most frequently mentioned
terms in Nature, Cell, and Science. Some terms appear frequently in all of
the major journals, like *exogen, apoptosis, mutagenesis, cytokinesis, antiporter
activity, DNA replication and phage assembly.* Some terms are mentioned
more often in one of the journals in comparison to others. E.g. Cell was
found to list articles on *transcription initiation, endoplasmic reticulum membrane, nucleosome, protein targeting, protein-ER targeting and regulation of
cell cycle* more frequently, which reflects its focus on molecular cell biology very
well. Science (Weekly) was found to list abstracts containing *T-cell activation,
nucleic acid transport, regulation of action potential, carbon dioxide transport*

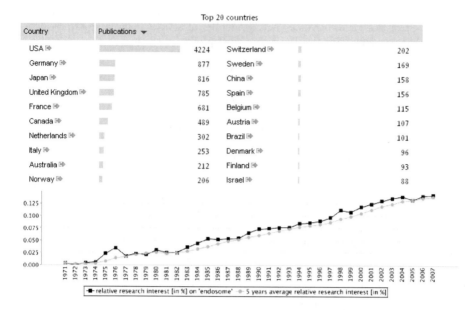

Fig. 8. Screenshot of GoPubMed statistics for the concept ENDOSOME. Top: Most prolific countries. Bottom: Articles on endosome over time.

and response to carbon dioxide more than other journals and Nature *nucleic acid transport, myosin, regulation of action potential and generation of action potential.*

7.3.5 20 Journals for the Molecular Biologist

The PubMed database is the main source for literature abstracts in the biomedical field covering thousands of journals, dating back to the 60s, including millions of authors, and several million abstracts. The GeneOntology on the other hand is a large vocabulary of over 24.000 terms covering many aspects of interest for a molecular biologist. Assuming that GO reflects the topics of interest for a molecular biologist, we wish to analyse how much of the PubMed literature might be actually of interest, which GO terms are mentioned most frequently, and which journals are the most relevant.

How many articles mention GO terms? PubMed is growing at a tremendous pace and in 2004 alone there were 598278 new abstracts registered. But how much of this is relevant to a molecular biologist as there are is also very general articles such as "Relative efficacy of the proposed Space Shuttle antimotion sickness medications", "Point-counterpoint: should physicians accept gifts from their patients? No: Gifts debase the true value of care", "The why and wherefore of empowerment: the key to job satisfaction and professional advancement". nonetheless, our analysis shows that nearly half (47.9%) of all articles in English mention at least one GO term. This figure is quite high, as e.g. Nature, which

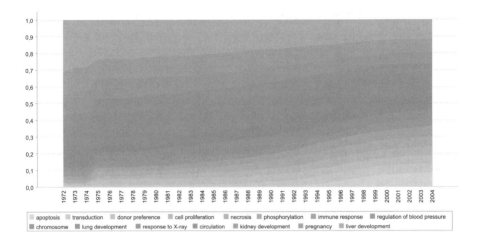

Fig. 9. Topic evolution for the most mention GO concepts since 1972. The list of GO concepts is ordered by the research interest over time. Topics like apoptosis, transduction, donor preference, cell proliferation and necrosis are of increasing research interest whereas liver development, pregnancy and kidney development show an relatively decreasing rate of mentions in PubMed abstracts.

is certainly the accepted journal for any molecular biologist, has only in 38.5% of its articles at least one GO term.

Which are the 10 most frequently used GO terms per year? A specialization level of greater than 5 was choosen in the Gene Ontology to compare research topics in two years 1972 and 2004. That means high-level ontology terms were ignored.

GO concepts mentioned in abstracts frequently in both year were: KIDNEY DEVELOPMENT, LUNG DEVELOPMENT, RESPONSE TO X-RAY, RESPONSE TO VIRUS and PHOSPHORYLATION. Articles from 1972 mentioned additionally following GO concepts more frequently: MICROSOME, CELLULAR RESPIRATION, PHAGE ASSEMBLY, ANTIPORTER ACTIVITY, CENTRAL NERVOUS SYSTEM DEVELOPMENT, DNA REPLICATION, ALKALINE PHOSPHATASE ACTIVITY, NUCLEIC ACID TRANSPORT, REGULATION OF BALANCE, RENIN ACTIVITY, LYSOSOME, SALIVARY GLAND (DETERMINATION/MORPHOGENESIS) and OVULATION.

In contrast to that the following new topics were subjects of research in 2004: APOPTOSIS, DONOR PREFERENCE, ENDOTHELIAL CELL (ACTIVATION/MORPHOGENESIS), EQUATOR SPECIFICATION, EXOGEN, ANGIOGENESIS, VISUAL PERCEPTION, INFLAMMATORY RESPONSE, INTERFERON-GAMMA BIOSYNTHESIS, RESPONSE TO REACTIVE OXYGEN SPECIES.

In figure 9 the evolution of the most frequently used GO concepts over time is shown. One can clearly see that topics like apoptosis, transduction, donor preference, cell proliferation and necrosis are of increasing research interest whereas liver development, pregnancy and kidney development show an relatively decreasing rate of mentions in PubMed abstracts.

Table 5. 20 journals for the molecular biologist

Pos. Journal
1. Biological chemistry
2. PNAS
3. Biochimica et biophysica acta
4. Immunology
5. Biochemical and biophysical research communications
6. American journal of physiology
7. Biochemistry
8. Biochemical journal
9. Brain research
10. Cancer research
11. Virology
12. FEBS letters
13. Bacteriology
14. Blood
15. Endocrinology
16. Cell biology
17. European journal of biochemistry / FEBS
18. Infection and immunity
19. Molecular and cellular biology
20. Nature

Which 20 journals mention the most GO terms? Finally, let us turn to the "20 journals for the molecular biologist", which are as partially shown in Table 5. Biological chemistry, PNAS, Biochimica et biophysica acta, Immunology, Biochemical and biophysical research communications, American journal of physiology, Biochemistry, Biochemical journal, Brain research, Cancer research, Virology, FEBS letters, Bacteriology, Blood, Endocrinology, Cell biology , European journal of biochemistry, FEBS, Infection and immunity, Molecular and cellular biology, and Nature. Other journals such as Science rank at position 32, EMBO at 26, Cell at 52. It is interesting that many biochemistry journal rank in the top positions.

This shows the high relevance of the selection of the journals. In the online version the user can browse all abstracts of a selected journal mentioning a previously selected GO term via GO.

7.4   Conclusion

GoPubMed was developed into a full application as part of the REWERSE project. The web based search engine is online at www.gopubmed.org since 2004 and serves thousands of users every day.

Contributions made during the course of the REWERSE project include a complete survey on state-of-the-art biomedical search engines. The systems were categorized according to their features supporting Information and Knowledge

Retrieval. One conclusion is that some systems semantically process documents but no system uses ontologies to organize search result.

Another contribution is a survey over 15 freely available annotation corpora. There is no corpus directly applicable for the evaluation of Concept Recognition algorithms for Gene Ontology terms. The BioCreAtIvE dataset is the best source for manually confirmed Gene Ontology annotations in full texts. However it is biased toward the annotation of selected proteins.

As part of REWERSE's output a new Concept Recognition pipeline was developed(15) which improved the previously reported (14) performance by 25,7% achieving 79,9% precision and 72,7% recall. To access the quality of the pipeline and to facilitate further training for ambiguous ontology labels a curations tool was designed. With these tools a new benchmark comprised of 689 PubMed abstracts and 18,356 curations, personally curated by the original authors of PubMed articles, was created.

(15) also summarizes the first ontology-based, large scale, online available, up-to-date bibliometric analysis for topics in molecular biology represented by Gene Ontology concepts is evaluated. It is shown that the method is in line with existing, but often out-dated, analyses.

The methods developed during the REWERSE project are not limited to the biomedical domain. Currently it is investigated how new ontologies can be generated by extending available domain ontologies or from the scratch. Ontology design is a labor intensive task. It is of great interest in how available literature corpora can be used to generate suggestions for ontology concept labels, synonyms, abbreviations, textual definitions and relations between concepts.

The induced ontologies are a rich source of information. However Concept Recognition methods can not be perfect. Even human experts disagree on a large number of concept mentions in natural language texts. Therefore the induced ontology will always carry irrelevant information and lack of some relevant information hidden in the texts. Recurrently stated knowledge is well represented in the induced ontology. It is of great interest how to identify under-represented or new knowledge in document collections. The difficulty here is to distinguish this knowledge from noise in the form of irrelevant markups.

References

[1] Ackoff, R.L.: From data to wisdom. Journal of Applies Systems Analysis 16, 3–9 (1989)
[2] Andreopoulos, B., Alexopoulou, D., Schroeder, M.: Word sense disambiguation in biomedical ontologies with term co-occurrence analysis and document clustering. Internation Journal of Data Mining and Bioinformatics (2008) (Special Issue on Text Mining and Information Retrieval)
[3] Apweiler, R., Bairoch, A., Wu, C., Barker, W., Boeckmann, B., Ferro, S., Gasteiger, E., Huang, H., Lopez, R., Magrane, M., Martin, M., Natale, D., ODonovan, C., Redaschi, N., Yeh, L.: UniProt: the universal protein knowledgebase. Nucleic Acids Res. 32(D), D115–D119 (2004)

[4] Ashburner, M., Ball, C.A., Blake, J.A., Botstein, D., Butler, H., Cherry, J.M., Davis, A.P., Dolinski, K., Dwight, S.S., Eppig, J.T., Harris, M.A., Hill, D.P., Issel-Tarver, L., Kasarskis, A., Lewis, S., Matese, J.C., Richardson, J.E., Ringwald, M., Rubin, G.M., Sherlock, G.: Gene ontology: tool for the unification of biology. The gene ontology consortium. Nat. Genet. 25(1), 25–29 (2000)

[5] Bellinger, G., Castro, D., Mills, A.: Data, Information, Knowledge, and Wisdom (2004), http://www.systems-thinking.org/dikw/dikw.htm

[6] Berman, H.M., Westbrook, J., Feng, Z., Gilliland, G., Bhat, T.N., Weissig, H., Shindyalov, I.N., Bourne, P.E.: The protein data bank. Nucleic Acids Res. 28(1), 235–242 (2000)

[7] Blaschke, C., Hirschman, L., Valencia, A.: Information extraction in molecular biology. Briefings in Bioinformatics 3, 154–165 (2002)

[8] Blaschke, C., Leon, E.A., Krallinger, M., Valencia, A.: Evaluation of biocreative assessment of task 2. BMC Bioinformatics 6(suppl. 1) (2005)

[9] Boyack, K.: Mapping knowledge domains: Characterizing PNAS. PNAS 101(1), 5192–5199 (2004)

[10] Börner, K., Mary, J., Goldstone, R.: The simultaneous evolution of author and paper networks. PNAS 101(1), 5266–5273 (2004)

[11] Brown, P.O., Botstein, D.: Exploring the new world of the genome with dna microarrays. Nat. Genet. 21(suppl. 1), 33–37 (1999)

[12] Chen, C.: Searching for intellectual turning points: Progressive knowledge domain visualization. PNAS 101(1), 5303–5318 (2004)

[13] Chen, H., Sharp, B.M.: Content-rich biological network constructed by mining PubMed abstracts. BMC Bioinformatics 5(1) (October 2004)

[14] Doms, A.: Using sequence alignment algorithms to extract gene ontology terms in biomedical literature abstracts. Diplomathesis, TU Dresden (2004)

[15] Doms, A.: GoPubMed: Ontology-based literature search for the life sciences. PhD thesis, Technical University of Dresden (2009)

[16] Doms, A., Schroeder, M.: GoPubMed: exploring PubMed with the Gene Ontology. Nucl. Acids Res. 33, W783–W786 (2005)

[17] Eaton, A.D.: Hubmed: a web-based biomedical literature search interface. Nucleic Acids Res. 34(Web Server issue) (July 2006)

[18] Fontelo, P., Liu, F., Leon, S., Anne, A., Ackerman, M.: PICO linguist and BabelMeSH: Development and partial evaluation of evidence-based multilanguage search tools for medline/pubmed. Stud. Health Technol. Inform. 129, 817–821 (2007)

[19] Fontelo, P., Liu, F., Ackerman, M.: Askmedline: a free-text, natural language query tool for medline/pubmed. BMC Med. Inform. Decis. Mak. 5(1) (March 2005)

[20] Forrey, A.W., McDonald, C.J., DeMoor, G., Huff, S.M., Leavelle, D., Leland, D., Fiers, T., Charles, L., Griffin, B., Stalling, F., Tullis, A., Hutchins, K., Baenziger, J.: Logical observation identifier names and codes (loinc) database: a public use set of codes and names for electronic reporting of clinical laboratory test results. Clin. Chem. 42(1), 81–90 (1996)

[21] Friedman, C., Kra, P., Yu, H., Krauthammer, M., Rzhetsky, A.: GENIES: a natural-language processing system for the extraction of molecular pathways from journal articles. In: Proceedings of the International Conference on Intelligent Systems for Molecular Biology, pp. 574–582 (2001)

[22] Fukuda, K., Tamura, A., Tsunoda, T., Takagi, T.: Toward information extraction: identifying protein names from biological papers. In: Pac. Symp. Biocomput., pp. 707–718 (1998)

[23] Garfield, E., Melino, G.: The growth of the cell death field: an analysis from the isi science citation index. Cell Death and Differentiation 4, 352–361 (1997)

[24] Gene Ontology Consortium: The Gene Ontology (GO) database and informatics resource. Nucleic Acids Res. 1(32), D258–D261 (2004)

[25] Goetz, T., von der Lieth, C.W.: Pubfinder: a tool for improving retrieval rate of relevant pubmed abstracts. Nucleic Acids Res. 33(Web Server issue) (July 2005)

[26] Gruber, T.R.: A translation approach to portable ontology specifications. Knowl. Acquis. 5(2), 199–220 (1993)

[27] Guimer, R., Uzzi, B., Spiro, J., Amaral, L.: Team assembly mechanisms determine collaboration network structure and team performance. Science 308(5722), 697–702 (2005)

[28] Hakenberg, J., Plake, C., Royer, L., Strobelt, H., Leser, U., Schroeder, M.: Gene mention normalization and interaction extraction with context models and sentence motifs. Genome Biology 9(suppl. 2) (2008)

[29] Hamosh, A., Scott, A.F., Amberger, J., Bocchini, C., Valle, D., McKusick, V.A.: Online mendelian inheritance in man (omim), a knowledgebase of human genes and genetic disorders. Nucleic Acids Res. 30(1), 52–55 (2002)

[30] Hersh, W., Cohen, A.M., Roberts, P., Rekapalli, H.K.: Overview of the TREC 2006 question answering track. In: The Fifteenth Text REtrieval Conference (TREC 2006) Proceedings (2006)

[31] Hirschman, L., Colosimo, M., Morgan, A., Yeh, A.: Overview of BioCreAtIvE task 1b: normalized gene lists. BMC Bioinformatics 6(1), S11 (2005)

[32] Hoffmann, R., Valencia, A.: A gene network for navigating the literature. Nature Genetics 36, 664 (2004)

[33] Hoffmann, R., Valencia, A.: A gene network for navigating the literature. Nat. Genet. 36(7) (2004)

[34] Jenssen, T.K., Laegreid, A., Komorowski, J., Hovig, E.: A literature network of human genes for high-throughput analysis of gene expression. Nat. Genet. 28(1), 21–28 (2001)

[35] Joachims, T.: A probabilistic analysis of the rocchio algorithm with tfidf for text categorization. In: Fisher, D.H. (ed.) Proceedings of ICML 1997, 14th International Conference on Machine Learning, Nashville, US, pp. 143–151. Morgan Kaufmann Publishers, San Francisco (1997)

[36] Kaenel, I.d., Iriarte, P.: Alternative interfaces for PubMed searches. In: European Association for Health Information & Libraries Workshop (2006)

[37] Koster, J.: PubMed Pubreminer: a tool for PubMed query building and literature mining (2007)

[38] Lambrix, P., Tan, H., Jakoniene, V., Strömbäck, L.: Biological Ontologies, pp. 85–99. Springer, Heidelberg (2007)

[39] Law, S., Jerzy, O., Dawid, S.: Lingo: Search results clustering algorithm based on singular value decomposition (2004)

[40] Lowe, H.J., Barnett, G.O.: Understanding and using the medical subject headings (mesh) vocabulary to perform literature searches. JAMA 271(14), 1103–1108 (1994)

[41] Müler, H.M., Kenny, E.E., Sternberg, P.W.: Textpresso: An ontology-based information retrieval and extraction system for biological literature. PLoS Biology 2(11) (2003)

[42] Muin, M., Fontelo, P.: Technical development of PubMed interact: an improved interface for Medline/PubMed searches. BMC Medical Informatics and Decision Making 6, 36+ (2006)

[43] Newman, M.: The structure of scientific collaboration networks. PNAS 98(2), 404–409 (2001)

[44] Newman, M.: Coauthorship networks and patterns of scientific collaboration. PNAS 101(1), 5200–5205 (2004)

[45] Patrick, J., Wang, Y., Budd, P.: An automated system for conversion of clinical notes into snomed clinical terminology. In: ACSW 2007: Proceedings of the fifth Australasian symposium on ACSW frontiers, Darlinghurst, Australia, pp. 219–226. Australian Computer Society, Inc., Australia (2007)

[46] Perez-Iratxeta, C., Perez, A., Bork, P., Andrade, M.: Update on XplorMed: A web server for exploring scientific literature. Nucleic Acids Res. 31(13), 3866–3868 (2003)

[47] Plake, C., Schiemann, T., Pankalla, M., Hakenberg, J., Leser, U.: ALIBABA: PubMed as a graph. Bioinformatics 22(19), 2444 (2006)

[48] Plikus, M.V., Zhang, Z., Chuong, C.M.: Pubfocus: Semantic medline/pubmed citations analytics through integration of controlled biomedical dictionaries and ranking algorithm. BMC Bioinformatics 7, 424 (2006)

[49] Price, D.: Networks of scientific papers. Science 30(149), 510–515 (1965)

[50] Quackenbush, J.: Genomics. microarrays–guilt by association. Science 302(5643), 240–241 (2003)

[51] Rebholz-Schuhmann, D., Arregui, M., Gaudan, S., Kirsch, H., Jimeno, A.: Text processing through web services: calling whatizit. Bioinformatics 24(2), 296–298 (2008)

[52] Rebholz-Schuhmann, D., Kirsch, H., Arregui, M., Gaudan, S., Riethoven, M., Stoehr, P.: EBIMed–text crunching to gather facts for proteins from medline. Bioinformatics 23(2), e237–e244 (2007)

[53] Sharma, N.: The origin of the data information knowledge wisdom hierarchy (February 2008) (unpublished)

[54] Siadaty, M.S., Shu, J., Knaus, W.A.: Relemed: Sentence-level search engine with relevance score for the medline database of biomedical articles. BMC Medical Informatics and Decision Making 7, 1+ (2007)

[55] Smith, T., Cleary, J.: Automatically linking medline abstracts to the geneontology. In: Proc. of the Sixth Annual Bio-Ontologies Meeting, Brisbane, Australia (2003)

[56] Taylor, D.P.: An integrated biomedical knowledge extraction and analysis platform: using federated search and document clustering technology. Methods Mol. Biol. 356, 293–300 (2006)

[57] Thomas, J., Milward, D., Ouzounis, C., Pulman, S., Carroll, M.: Automatic extraction of protein interactions from scientific abstracts. In: Proc. of the Pacific Symp. on Biocomputing, pp. 538–549 (2002)

[58] Tyers, M., Mann, M.: From genomics to proteomics. Nature (London) 422, 193–197 (2003)

[59] Yao, Y., Zeng, Y., Zhong, N., Huang, X.: Knowledge retrieval (KR). In: Proceedings of the IEEE/WIC/ACM International Conference on Web Intelligence (2007)

[60] Yeh, A., Morgan, A., Colosimo, M., Hirschman, L.: BioCreAtIvE task 1a: gene mention finding evaluation. BMC Bioinformatics 6(1), S2 (2005)

Chapter 8
Information Integration in Bioinformatics with Ontologies and Standards

Patrick Lambrix, Lena Strömbäck, and He Tan

Department of Computer and Information Science
Linköpings universitet
581 83 Linköping, Sweden

Abstract. New experimental methods allow researchers within molecular and systems biology to rapidly generate larger and larger amounts of data. This data is often made publicly available on the Internet and although this data is extremely useful, we are not using its full capacity. One important reason is that we still lack good ways to connect or integrate information from different resources.

One kind of resource is the over 1000 data sources freely available on the Web. As most data sources are developed and maintained independently, they are highly heterogeneous. Information is also updated frequently. Other kinds of resources that are not so well-known or commonly used yet are the ontologies and the standards. Ontologies aim to define a common terminology for a domain of interest. Standards provide a way to exchange data between data sources and tools, even if the internal representations of the data in the resources and tools are different.

In this chapter we argue that ontological knowledge and standards should be used for integration of data. We describe properties of the different types of data sources, ontological knowledge and standards that are available on the Web and discuss how this knowledge can be used to support integrated access to multiple biological data sources. Further, we present an integration approach that combines the identified ontological knowledge and standards with traditional information integration techniques. Current integration approaches only cover parts of the suggested approach. We also discuss the components in the model on which much recent work has been done in more detail: ontology-based data source integration, ontology alignment and integration using standards.

Although many of our discussions in this chapter are general we exemplify mainly using work done within the REWERSE[1] working group on Adding Semantics to the Bioinformatics Web.

8.1 Introduction

New experimental methods allow researchers within molecular and systems biology to rapidly generate larger and larger amounts of data. This data is often made publicly available on the Internet. However, although this data is extremely useful, we are not using its full capacity. One important reason is that we still lack good ways to connect

---

[1] The work on this chapter and on the articles by this chapter's authors that are referenced in this chapter was performed in the context of REWERSE.

F. Bry and J. Maluszynski (Eds.): Semantic Techniques for the Web, LNCS 5500, pp. 343–376, 2009.
© Springer-Verlag Berlin Heidelberg 2009

or integrate information from different resources. Semantic Web technology can play an important role to alleviate this problem.

Assume a future crisis, where a new virus creates an epidemic of fatal disease. Researchers manage to isolate the genetic material of the virus, determine the sequence and want to create an antidote. Publicly available resources and specialized computer programs can be used to determine relationships with previously studied viruses and to determine proteins that are possible targets of antidotes. Additional useful information such as sequences, structure, functionality and interactions of the protein can be found and compared with data from other viruses to more rapidly give information on what antidotes to create and what further experiments to perform.

There are several kinds of resources that contain biological information which are relevant for the scenario. The most common and well-known resources are the data sources. They may contain, for instance, information on gene maps, protein structures, molecular interactions, and models of metabolic and other kinds of pathways. The 2007 Database issue of the Nucleic Acids Research journal listed 968 data sources freely available on the Web and the 2008 issue introduced 98 new data sources. As most data sources are developed and maintained independently, they are highly heterogeneous. They vary in the type of the stored data, the data format, and access methods. In addition, there is a terminology discrepancy at the data level and at the schema level, which even more complicates the data retrieval process. Also, information is frequently updated and new information is frequently added.

Other kinds of resources, i.e. ontologies and standards, are not so well-known or commonly used yet. Ontologies aim to define the basic terms and relations of a domain of interest, as well as the rules for combining these terms and relations. In the last decade much research has been performed on defining a common terminology through the development of ontologies. To describe data items and relations between data items each data source has its own internal data model. Today, a number of standardized export formats for data have been developed. These standards provide a way to exchange data between data sources and tools, even if the internal representations of the data in the resources and tools are different. The latest development is to define meta standards, which determine minimum requirements for what should be included in a standard or representation format.

In order to find all relevant data for our scenario, information from all these resources needs to be connected and integrated. For instance, to find information about a particular protein, we may have to access different data sources, such as UniProt for sequence information and PDB for structure information. Therefore, we need to find out how proteins are modeled in the two sources (schema) and we need a way to know when data items in the two sources represent the same real-world entity such as a particular protein. These tasks are further complicated by the fact that different resources may use different terminology, possibly based on different ontologies. Further, data needs to be transferred between data sources (e.g. an identifier for a protein in UniProt is used to query PDB) or tools (e.g. protein data is used in a reaction network model building or analysis tool). To be able to integrate all this information we need knowledge about the relationships between the different resources.

In this chapter we describe the different kinds of resources for bioinformatics mentioned above (section 8.2). We describe data sources (section 8.2.1), ontologies (section 8.2.2) and standards (section 8.2.3). In section 8.3 we deal with integration of the resources. First, we describe an integration model in section 8.3.1 and show that not only the resources, but also the connections between the resources are important for integration. We then briefly describe integration approaches for data sources (section 8.3.2) and the connection between data sources and ontologies (section 8.3.3) and focus in more details on recent work on ontology alignment (section 8.3.4), integration of standards (section 8.3.5) and the connection between standards and other resources (section 8.3.6).

8.2 Resources

8.2.1 Data Sources

There are over 1000 data sources that store information related to biological data freely available on the Web and they are highly heterogeneous in different ways. We describe here the properties of these data sources.

Data source content. The stored biological data ranges from experimental results, DNA and protein sequences, to three-dimensional molecule structures and networks representing interactions between molecules. As researchers worked independently and collected biological data relevant to their own research issues in parallel, the available data is spread over a large number of data sources, the data sources differ in their focus, but they often store data that is highly related to each other. To facilitate the discovery of relevant information, links between relevant entries at different data sources are usually stored explicitly. Links may differ in their quality and semantics [3] and not all possible relationships are explicitly stated.

Data quality. The data sources differ in their data quality. The source of the data can be researchers (that submit their data), literature (data from published articles) and other data sources. The collected data may be checked, modified or appended by the local curators. Some data sources use tools (instead of curators) to gather information from different sources. This means that as the original data sources change, the data in the secondary data sources should also change.

Data updates. The frequency of data updates differs among the data sources. The user may receive the source releases in periods of two or four months, the data updates between one day or two weeks, or the period can be irregular.

Inconsistency. Many inconsistencies appear in biological data sources. There are several reasons for this: data is merged from different communities, data is submitted in a flexible way, there are errors in annotations and the area is inherently dynamic. In contrast to traditional database approaches, constraints are often not used to check the validity of the submitted data. To speed up publishing the discovered knowledge, correctness of the submitted data may not always be checked. In [14] it was illustrated that errors in predicted annotations in a genome may reach up to 40%. Data sources frequently change and evolve as new approaches and tools are developed that generate new types of data. This causes irregularity of data structure at the data sources. Another source of data inconsistencies is non-synchronized frequency of updates at the data sources. For example, as explicitly specified links between data sources may not be updated, or as biological objects evolve over time and their identifiers change, some data sources may still use old identifiers to refer to the data.

Semantic heterogeneity. In addition to syntactic heterogeneity biological data sources display a high level of semantic heterogeneity. Since there is no agreed terminology in the area of bioinformatics, we encounter cases where the same representation is implicitly assigned to different definitions or different representations are used to refer to the same concept. For instance, the concept gene is a DNA sequence fragment that either encodes a protein (e.g. UniProt) or that carries some information of biological interest (e.g. GenBank). Different identifiers and names can be used to refer to the same biological object. For example, a UniProt data entry describing Pancreas/duodenum homeobox protein 1 (short name: PDX-1) has P52945 as the entry identifier (and previously O60594 as a withdrawn entry identifier in SwissProt which is now a part of UniProt). This protein can also be referred to as Insulin promoter factor 1 (IPF-1), Islet/duodenum homeobox-1 (IDX-1), Somatostatin-transactivating factor 1 (STF-1), Insulin upstream factor 1 (IUF-1) and Glucose-sensitive factor (GSF).

Data models. Different data models (from flat files to relational and object models) are used to represent biological data and different data management systems are used to manage the data. For instance, flat files were selected at the beginning of the 1990s, as a simple, flexible and working solution for storing biological data. Currently, relational or object-oriented database techniques are also used. Regardless of the type of the underlying data model, the data can usually be exported to the user as flat files or in XML[2] format.

Data retrieval possibilities. Regardless of the expressivity of the underlying data model, the users are provided with data retrieval interfaces having limited query capabilities. Most systems support boolean queries (supporting the AND, OR and NOT operators) as well as wildcards in the text strings. The systems allow search for the given strings within a data entry (full-text search) or within selected predefined fields. The retrieval interface specifies which fields are searchable. The queries can be formulated using a form-based query interface or entered into a command line (sometimes expressed as URLs). The fields that can be searched on differ in the form-based and the command-line interfaces.

8.2.2 Ontologies

A second important source of biological information is the ontologies. Intuitively, ontologies can be seen as defining the basic terms and relations of a domain of interest, as well as the rules for combining these terms and relations [63]. In recent years many biomedical ontologies (e.g [49]) have been developed. They are a key technology for the Semantic Web [71,43]. The benefits of using ontologies include reuse, sharing and portability of knowledge across platforms, and improved documentation, maintenance, and reliability. Ontologies lead to a better understanding of a field and to more effective and efficient handling of information in that field. The work on ontologies is recognized as essential in some of the grand challenges of genomics research [10] and there is much international research cooperation for the development of ontologies. The number of researchers working on methods and tools for supporting ontology engineering

---

[2] See chapter 2.

is constantly growing and more and more researchers and companies use ontologies in their daily work.

The use of biological ontologies has grown drastically since database builders concerned with developing systems for different (model) organisms joined to create the Gene Ontology (GO) Consortium in 1998 [11]. The goal of GO was and still is to produce a structured, precisely defined, common and dynamic controlled vocabulary that describes the roles of genes and proteins in all organisms. The GO ontologies are a de facto standard and many biological data sources are today annotated with GO terms. The terms in GO are arranged as nodes in a directed acyclic graph, where multiple inheritance is allowed.

Another milestone was the start of Open Biomedical Ontologies as an umbrella Web address for ontologies for use within the genomics and proteomics domains [70]. In addition to being a common portal for ontologies, OBO also promotes a scientific method for ontology development. For an ontology to become a member ontology originally five requirements needed to be fulfilled: the ontologies are required to be open, to be written in a common shared syntax, to have a delineated content (be orthogonal to other ontologies), to share a unique identifier space and to include textual definitions. In 2006 five new requirements were added for ontologies to be part of the OBO Foundry: procedures for identifying distinct successive versions need to be provided, relations in the ontologies follow the OBO Relation Ontology [86], the ontology is well documented, the ontology has a plurality of independent users, and the ontology is developed collaboratively with other OBO Foundry members. Many biological ontologies, including GO, are already available via OBO.

The field has also matured enough to start talking about standards. An example of this is the organization of the first conference on Standards and Ontologies for Functional Genomics (SOFG) in 2002 and the development of the SOFG resource on ontologies. Another example is the start of the development of the Common Anatomy Reference Ontology [7] to facilitate interoperability between existing anatomy ontologies for different species. Further, in systems biology ontologies are used more and more, for instance, in the definition of standards for representation and exchange of molecular interaction data.

Ontologies differ regarding the kind of information they can represent. From a knowledge representation point of view ontologies can have the following components (e.g. [88,49]). Concepts represent sets or classes of entities in a domain. For instance, in figure 1 nose represents all noses. The concepts may be organized in taxonomies, often based on the is-a relation (e.g. nose is-a sensory organ in figure 1) or the part-of relation (e.g. nose part-of respiratory system in figure 1). Instances represent the actual entities. They are, however, often not represented in ontologies. Further, there are many types of relations (e.g. chromosone has-sub-cellular-location nucleus). Finally, axioms represent facts that are always true in the topic area of the ontology. These can be such things as domain restrictions (e.g. the origin of a protein is always of the type gene coding origin type), cardinality restrictions (e.g. each protein has at least one source), or disjointness restrictions (e.g. a helix can never be a sheet and vice versa). Ontologies can be classified according to the components and the information regarding the components they contain. A simple type of ontology is the controlled vocabulary. These are essentially

```
[Term]
id: MA:0000281
name: nose
is_a: MA:0000017 ! sensory organ
is_a: MA:0000581 ! head organ
relationship: part_of MA:0000327 ! respiratory system
relationship: part_of MA:0002445 ! olfactory system
relationship: part_of MA:0002473 ! face
```

Fig. 1. Example concept from the Adult Mouse Anatomy ontology (available from OBO)

lists of concepts. When these concepts are organized in an is-a hierarchy, we obtain a taxonomy. A slightly more complex kind of ontology is the thesaurus. In this case the concepts are organized in a graph. The arcs in the graph represent a fixed set of relations, such as synonym, narrower term, broader term, similar term. The data models allow for defining a hierarchy of classes (concepts), attributes (properties of the entities belonging to the classes, functional relations), relations and a limited form of axioms. The knowledge bases are often based on a logic. They can contain all types of components and provide reasoning services such as checking the consistency of the ontology. An ontology and its components can be represented in a spectrum of representation formalisms ranging from very informal to strictly formal [32]. In general, the more formal the used representation language, the less ambiguity there is in the ontology. Formal languages are also more likely to implement correct functionality. Furthermore, the chance for interoperation is higher. In the informal languages the ontology content is hard-wired in the application. This is not the case for the formal languages as they have a well-defined semantics. However, building ontologies using formal languages is not an easy task. In practice, biological ontologies have often started out as controlled vocabularies. This allowed the ontology builders, which were domain experts, but not necessarily experts in knowledge representation, to focus on the gathering of knowledge and the agreeing upon definitions. More advanced representation and functionality was a secondary requirement and was left as future work. However, some of the biological ontologies have reached a high level of maturity and stability regarding the ontology engineering process and their developers have now started investigating how the usefulness of the ontologies can be augmented using more advanced representation formalisms and added functionality. Moreover, some recent efforts have started out immediately as knowledge bases.

8.2.3 Standards

The third important source for biological information is the standards. These were developed for exchange and integration of data.

Data can originate from experimental results or created models and researchers often want to submit them to databases. Further, often several systems and data sources are used when analyzing biological data, and data and results are exported from one system or data source to the next. Earlier, the task of transferring data from one system to the next and translating the output format from the first to the input format of the second, was often done on case-to-case base and was therefore a time-consuming and error-prone

Table 1. Standards for molecular interaction data. (Information from [90]).

Name	Substances			Interactions	Pathways	Compartments	Organism	Experiments
	DNA, RNA	Protein	Other					
SBML	X	X	X	X	X	X		
PSI MI	X	X	X	X		X	X	X
BioPAX	X	X	X	X	X		X	X
CellML	X	X	X	X	X	X	X	
CML			X	X				
EMBLxml	X	X					X	
INSDseq	X	X					X	
Seqentry	X	X					X	
BSML	X	X					X	X
HUP-ML	X	X					X	X
MAGE-ML	X	X					X	X
mzXML								X
mzData								X
AGML							X	X

task. To avoid this we need common data models, i.e. descriptions of the different kinds of data, their representations and how the different pieces of data relate to each other. This is often done by the definition of standards.

One important decision when defining a standard for data representation is the choice of underlying representation language. Previous evaluations [1,58] have shown that XML is a suitable representation language for bioinformatics data. Moreover, an overview of existing standards shows that within bioinformatics and systems biology most standards use XML or XML-based representation formats such as RDF and OWL.[3] For instance, a search for XML-based standards within systems biology provided 85 standards of varying levels of interest [90]. We limited the scope to standards for molecular interaction and signaling pathways, and standards for describing the basic entities often included in standards for signaling pathways, that is, standards for describing proteins, DNA, genes or other substances, compartments, and experimental results. We further limited the result list by requiring that the standards must have been under recent development or use, they are referred to in more than one source, there is data available that uses the standard for representation or there are tools available for manipulation of data in the standard. With these restrictions we still found 14 standards of higher interest.

An overview of these standards is given in table 1. We show whether the standards contain information on substances, interactions, pathways, compartments, organisms and experiments. The standards fall into three categories. First, there are the standards whose aim is to represent some aspects of molecular interactions or pathways. In this category SBML ([31], http://www.sbml.org/) and CellML ([24], http://www.cellml.org/) are tuned towards simulation, PSI MI ([29], http://www.psidev.info/) towards description of experiments, and BioPAX ([108],

---

[3] For an introduction to XML and RDF see chapters 1 and 2.

http://www.biopax.org) has a more general scope. The second group focuses on describing genes and proteins and the third group models results of experiments. Considering our example in section 8.1, the kind of information described in these standards is useful when we have identified a number of proteins similar to the ones appearing in our virus. The information described in the standards can give useful information on their interactions and functionality. This information could be very valuable to determine the probable functionality of our unknown virus and to suggest experiments to verify these properties.

In previous studies ([92], updated in [91]) we have put further emphasis on standards for molecular interactions by comparing the three standards: SBML, PSI MI

```xml
<?xml version="1.0" encoding="UTF-8"?>
<sbml xmlns="http://www.sbml.org/sbml/level2" level="2" version="1">
  <model id="Tyson1991CellModel_6" name="Tyson1991_CellCycle_6var">
    + <annotation>
    <listOfSpecies>
      <species id="C2" name="cdc2k" compartment="cell">
        + <annotation>
      </species>
      + <species id="M" name="p-cyclin_cdc2" compartment="cell">
      + <species id="YP" name="p-cyclin" compartment="cell">
      ... more species
    </listOfSpecies>
    <listOfReactions>
      <reaction id="Reaction1" name="cyclin_cdc2k dissociation">
        <annotation>
          <rdf:li rdf:resource="http://www.reactome.org/#REACT_6308"/>
          <rdf:li rdf:resource="http://www.geneontology.org/#GO:0000079"/>
        </annotation>
        <listOfReactants>
          <speciesReference species="M"/>
        </listOfReactants>
        <listOfProducts>
          <speciesReference species="C2"/>
          <speciesReference species="YP"/>
        </listOfProducts>
        <kineticLaw>
          <math xmlns="http://www.w3.org/1998/Math/MathML">
            <apply> <times/> <ci> k6 </ci> <ci> M </ci>
            </apply></math>
          <listOfParameters>
            <parameter id="k6" value="1"/>
          </listOfParameters>
        </kineticLaw>
      </reaction>
      + <reaction id="Reaction2" name="cdc2k phosphorylation">

      ... more reactions

    </listOfReactions>
  </model>
</sbml>
```

Fig. 2. An SBML example

and BioPAX. An example of an SBML description fetched from the Biomodels [64] database is given in figure 2. The figure is shortened to improve readability. The figure shows an SBML representation of the Tyson cell model [101] which describes one of the basic functions in the cell life cycle. In the example we can see the interacting molecules (the species) and where these molecules exist. The reaction part of the example shows how these molecules interact to provide the studied functionality.

In general, the aim of SBML is to represent several kinds of pathways, biochemical reactions and gene regulation. The main concepts in SBML are the interacting substances (Species), how these substances interact (Reaction) and where the reaction takes place (Compartment). In addition, the user can specify mathematical properties describing the reaction's behavior, sizes of compartment, concentrations of substances and similar information. SBML contains several features. One is a framework for linking SBML descriptions to complementary information about the objects in available data sources. Another addition is the ability to place restrictions on the type of objects. For some of the main concepts, such as Reactions and participants in the reactions, the user can refer to controlled vocabularies, thereby providing a more detailed specification of the concept. These vocabularies are provided as a controlled vocabulary, included in OBO [70]. For Species and Compartments, there is another solution. Here the user can group concepts by a type specification that is specific to each model. In the proposal for level 3, future versions of the standard will enable the encoding of protein states from protein structure.

The main aim for PSI MI is to provide a mean for representation and exchange of data from experiments. The main objects in PSI MI are Interactors (in SBML: species), Interactions (in SBML: reactions) and Experiments. In addition, information about the type of experiment, methods for detecting a substance, statistical evidence for an interaction and the participating Interactors can be stored. Later developments of PSI MI provide a means for a more fine-grained representation of Interactor and Interactions. It also allows representation of both the biological and the experimental role of a participant in a detected interaction. There are also additions to Interaction allowing the user to represent deduced interactions and experiments made on species other than the one the interaction is reported from. Another feature is the use of ontologies, providing means of referring to Interactortypes, Interactiontypes, Experimenttypes and different kinds of experimental methods in a consistent way. As with SBML, these vocabularies are part of OBO [70].

The main aim of the BioPAX standard is to define a unified framework for sharing pathway information. It uses OWL as the representation format. In BioPAX information is centered around substances, called Physical Entities, and Interactions. For each of these main concepts a number of subclasses are defined specifying many types of substances, such as proteins and DNA, together with different kinds of interactions. BioPAX also includes a means for the user to combine single Interactions into Pathways in various ways. BioPAX also provides an import of the PSI MI features for representation of sequences for proteins, DNA and RNA. BioPAX is adapted towards experimental data by the ability to represent information about experimental evidence of an interaction.

This short description gives an idea about the similarities and differences between the different standards. In practice, all the standards studied in [90] show some overlap to other XML standards. In some cases, there are competitive standards, i.e. two suggestions on how to standardize the same kind of information. In most cases, however, there are differences in scope and aim of the standards which makes this large amount of co-existing standards valuable to the community. On the other hand, the existence of many standards is a hinder for data integration, since the information exists in several formats.

To alleviate the problem that many standards are being developed covering much overlapping information, the latest development within this field includes efforts to determine minimum requirements for a standard. For instance, MIAME (Minimum Information About a Microarray Experiment) [5] defines minimum requirements for microarray data and within the genomic technology society, several minimum requirements have been developed, such as MIAPE (The Minimum Information About a Proteomics Experiment) [100] for proteomics data and MIRIAM (Minimum Information Requested In the Annotation of biochemical Models) [65] for models in systems biology. One common theme among these requirements is a link to ontologies by the recommendation to store metadata according to controlled vocabularies instead of free text. Other important requirements are that information about participating substances is included as well as information on which organisms they are collected from, and references to sources in the literature.

Important for integration of information is the ability to link data and information between resources. Many XML-based standards contain ontology information or links to external sources. This is often in accordance with the recommendations in the specifications of minimal information for a standard. For instance, MIAME recommends the use of ontologies to represent data where such ontologies exist. MIAME uses the notion of qualifier, value, source-triplets to refer to external knowledge. The source can be defined either by the user or can refer to an external ontology. Also MIRIAM has an annotation scheme for external resources that requires the use of unique resources identifiers (URIs) to identify model constituents, such as model, compartments, reacting entity or reaction. These URIs are unique, permanent references to information about the particular object in that data source or ontology, that are built up so they do not necessarily reflect the current server address or entry name but contain information to identify organization, data source and accession code. The standards listed in table 1 all have some variant of this feature.

8.3   Integrating the Resources

8.3.1   Integration Model

Figure 3 [46] represents the different kinds of resources as well as the different kinds of relationships among the (components) of resources.

Within a data source there is usually[4] a well-defined relationship between the data items and the schema (1). When integrating information from two data sources (e.g. we

---

[4] This is especially true for the structured data sources such as relational databases.

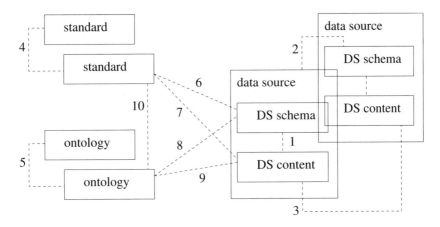

Fig. 3. Integration Model [46]

need information about the sequence and structure about proteins for our example in section 8.1 and need to integrate data from UniProt, a protein data source, and PDB, a data source containing information about the 3-dimensional structure of proteins), we need to find out how similar data items are represented. This is usually done by providing a mapping between the schemas (2). Several information integration systems in this field define an integrated schema (not shown in figure) which is mapped to the schemas of the data sources that are integrated. Many of these information integration systems are still academic. Further, it is necessary to integrate data items from different sources (3). In the current systems this is often done by explicitly linking data items. Popular integration systems such as SRS and Entrez, rely heavily on this. Some duplicate detection or instance matching mechanisms are also used. Further, many data sources already provide the possibility to export their data in standardized formats (6 and 7).

Some integration systems use ontologies as an integrated schema (8). However, the main use of ontologies is currently for annotation of data sources (8 and 9). For instance, terms from the GO molecular function ontology are used to describe gene and protein functions in many data sources. The use of annotations reduces the terminology discrepancy and supports finding similar data items in one or more data sources.

One problem, however, is the fact that many ontologies with overlapping information are being developed. For instance, OBO lists 26 anatomy ontologies (October 2008). Finding the relationships between these ontologies would give valuable information for information integration (5). Currently, a number of ontology alignment systems have been developed to deal with this issue. Similarly, standards have been developed with different focus or application area, but with overlapping information, and it is only recently that work has started on finding this overlap (4). Some of the current standards are also based on or connected to ontologies (10).

Integrating information from different resources is not easy due to the different levels of heterogeneity. To overcome these difficulties and obtain higher quality search results, knowledge about relationships between entities in the different resources should be used. While integration of data sources is reasonably well understood, still not so many systems exist. Work on dealing with integration of ontologies has started and good results have already been obtained, while integration of standards is still in its infancy. However, with the recent interest in developing a Semantic Web, both these issues attract more and more research.

In the remainder of this section we discuss some of the connections in figure 3. We briefly describe integration approaches for data sources and the connection between data sources and ontologies. Then, we focus in more details on recent work on ontology alignment, integration of standards and the connection between standards and other resources.

8.3.2 Integration of Data Sources

The traditional database integration solutions cannot be used in a straightforward way to meet the requirements for information integration in bioinformatics. These approaches usually lack the power and flexibility to cope with the heterogeneity of the environment and the user needs that evolve over time. As noted in [41], traditional database systems rely on data structures and object identities that are predefined and do not change over time, lack the flexibility to represent similar data and limit users to data manipulation queries. Existing information integration systems can be grouped into virtual and materialized systems depending on whether they preserve the autonomy of data sources or whether they need to be downloaded and processed locally. As discussed in [38], data warehouses - materialized systems - do not guarantee that accessed information is up-to-date, provide access to a limited number of data sources and do not cope well with changes in the data sources. Virtual information integration systems retrieve data on demand, i.e. query results are up-to-date. With respect to the transparency, information integration systems can be grouped into tightly coupled and loosely coupled federations. Tightly coupled federations hide the integrated data sources from the user and select the relevant data sources for the query processing. Loosely coupled federations expose the integrated data sources to the user. In this approach, the user is responsible for selecting data sources that are relevant to a query. Loosely coupled federations count on user knowledge about existing data sources and ways to link them. In [78] it is emphasized that tightly coupled federated database systems are not practical since the required level of integrated data management is too high. In many cases, the simplifying assumptions made by the available integration systems are not appropriate for biological data integration that relates to management of scientific data on the Web [41]. For instance, systems developed to integrate Web sources assume that conjunctive queries are enough to retrieve data of interest. The systems deal with limited data source query capabilities but do not consider multiple capabilities of the data source [17].

Within the field of bioinformatics several integration approaches have been proposed and systems have been implemented. Different systems have focused on different issues and requirements for information integration systems. The earliest and currently most widely used systems are index-based systems such as Entrez [19] and SRS (Sequence Retrieval System, [20]). An Entrez data source stores information collected from different other data sources. Before adding new information into the integrated data source, it is modified, e.g. assigned a unique identifier, converted into a common data representation, validated and matched to literature and taxonomy data sources. Data sources accessed through Entrez are interlinked between each other by cross-references. Entrez also introduces links between entries at a data source. For each entry, a list of neighbors, i.e. similar entries, is assigned. SRS is a loosely coupled federation supporting a uniform interface and a query language to retrieve data. Most of the other (more recent) systems have been used in field trials or have reached the stage of prototype, but they often have not yet been used to the same extent as Entrez and SRS.

The information integration systems differ with respect to the type of the stored knowledge and the formalisms used to represent the knowledge. The data models for representing the integrated schemas can be separated into two groups: data models established in the database community, i.e. relational, object-oriented and functional data models, and data models used in the knowledge representation community, mainly logic-based models. Some systems emphasize the support of web technologies and suggest to use data models specialized for the Web (e.g. TAMBIS [89,9]). Some systems support multiple integrated schemas (e.g. BioMediator [83]). The systems maintaining integrated schemas differ in expressivity of the modeled mapping rules between the integrated schemas and the source schemas. Mapping rules are relations between terms in an integrated schema and terms in the data source schemas. These rules specify how queries expressed in terms of an integrated schema can be rewritten into the queries referring to the data sources. Two important kinds of mapping rules are local-as-view and global-as-view mapping rules (e.g. [53]). Both kinds of mapping rules are used. For instance, BACIIS [60] uses local-as-view mapping rules, K2 [99] uses global-as-view, while KIND [55] supports both. In addition to schemas and mapping rules, the systems may use other types of knowledge to improve query processing such as capabilities of data sources (e.g. BioFAST [17]), domain knowledge (e.g. KIND [55]), links between data sources (BioMediator [83]) and statistical information on the data and data sources (e.g. BioNavigation [42]).

Other types of information integration systems are also used in bioinformatics. For instance, [37] explores the use of agent ontology for information integration in bioinformatics and grid technology is used in, for instance, the Cancer Bioinformatics Grid CaBIG [73].

8.3.3   Data Sources and Ontologies

The main use of ontologies in bioinformatics is currently for annotation of data sources. For instance, [2] describes the use of GO for Mouse Genome Informatics (MGI). MGI integrates genetic and genomic data about the mouse. GO is used to assign consistent functional annotations for data that is gathered from different other data sources and literature. The use of consistent annotations reduces the terminology discrepancy and supports finding similar data items in different data sources. The GO annotation sets are also used in MGI for computing initial functional annotations for previously uncharacterized genes.

Two other advantages of using ontologies which can be used in integration, are the use of the ontologies for query expansion and the use of ontologies for reasoning [81]. In both cases new relationships may become apparent and used for integration of information.

Some of the information integration systems described in section 8.3.2 use Semantic Web technologies based on the use of ontologies together with reasoning mechanisms over the ontologies for dealing with data on the Web. BACIIS and TAMBIS use a domain ontology to represent the integrated schema, to specify mapping rules, to guide rewriting of a user query into data source terms and to structure the retrieved results. In addition, TAMBIS uses the domain ontology to guide a user during query formulation and to semantically optimize queries. KIND uses the domain ontology to bridge the gap between the data sources to integrate, i.e. to model and identify the relationships between data objects that come from domains that are not directly related. Views are formulated over the data sources and the domain ontology. In addition to the terminological ontologies, the system uses a special formalism to model ontologies describing processes. To enable automatic inclusion of the newly integrated data sources to the previously defined views and to organize the retrieved data, similarly to BACIIS and TAMBIS, KIND maps local data source ontology terms to the domain ontology.

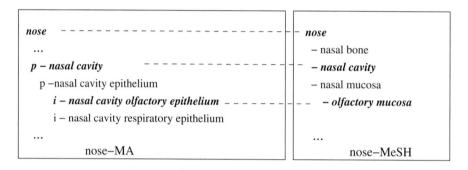

Fig. 4. Example of overlapping ontologies

8.3.4   Ontology Alignment

As stated before, many ontologies with overlapping information are being developed. For instance, figure 4 shows a piece of the Adult Mouse Anatomy (MA) ontology and a piece of the Medical Subject Headings (MeSH) ontology both representing nose. In MA is-a (i) is used for specialization relationships (e.g. nasal cavity olfactory epithelium is-a nasal cavity epithelium) and part-of (p) for partitive relationships (nasal cavity part-of nose). MeSH uses the same relationship (-) for both these types of relationships. There is overlap between these pieces. We know that nose, nasal cavity and nasal cavity olfactory epithelium in MA represent the same concepts as nose, nasal cavity and olfactory mucosa, respectively, in MeSH.

Ontology alignment[5] is recognized as an important step in ontology engineering that needs more extensive research and during the last years a large number of ontology alignment systems have been developed. Examples of such systems can be found in tables 2, 3, 4 and 5. More information can also be found in review papers (e.g.

---

[5] See also chapter 6, section on ontology modularization, for an application in component models for semantic web languages.

[47,84,66,36]), the book [21] on ontology matching, and on the ontology matching web site at http://www.ontologymatching.org/.

Ontology alignment framework. Many ontology alignment systems are based on the computation of similarity values between terms in different ontologies and can be described as instantiations of the general framework shown in figure 5[6]. The framework consists of two parts. The first part (*I* in figure 5) computes mapping suggestions. The second part (*II*) interacts with the user to decide on the final mappings.[7] An alignment algorithm receives as input two source ontologies. The ontologies can be preprocessed, for instance, to select pieces of the ontologies that are likely to contain matching terms. The algorithm includes one or several matchers, which calculate similarity values between the terms from the different source ontologies and can be based on knowledge about the linguistic elements, structure, constraints and instances of the ontology. Also auxiliary information can be used. Mapping suggestions are then determined by combining and filtering the results generated by one or more matchers. By using different matchers and combining and filtering the results in different ways we obtain different alignment strategies. The suggestions are then presented to the user who accepts or rejects them. The acceptance and rejection of a suggestion may influence further suggestions. Further, a conflict checker is used to avoid conflicts introduced by the mapping relationships. The output of the ontology alignment system is an alignment which is a set of mappings between terms from the source ontologies.

Strategies. The matchers use different strategies to calculate similarities between the terms from the different source ontologies. They use different kinds of knowledge that can be exploited during the alignment process to enhance the effectiveness and efficiency. Some of the approaches use information inherent in the ontologies. Other approaches require the use of external sources. We describe the types of strategies that are used by current ontology alignment systems and in tables 2, 3, 4 and 5[8] we give an overview of the used knowledge per system. The information in table 2 stems from the study in [47]. The systems in tables 3, 4 and 5 participated in the Ontology Alignment Evaluative Initiative in 2007 and/or 2008.

- *Strategies based on linguistic matching.* These approaches make use of textual descriptions of the concepts and relations such as names, synonyms and definitions. The similarity measure between concepts is based on comparisons of the textual descriptions. Simple string matching approaches and information retrieval approaches

---

[6] The framework in figure 5 is an extension of the framework defined in [47]. The framework is further extended in [45] where we investigated the use of partial reference alignments in ontology alignment.

[7] Some systems are completely automatic (only part I). Other systems have a completely manual mode where a user can manually align ontologies without receiving suggestions from the system (only part II). Several systems implement the complete framework (parts I and II) and allow the user to add own mapping relationships as well.

[8] Also the approaches that are not based on the computation of similarity values may use these types of knowledge and are therefore included in the table.

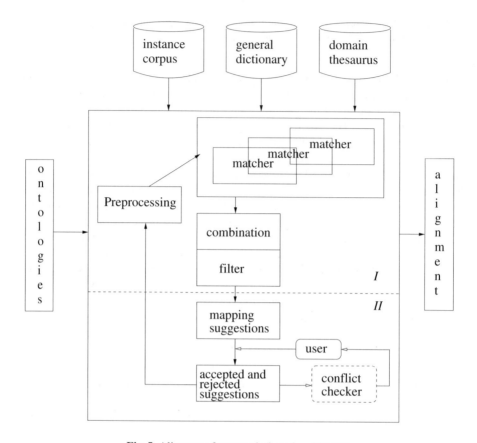

Fig. 5. Alignment framework (based on [47,45])

(e.g. based on frequency counting) may be used. Most systems use this kind of strategies.

– *Structure-based strategies.* These approaches use the structure of the ontologies to provide suggestions. Typically, a graph structure over the concepts is provided through is-a, part-of or other relations. The similarity of concepts is based on their environment. An environment can be defined in different ways. For instance, using the is-a relation an environment could be defined using the parents (or ancestors) and the children (or descendants) of a concept.

– *Constraint-based approaches.* In this case the axioms are used to provide suggestions. For instance, knowing that the range and domain of two relations are the same, may be an indication that there is a relationship between the relations. Constraint-based approaches are currently used by only a few systems.

– *Instance-based strategies.* In some cases instances are available directly or can be obtained. For instance, the entries in biological data sources that are annotated with GO terms, can be seen as instances for these GO terms. When instances are available, they may be used for defining similarities between concepts.

Table 2. Knowledge used by (earlier) alignment systems [47]

	linguistic	structure	constraints	instances	auxiliary
ArtGen [74]	name	parents, children		domain-specific documents	WordNet
ASCO [52]	name, label, description	parents, children, siblings, path from root			WordNet
Chimaera [59]	name	parents, children			
FCA-Merge [93]	name			domain-specific documents	
FOAM [18]	name, label	parents, children	equivalence		
GLUE [15]	name	neighborhood		instances	
HCONE [40]	name	parents, children			WordNet
IF-Map [35]				instances	a reference ontology
iMapper [94]		leaf, non-leaf, children, related node	domain, range	instances	WordNet
OntoMapper [75]	name	parents, children		documents	
(Anchor-) PROMPT [68,69]	name	direct graphs			
S-Match [25]	label	path from root	semantic relations in text		WordNet

– *Use of auxiliary information.* Dictionaries and thesauri representing general or domain knowledge, or intermediate ontologies may be used to enhance the alignment
process. Also information about previously aligned or merged ontologies may be
used. Many systems use auxiliary information.

An ontology alignment tool. As an example of an ontology alignment tool and its use,
we briefly discuss SAMBO[9] [47]. SAMBO is developed according to the framework
described above. The system separates the process into two steps: aligning relations
and aligning concepts. The second step can be started after the first step is finished.
In the suggestion mode several kinds of matchers can be used and combined. The
implemented matchers are a terminological matcher (TermBasic), the terminological
matcher using WordNet (TermWN), a structure-based matcher (Hierarchy), a matcher
(UMLSKSearch) using domain knowledge in the form of the Unified Medical Language System (UMLS) of the U.S. National Library of Medicine [103] and an instance-
based matcher (BayesLearning). TermBasic contains matching algorithms based on the

---

[9] System for Aligning and Merging Biomedical Ontologies.

Table 3. Knowledge used by alignment systems participating in both OAEI 2007 and 2008

	linguistic	structure	constraints	instances	auxiliary
ASMOV [33,34]	textual descriptions	parents, children	property		WordNet, UMLS
DSSim [61,62]	textual descriptions		semantic relations in text		WordNet
Lily [105,106]	textual descriptions	hierarchy information	property	web knowledge, instances	
RiMOM [54,113]	label, comment	ancestors		instances	WordNet
SAMBO(dtf) [98,50]	name, synonym	is-a and part-of, descendants and ancestors		domain-specific documents	WordNet, UMLS
TaxoMap [110,28]	label	children	property		

Table 4. Knowledge used by alignment systems participating in OAEI 2007 but not in 2008

	linguistic	structure	constraints	instances	auxiliary
Agreement-Maker [95]	labels path to root	parents			
AOAS [112]	textual descriptions	is-a, part-of relations			UMLS a reference ontology
Falcon-AO [30]	textual descriptions	structural relations			
OLA [23]	textual descriptions	structural relations	domain, range property		
OWL-CM [109]	label	children	equivalence		WordNet
Prior+ [56]	name	sub-elements	property		
SEMA [87]	label, comment	parents, children	domain,range property	instances	WordNet
SILAS [72]	name			related text	
SODA [111]	label, comment	neighborhood			
X-SOM [12]	textual descriptions	neighborhood		google search	WordNet

names and synonyms of concepts and relations. The matcher is a combination matcher based on two approximate string matching algorithms (n-gram and edit distance) and a linguistic algorithm. In TermWN a general thesaurus, WordNet [107], is used to enhance the similarity measure by using the hypernym relationships in WordNet. The structure-based algorithm requires as input a list of mapping relationships and

Table 5. Knowledge used by alignment systems participating in OAEI 2008, but not in 2007

	linguistic	structure	constraints	instances	auxiliary
Anchor-Flood [82]	lexical info	parents, children neighborhood	property		
AROMA [13]	textual descriptions		subsumption relations in text	instances	
CIDER [27]	textual descriptions	structural relations			
GeRoMe [77]	textual descriptions	paths, children			WordNet
MapPSO [4]	name, label	parents	property		WordNet
SPIDER [80]	textual descriptions		semantic relations in text		WordNet

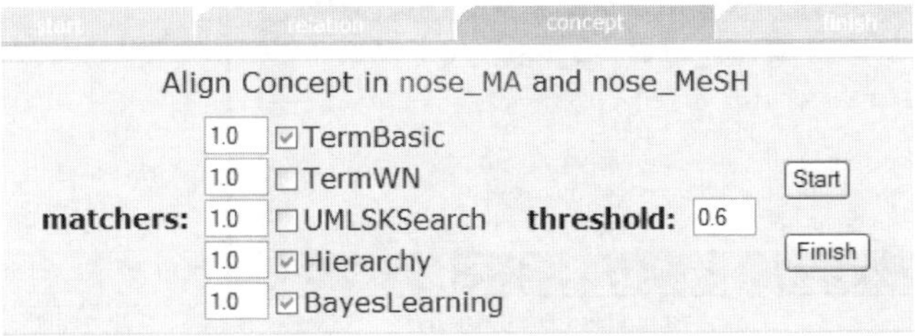

Fig. 6. Combination and filtering

similarity values and can therefore not be used in isolation. The intuition behind the algorithm is that if two concepts lie in similar positions with respect to is-a or part-of hierarchies relative to already aligned concepts in the two ontologies, then they are likely to be similar as well. UMLSKSearch uses the Metathesaurus in the UMLS which contains more than 100 biomedical and health-related vocabularies. The Metathesaurus is organized using concepts. The concepts may have synonyms which are the terms in the different vocabularies in the Metathesaurus that have the same intended meaning. The similarity of two terms in the source ontologies is determined by their relationship in UMLS. BayesLearning makes use of life science literature that is related to the concepts in the ontologies. It is based on the intuition that a similarity measure between concepts in different ontologies can be defined based on the probability that documents about one concept are also about the other concept and vice versa. For more detailed information about these matchers we refer to [47]. In addition to these techniques we have also experimented with other matchers [51,96,104]. The combination algorithm in SAMBO is a weighted sum strategy. Figure 6 shows how different matchers can be chosen and weights can be assigned to these matchers.

Filtering is performed using a threshold value. The pairs of terms with a similarity value above this value are shown to the user as mapping suggestions. We have also developed the double threshold filtering method [8] and implemented in the

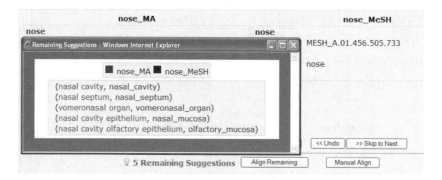

Fig. 7. Mapping suggestion

Fig. 8. Information about the remaining suggestions

SAMBOdtf system, an extension of SAMBO. The double threshold filtering approach uses the structure of the ontologies. It is based on the observation that (for the different approaches in the evaluation in [47]) for single threshold filtering the precision of the results is decreasing and the recall is increasing when the thresholds are decreasing.[10] Therefore, we propose to use two thresholds. Pairs with similarity values equal or higher than the upper threshold are retained as suggestions. The intuition is that this gives suggestions with a high precision. Further, pairs with similarity values between the lower and the upper threshold are filtered using structural information and the rest is discarded. We require that the pairs with similarity values between the two thresholds are 'reasonable' from a structural point of view. The intuition is that the recall is augmented by adding new suggestions, while at the same time the precision stays high because only structurally reasonable suggestions are added. For details we refer to [8].

An example mapping suggestion is given in figure 7. The system displays information (definition/identifier, synonyms, relations) about the source ontology terms in the suggestion. For each mapping suggestion the user can decide whether the terms are equivalent, whether there is an is-a relation between the terms, or whether the suggestion should be rejected. If the user decides that the terms are equivalent, a new name

---

[10] Recall is the number of correct suggestions divided by the number of correct mapping relationships. Precision is the number of correct suggestions divided by the number of suggestions.

```
                    nose_MA                                          nose_MeSH
M nose (nose)                                        M nose (nose)
   ⊢ O nasal turbinate                                  ⊢ M nasal_cavity (nasal cavity)
   ⊢ M nasal cavity (nasal_cavity)                      ⊢ O nasal_mucosa
      ⊢ O nasal cavity epithelium                          ⊢ O olfactory_mucosa
         ⊢ O nasal cavity olfactory epithelium              ⊢ O olfactory_receptor_neuron
         ⊢ O nasal cavity respiratory epithelium            ⊢ O goblet_cell
   ⊢ O naris                                            ⊢ O nasal_bone
      ⊢ O external naris                                ⊢ O paranasal_sinus
         ⊢ O primary choana                                ⊢ O maxillary_sinus
         ⊢ O external naris epithelium                     ⊢ O ethmoid_sinus
      ⊢ O internal naris                                   ⊢ O sphenoid_sinus
   ⊢ M nasal septum (nasal_septum)                         ⊢ O frontal_sinus
      ⊢ O septal olfactory organ                        ⊢ M nasal_septum (nasal septum)
   ⊢ M vomeronasal organ (vomeronasal_organ)              ⊢ M vomeronasal_organ (vomeronasal organ)
      ⊢ O vomeronasal organ sensory epithelium          ⊢ O turbinate
   ⊢ O olfactory nerves
   ⊢ O nasal capsule
   ⊢ O olfactory gland
```

```
   comment on the alignment

                                              Concept Name:                        in  nose_MA ▾    search

            << Undo      = Equiv. Concepts    ≤ Sub-Concept    ≥ Super-Concept

                              Suggestion Align
```

Fig. 9. Manual mode

for the term can be given as well. Upon an action of the user, the suggestion list is up-dated. If the user rejects a suggestion where two different terms have the same name, she is required to rename at least one of the terms. The user can also add comments on a mapping relationship. At each point in time during the alignment process the user can view the ontologies represented in trees with the information on which actions have been performed, and she can check how many suggestions still need to be processed. Figure 8 shows the remaining suggestions for a particular alignment process. A similar list can be obtained to view the previously accepted mapping suggestions. In addition to the suggestion mode, the system also has a manual mode in which the user can view the ontologies and manually map terms (figure 9). The source ontologies are illustrated us-ing is-a and part-of hierarchies (i and p icons, respectively). The user can choose terms from the ontologies and then specify an alignment operation. Previously aligned terms are identified by different icons. For instance, the M icons in front of 'nasal_cavity' in the two ontologies in figure 9 show that these were aligned using an equivalence relationship. There is also a search functionality to find specific terms more easily in the hierarchy. The suggestion and manual modes can be interleaved. The suggestion mode can also be repeated several times, and take into account the previously performed operations.

After the user accomplishes the alignment process, the system receives the final map-ping list and can be asked to create the new ontology. The system merges the terms in the mapping list, computes the consequences, makes the additional changes that follow from the operations, and finally copies the other terms to the new ontology. Furthermore,

SAMBO uses a DIG description logic reasoner to provide a number of reasoning services. The user can ask the system whether the new ontology is consistent and can ask for information about unsatisfiable concepts and cycles in the ontology.

Evaluation of ontology alignment strategies for life sciences. Considering the fact that many strategies and systems are being developed, it becomes increasingly difficult to choose what techniques to use for aligning ontologies. However, the study of the properties, and the evaluation and comparison of the alignment strategies and their combinations, can give us valuable insight in how the strategies could be used in the best way. It would also lead to recommendations on how to improve the alignment techniques. There are two main evaluations of the performance of ontology alignment strategies for the life sciences in terms of the quality of the mapping suggestions. In [47,96,104,51] different *matchers* were evaluated on five smaller test cases. Two cases used a part of a GO ontology together with a part of SigO. The other test cases were based on MeSH (anatomy category) and Adult Mouse Anatomy. An analysis of some of the results using the KitAMO environment is given in [48].

The largest evaluation is performed within the Ontology Alignment Evaluation Initiative (OAEI, http://oaei.ontologymatching.org/). This is a yearly initiative that was started in 2004. The goals are, among others, to assess the strengths and weaknesses of alignment *systems*, to compare different techniques and to improve evaluation techniques. This is to be achieved through controlled experimental evaluation. For this purpose OAEI publishes different cases of ontology alignment problems, some of which are open (reference alignment is known beforehand), but most are blind (reference alignment is not known - participants send their mapping suggestions to organizers who evaluate the performance). OAEI currently only evaluates the non-interactive part of ontology alignment systems. The case that is related to the life sciences is the anatomy case. In 2008 participants were required to align the Adult Mouse Anatomy (2744 concepts) and the NCI Thesaurus - anatomy (3304 concepts). The case is divided into 4 tasks (of which task 4 was new for 2008). The anatomy case is a blind case. The reference alignment contains only equivalence correspondences between concepts. In all tasks the two ontologies should be aligned. In task 1 the system should be tuned to optimize the f-measure[11]. This means that both precision and recall are important. The systems are compared with respect to precision, recall, f-measure and recall+ (recall with respect to non-trivial mappings). 9 systems participated in task 1. In tasks 2 and 3, in which 4 systems participated, the system should be optimized with respect to precision and recall, respectively. In task 4[12], in which 4 systems participated, a partial reference alignment is given which can be used during the computation of mapping suggestions. It contains all trivial and some non-trivial mappings. The best results in tasks 1 and 4 were obtained by SAMBO with SAMBOdtf in second place. The version of SAMBO for OAEI used a maximum-based combination of TermWN and UMLSKSearch. This suggests that domain knowledge is important to obtain good results for this task.

---

[11] F-measure is the weighted harmonic mean of precision and recall. For task 1 precision and recall are weighted evenly.

[12] As a follow-up on task 4, we have studied the use of partial reference alignments in the different components of an ontology alignment system. For details we refer to [45].

However, as the recall+ of the best system is still around 0.6, work still needs to be done to find non-trivial mappings. Also, the best systems that did not not use domain knowledge managed to find non-trivial mappings that were not found by any system using domain knowledge. As [6] suggests, a combination of different strategies may improve the results. Taking the union of the SAMBO results with the results of the Ri-MOM and Lily systems gave a recall of 0.922 and a recall+ of 0.8. RiMOM and Lily use linguistic and structure-based approaches. The 4 systems participating in task 4 all managed to improve their f-measure on the non-given part of the reference alignment. However, only SAMBOdtf managed to improve both precision and recall. The best system for task 2[13] (RiMOM) obtained a precision of 0.964 (with a recall of 0.677) and the best system for task 3 (RiMOM) obtained a recall of 0.808 (with a precision of 0.450 and a recall+ of 0.538). We note that the best recall is lower than the recall for SAMBO and SAMBOdtf in task 1.

Current challenges. [85] lists ten challenges for ontology alignment systems. A first challenge is more large-scale evaluation. This includes larger test sets than the current OAEI provides and evaluation measures. Further, there is an issue of performance. When mappings are generated off-line and not very often, this may not be a major factor, but when mappings need to be generated on-line in semantic web applications, this may become a bottleneck. When ontologies are being developed, it is usually within a certain context and with certain background knowledge. This is not always explicitly represented in the ontologies and therefore makes the alignment task harder. Other challenges are uncertainty in ontology alignment and reasoning with mappings. Also, not so much work has been done on user involvement, user interfaces and ontology and ontology alignment visualization [44,22]. To help users in deciding whether to accept or reject mapping suggestions, we may need to explain why the systems propose the suggestions. Further, as ontology alignment may be a large task, one may want to involve a community of users that participate in social and collaborative ontology alignment. There is currently not much infrastructure for alignment management. An ontology alignment management system may, in addition to components for aligning ontologies, also provide functionality for storing, retrieving and searching the results. Some initial ideas are realized in the KitAMO environment [48] and the BioPortal [67]. Finally, as many alignment strategies (matchers, combinations and filters) have been developed, it is not always easy to choose the best strategy for aligning two ontologies. One way to address this issue is to develop systems that recommend strategies (e.g. [97]).

8.3.5 Integrating Standards

As for ontologies, although different standards are developed with different aims, the fact that there exists so many and diverse standards gives problems for the community. The different standards overlap partially and have their own specific strengths, e.g. PSI MI [29] is designed for the description of molecular interactions while SBML [31] and CellML [24] are designed for describing pathways and simulation models.

In parallel to this, the importance of data integration becomes more and more obvious as the number of interdisciplinary research projects increases constantly [79,102].

---

[13] SAMBO and SAMBOdtf did not participate in tasks 2 and 3.

When the scope of the projects increases there is a need to model different kinds of data and several of the standards are likely to be used within the same project. To provide an interface between the standards, transformations between the standards are needed, i.e. connection 4 in figure 3. For instance, of the variety of software tools being available for the work on biochemical models, many only support a limited number of formats and develop their own specific converters for the standards they support [26,64]. Some converters are provided by the communities, for example on www.sbml.org for SBML models. Those are mostly XSLT scripts that convert versions of a standard into versions of another standard. The main drawback of this solution is that the transformation is hard coded, and thus as soon as a new version of one of the participating standards becomes available, the converter has to be rewritten. This leads to a situation where a conversion expert is needed in order to maintain the facilities.

Another problem is that a great number of converters is needed for, for instance, all the combinations of pathway standards that a community wants to support. One solution is to use tools for integration of XML that work on the XML schema, e.g. COMA++ [57]. Another solution that allows integration of XML-based and OWL-based standards is proposed in [39]. The main architecture of this solution is described in figure 10. The task is to convert XML or OWL data from one standardized format to another. The main work is done on the schema level. For cases where the standard is described as an XML schema the XML schema is converted to an OWL description. With this transformation the integration can be done, semi-automatically, with a tool for integration of OWL ontologies such as SAMBO. The alignment results can then be used for transforming the original XML files from one format to the other. The proposed architecture provides a general and semi-automatic solution for integrating standards in order to support reusability and comparability of biochemical models defined in XML-based and OWL-based formats. XML Schema lacks the formal foundations of OWL and thus a lossless transformation of OWL relations into XML Schema structures is not possible.

The general architecture can be divided into the following steps:

1. Provide a schema definition (e. g. the SBML Schema), if the starting point was an instance file (e. g. an SBML model).
2. Transform the XML Schema into an OWL model representation.
3. Repeat 1 and 2 for all standards that should be compared.
4. Match the (created) OWL models on OWL level.
5. Use the matching correspondences to either form a joint format containing all the information of both starting schemas, or to assign data of the source document to the target document.

During the transformation step, an existing XML Schema is transformed into an equivalent OWL model. The transformation is focused on keeping the naming and structure of the original models, as those are of great importance for the success of a matching process. This especially means that the hierarchy of the XML Schema has to be kept unchanged. Notions such as cardinalities or data models (differences between choice, sequence and all) are of minor interest for the matching process. No additional names or identifiers should be added in order to avoid a distortion of matching results later on. The result of the transformation is a valid OWL model which can be read by existing OWL tools.

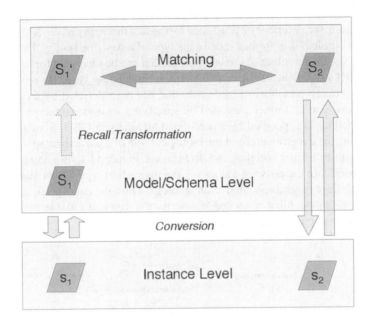

Fig. 10. An architecture for integration of XML standards [39]

To get back from the created OWL model to the XML based representation, the whole transformation from XML Schema to OWL model is recorded (recall transformation) using XML Path Language (XPath). XPath can address parts of an XML Schema unambiguously and by storing the XPath expressions during the transformation step, the XML Schema can be recreated from the generated OWL model. It has been shown that the backwards transformation from OWL model to XML Schema keeps all the information needed for the transformation tool, and that it is possible to unambiguously identify each part of the created OWL model and its equivalent in the according XML Schema.

During a matching step, the created and/or original models are matched. A matching takes two schemas S1 and S2 as input and returns a mapping between those two schemas as output. The resulting mapping then defines relations between the two schemas and therefore allows for a comparison of common parts in both schemas. As described in section 8.3.4 there exist a number of matching algorithms and we have currently evaluated SAMBO, PROMPT and COMA++.

The described architecture could, if fully realized, be a tool that provides semi-automatic conversion between formats. We also see the possibility of reusing old conversions when new versions of a representation format appear. This is needed but not available in current technology.

8.3.6   Connection between Standards and Other Resources

We have already mentioned in section 8.2.3 that connections between standards and ontologies have been required by several of the efforts defining minimum requirements for standards.

This relation between standards, described in XML format, and ontologies (connection 10 in figure 3) can further be illustrated by the fact that many of the XML standards make use of ontologies to further specify the type of a described entity. Here we illustrate this by showing a subset of the information that can be identified for an interaction in the PSI MI [29] standard. In this standard an interaction typically represents any interaction between a number of molecules that has been detected by an experiment. The interaction can be further classified by specifying an interactiontype. The values of the interactions are specified by a predefined hierarchy in OBO. Similarly, the values specifying the experimental and the biological role of a participant molecule of the interaction can be further specified as OBO classes. Figure 11 shows an excerpt of the ontology specifying the possible values for the interaction type. Other standards, e.g. BioPAX [108], are instead specified as an ontology. Thus, the concepts in the hierarchy build up the data model that is needed to describe the data and suitable parameters can be connected to a specific class in the ontology.

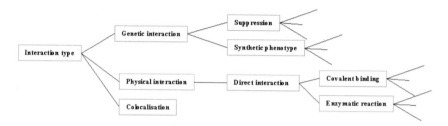

Fig. 11. The PSI MI interaction ontology

The connection between the data model of an internal data source and a standard is kept by special implementations that allow a data source to import and export data in any of the standardized formats (connection 6 in figure 3). However, in addition to this, the specification of the standards tries to support integration between data sources (connections 7 and 3 in figure 3). One example of this is the xref attribute used in the PSI MI standard (see table 6). The standard supports links between different data sources, and by following the given links it is possible to find related interactions in other databases.

The MIRIAM [65] standard has put further effort on this. MIRIAM is a meta standard and the aim is to define the minimal amount of information that must be included in data to make it reusable by other sources. One important issue is to specify how the objects in a description could be annotated, i.e. how the references to other data sources should be defined. For this purpose MIRIAM contains both a database of different web resources and a specification of how these should be referred. Examples of how this is used for the Tyson Cell model in the Biomodels database are shown in table 7. The table shows that the Tyson model is considered to be part of the Fungi Metozoa group as specified in the UniProt database. There are also versions of this model in other databases and ontologies: KEGG, GO and Reactome. Similarly, the bottom part of the table shows where more information on one of the first reactions of this model can be found. Here the information is linked to the Enzyme nomenclature, GO and Reactome.

Table 6. PSI MI interaction

PSI MI: Interaction

id	identifier
xref	reference to identifier for this interaction in other databases
interactiontype	further specification of kind of interaction (OBO class)
experimentList	experiments to verify this interaction
participantList	molecules participating in the interaction
id	identifier for the participating molecule
names	name of the participating molecule
experimental-role	the molecules role in the experiment (OBO-class)
biological-role	the molecules biological role in the interaction (OBO-class)

Table 7. Tyson model

Tyson model	bqmodel:is	Taxonomy: Fungi/Metazoa group
	bqbiol:isVersionOf	KEGG Pathway: sce04111
		Gene Ontology: mitotic cell cycle
	bqbiol:hasVersion	Reactome REACT_152
Deactivation of cdc-2 kinase	bqbiol:isVersionOf	EnzymeNomenclature: 2.7.10.2
		Gene Ontology: protein amino acid phosphorylation
		Gene Ontology: negative regulation of
		cyclin-dependent protein kinase activity
	bqbiol:hasVersion	Reactome: REACT_3178
		Reactome: REACT_6327

The qualifier isVersionOf in table 7 specifies that this model is a version or instance of the more general model or concept defined by, in this case KEGG and GO. The qualifier hasVersion, on the other hand, specifies that models or concepts, in this case in KEGG, are a version of this model.

8.4   Conclusion

In this chapter we have briefly reviewed different kinds of resources of biological information that are publicly available on the Web. Researchers in the field need information from all these different kinds of resources. However, integrating this information is not an easy task. We described a model for integration that not only makes use of the resources, but also of the connections between the resources. Further, we described recent work on different components in the model.

A resource of biological information that we did not address in this chapter is the large amount of scientific literature. Many biological data sources have curators that collect data from literature and add the data to the data source. Other biological data sources use text mining approaches to gather data. Also here there is more and more work on using ontologies for annotation and search. For instance, all articles in MEDLINE, the main component of PubMed [76], are annotated with MeSH terms. An example of a literature search engine that uses ontologies is GoPubMed [16], which is the topic of chapter 7.

Acknowledgements

All authors of this chapter were members of the EU Network of Excellence REWERSE (Sixth Framework Programme project 506779), working group on Adding Semantics to the Bioinformatics Web.

A large part of this chapter is based on previous work. We thank our collaborators: Vaida Jakonienė for work on data sources, David Hall and Dagmar Köhn for work on standards, and Chen Bi, Anna Edberg and Wei Xu for work on the SAMBO system. Further, we thank Bo Servenius for many valuable discussions over the last years.

We acknowledge the financial support of the Swedish Research Council (Vetenskapsrådet), the Center for Industrial Information Technology (CENIIT), and the Swedish national graduate school in computer science (CUGS).

References

1. Achard, F., Vaysseix, G., Barillot, E.: XML, bioinformatics and data integration. Bioinformatics 17(2), 115–125 (2001)
2. Blake, J., Bult, C.: Beyond the data deluge: data integration and bio-ontologies. Journal of Biomedical Informatics 39, 314–320 (2006)
3. Bleiholder, J., Lacroix, Z., Murthy, L., Naumann, F., Raschid, L., Vidal, M.E.: BioFast: Challenges in exploring linked life science sources. SIGMOD Record 33(2), 72–77 (2004)
4. Bock, J., Hettenhausen, J.: MapPSO results for OAEI 2008. In: Proceedings of the Third International Workshop on Ontology Matching, pp. 176–181 (2008)
5. Brazma, A., Hingamp, P., Quackenbush, J., Sherlock, G., Spellman, P., Stoeckert, C., Aach, J., Ansorge, W., Ball, C.A., Causton, H.C., Gaasterland, T., Glenisson, P., Holstege, F.C.P., Kim, I.F., Markowitz, V., Matese, J.C., Parkinson, H., Robinson, A., Sarkans, U., Schulze-Kremer, S., Stewart, J., Taylor, R., Vilo, J., Vingron, M.: Minimum information about a microarray experiment (MIAME) - toward standards for microarray data. Nature Genetics 29, 365–371 (2001)
6. Caracciolo, C., Euzenat, J., Hollink, L., Ichise, R., Isaac, A., Malaisé, V., Meilicke, C., Pane, J., Shvaiko, P., Stuckenschmidt, H., Svab-Zamazal, O., Svatek, V.: Results of the ontology alignment evaluation initiative 2008. In: Proceedings of the Third International Workshop on Ontology Matching, pp. 73–119 (2008)
7. CARO. Common anatomy reference ontology,
   http://www.bioontology.org/-wiki/index.php/CARO:Main_Page
8. Chen, B., Tan, H., Lambrix, P.: Structure-based filtering for ontology alignment. In: Proceedings of the IEEE WETICE Workshop on Semantic Technologies in Collaborative Applications, pp. 364–369 (2006)
9. Cheung, K.-H., Smith, A.K., Yip, K.Y.L., Baker, C.J.O., Gerstein, M.B.: Semantic web approach to database integration in the life sciences. In: Baker, Cheung (eds.) Semantic Web: Revolutionizing Knowledge Discovery in the Life Sciences, pp. 11–30. Springer, Heidelberg (2007)
10. Collins, F., Green, E., Guttmacher, A., Guyer, M.: A vision for the future of genomics research. Nature 422, 835–847 (2003)
11. The Gene Ontology Consortium. Gene ontology: tool for the unification of biology. Nature Genetics 25(1), 25–29 (2000)
12. Curino, C., Orsi, G., Tanca, L.: X-SOM results for OAEI 2007. In: Proceedings of the Second International Workshop on Ontology Matching, pp. 276–285 (2007)

13. David, J.: AROMA results for OAEI 2008. In: Proceedings of the Third International Workshop on Ontology Matching, pp. 128–131 (2008)
14. Devos, D., Valencia, A.: Intrinsic errors in genome annotation. Trends in Genetics 17(8), 429–431 (2001)
15. Doan, A., Madhavan, J., Domingos, P., Halevy, A.: Ontology matching: A machine learning approach. In: Staab, Studer (eds.) Handbook on Ontologies in Information Systems, pp. 397–416. Springer, Heidelberg (2003)
16. Doms, A., Schroeder, M.: GoPubMed: exploring PubMed with the Gene Ontology. Nucleic Acids Research 33, W783–W786 (2005)
17. Eckman, B., Lacroix, Z., Raschid, L.: Exploiting biomolecular source capabilities for query optimization. In: Proceedings of the IEEE Symposium on Bioinformatics and Bioengineering, pp. 23–32 (2001)
18. Ehrig, M., Haase, P., Stojanovic, N., Hefke, M.: Similarity for ontologies - a comprehensive framework. In: Proceedings of the 13th European Conference on Information Systems (2005)
19. Entrez. The life sciences search engine at the national center for biotechnology information, http://www.ncbi.nlm.nih.gov/Entrez/
20. Etzold, T., Ulyanov, A., Argos, P.: SRS: Information retrieval system for molecular biology data banks. Methods in Enzymology 266, 114–128 (1996)
21. Euzenat, J., Shvaiko, P.: Ontology Matching. Springer, Heidelberg (2007)
22. Falconer, S., Storey, M.-A.: A cognitive support framework for ontology mapping. In: Aberer, K., Choi, K.-S., Noy, N., Allemang, D., Lee, K.-I., Nixon, L.J.B., Golbeck, J., Mika, P., Maynard, D., Mizoguchi, R., Schreiber, G., Cudré-Mauroux, P. (eds.) ASWC 2007 and ISWC 2007. LNCS, vol. 4825, pp. 114–127. Springer, Heidelberg (2007)
23. Francois, J., Kengue, D., Euzenat, J., Valtchev, P.: OLA in the OAEI 2007 evaluation contest. In: Proceedings of the Second International Workshop on Ontology Matching, pp. 188–195 (2007)
24. Garny, A., Nickerson, D., Cooper, J., Weber dos Santos, R., Miller, A.K., McKeever, S., Nielsen, P., Hunter, P.J.: CellML and associated tools and techniques. Philosophical Transactions of the Royal Society A 366(1878), 3017–3043 (2008)
25. Giunchiglia, F., Shvaiko, P., Yatskevich, M.: S-Match: an algorithm and an implementation of semantic matching. In: Bussler, C.J., Davies, J., Fensel, D., Studer, R. (eds.) ESWS 2004. LNCS, vol. 3053, pp. 61–75. Springer, Heidelberg (2004)
26. Goto, S., Okuno, Y., Hattori, M., Nishioka, T., Kanehisa, M.: Ligand: database of chemical compounds and reactions in biological pathways. Nucleic Acids Research 30(1), 402–404 (2002)
27. Gracia, J., Mena, E.: Ontology matching with CIDER: evaluation report for the OAEI 2008. In: Proceedings of the Third International Workshop on Ontology Matching, pp. 140–146 (2008)
28. Hamdi, F., Zargayouna, H., Safar, B., Reynaud, C.: TaxoMap in the OAEI 2008 alignment contest. In: Proceedings of the Third International Workshop on Ontology Matching, pp. 206–213 (2008)
29. Hermjakob, H., Montecchi-Palazzi, L., Bader, G., Wojcik, J., Salwinski, L., Ceol, A., Moore, S., Orchard, S., Sarkans, U., von Mering, C., Roechert, B., Poux, S., Jung, E., Mersch, H., Kersey, P., Lappe, M., Li, Y., Zeng, R., Rana, D., Nikolski, M., Husi, H., Brun, C., Shanker, K., Grant, S.G.N., Sander, C., Bork, P., Zhu, W., Pandey, A., Brazma, A., Jacq, B., Vidal, M., Sherman, D., Legrain, P., Cesareni, G., Xenarios, I., Eisenberg, D., Steipe, B., Hogue, C., Apweiler, R.: The HUPO PSI's molecular interaction format - a community standard for the representation of protein interaction data. Nature Biotechnology 22(2), 177–183 (2004)

30. Hu, W., Zhao, Y., Li, D., Cheng, G., Wu, H., Qu, Y.: Falcon-AO: results for OAEI 2007. In: Proceedings of the Second International Workshop on Ontology Matching, pp. 170–178 (2007)

31. Hucka, M., Finney, A., Sauro, H.M., Bolouri, H., Doyle, J.C., Kitano, H., Arkin, A.P., Bornstein, B.J., Bray, D., Cornish-Bowden, A., Cuellar, A.A., Dronov, S., Gilles, E.D., Ginkel, M., Gor, V., Goryanin II, Hedley, W.J., Hodgman, T.C., Hofmeyr, J.-H., Hunter, P.J., Juty, N.S., Kasberger, J.L., Kremling, A., Kummer, U., Le Novère, N., Loew, L.M., Lucio, D., Mendes, P., Minch, E., Mjolsness, E.D., Nakayama, Y., Nelson, M.R., Nielsen, P.F., Sakurada, T., Schaff, J.C., Shapiro, B.E., Shimizu, T.S., Spence, H.D., Stelling, J., Takahashi, K., Tomita, M., Wagner, J., Wang, J.: The systems biology markup language (SBML): a medium for representation and exchange of biochemical network models. Bioinformatics 19(4), 524–531 (2003)

32. Jasper, R., Uschold, M.: A framework for understanding and classifying ontology applications. In: Proceedings of the IJCAI Workshop on Ontologies and Problem-Solving Methods: Lessons Learned and Future Trends (1999)

33. Jean-Mary, Y.R., Kabuka, M.R.: ASMOV results for OAEI 2007. In: Proceedings of the Second International Workshop on Ontology Matching, pp. 150–159 (2007)

34. Jean-Mary, Y.R., Kabuka, M.R.: ASMOV results for OAEI 2008. In: Proceedings of the Third International Workshop on Ontology Matching, pp. 132–139 (2008)

35. Kalfoglou, Y., Schorlemmer, M.: IF-Map: an ontology mapping method based on information flow theory. In: Spaccapietra, S., March, S., Aberer, K. (eds.) Journal on Data Semantics I. LNCS, vol. 2800, pp. 98–127. Springer, Heidelberg (2003)

36. Kalfoglou, Y., Schorlemmer, M.: Ontology mapping: the state of the art. The Knowledge Engineering Review 18(1), 1–31 (2003)

37. Karasavvas, K.A., Baldock, R., Burger, A.: Bioinformatics integration and agent technology. Journal of Biomedical Informatics 37, 205–219 (2004)

38. Karp, P.: A strategy for database interoperation. Journal of Computational Biology 2(4), 573–586 (1996)

39. Köhn, D., Strömbäck, L.: A method for semi-automatic integration of standards in systems biology. In: Bhowmick, S.S., Küng, J., Wagner, R. (eds.) DEXA 2008. LNCS, vol. 5181, pp. 745–752. Springer, Heidelberg (2008)

40. Kotis, K., Vouros, G.A.: The HCONE approach to ontology merging. In: Bussler, C.J., Davies, J., Fensel, D., Studer, R. (eds.) ESWS 2004. LNCS, vol. 3053, pp. 137–151. Springer, Heidelberg (2004)

41. Lacroix, Z.: Issues to address while designing a biological information system. In: Lacroix, Critchlow (eds.) Bioinformatics: Managing Scientific data, pp. 75–108. Kaufmann, San Francisco (2003)

42. Lacroix, Z., Morris, T., Parekh, K., Raschid, L., Vidal, M.-E.: Exploiting multiple paths to express scientific queries. In: Proceedings of the International Conference on Scientific and Statistical Database Management (2004)

43. Lambrix, P.: Towards a semantic web for bioinformatics using ontology-based annotation. In: Proceedings of the 14th IEEE International Workshops on Enabling Technologies: Infrastructures for Collaborative Enterprises, pp. 3–7 (2005) (invited talk)

44. Lambrix, P., Edberg, A.: Evaluation of ontology merging tools in bioinformatics. In: Proceedings of the Pacific Symposium on Biocomputing, pp. 589–600 (2003)

45. Lambrix, P., Liu, Q.: Using partial reference alignments to align ontologies. In: Aroyo, L., et al. (eds.) ESWC 2009. LNCS, vol. 5554, pp. 188–202. Springer, Heidelberg (2009)

46. Lambrix, P., Strömbäck, L.: Where is my protein? - issues in information integration. BIOforum Europe 11(7-8), 24–26 (2007) (invited contribution)

47. Lambrix, P., Tan, H.: SAMBO - a system for aligning and merging biomedical ontologies. Journal of Web Semantics, Special issue on Semantic Web for the Life Sciences 4(3), 196–206 (2006)
48. Lambrix, P., Tan, H.: A tool for evaluating ontology alignment strategies. In: Spaccapietra, S., Atzeni, P., Fages, F., Hacid, M.-S., Kifer, M., Mylopoulos, J., Pernici, B., Shvaiko, P., Trujillo, J., Zaihrayeu, I. (eds.) Journal on Data Semantics VIII. LNCS, vol. 4380, pp. 182–202. Springer, Heidelberg (2007)
49. Lambrix, P., Tan, H., Jakonienė, V., Strömbäck, L.: Biological ontologies. In: Baker, Cheung (eds.) Semantic Web: Revolutionizing Knowledge Discovery in the Life Sciences, pp. 85–99. Springer, Heidelberg (2007)
50. Lambrix, P., Tan, H., Liu, Q.: SAMBO and SAMBOdtf results for the ontology alignment evaluation initiative 2008. In: Proceedings of the Third International Workshop on Ontology Matching, pp. 190–198 (2008)
51. Lambrix, P., Tan, H., Xu, W.: Literature-based alignment of ontologies. In: Proceedings of the Third International Workshop on Ontology Matching, pp. 219–223 (2008)
52. Le, B.T., Dieng-Kuntz, R., Gandon, F.: On ontology matching problem (for building a corporate semantic web in a multi-communities organization). In: Proceedings of 6th International Conference on Enterprise Information Systems (2004)
53. Lenzerini, M.: Data integration: a theoretical perspective. In: Proceedings of the twenty-first ACM SIGMOD-SIGACT-SIGART Symposium on Principles of Database Systems, pp. 233–246 (2002)
54. Li, Y., Zhong, Q., Li, J., Tang, J.: Result of ontology alignment with RiMOM at OAEI 2007. In: Proceedings of the Second International Workshop on Ontology Matching, pp. 227–235 (2007)
55. Ludäscher, B., Gupta, A., Martone, M.: A model-based mediator system for scientific data management. In: Lacroix, Critchlow (eds.) Bioinformatics: Managing Scientific data, pp. 335–370. Kaufmann, San Francisco (2003)
56. Mao, M., Peng, Y.: The PRIOR+: results for OAEI campaign 2007. In: Proceedings of the Second International Workshop on Ontology Matching, pp. 219–226 (2007)
57. Massmann, S., Engmann, D., Rahm, E.: COMA++: Results for the ontology alignment contest OAEI 2006. In: Proceedings of the First International Workshop on Ontology Matching (2006)
58. McEntire, R., Karp, P., Abrenethy, N., Benton, D., Helt, G., DeJongh, M., Kent, R., Kosky, A., Lewis, S., Hodnett, D., Neumann, E., Olken, F., Pathak, D., Tarzy-Hornoch, P., Tolda, L.L., Topaloglou, T.: An evaluation of ontology exchange languages for bioinformatics. In: International Conference on Intelligent Systems for Molecular Biology, pp. 239–250 (2000)
59. McGuinness, D., Fikes, R., Rice, J., Wilder, S.: An environment for merging and testing large ontologies. In: Proceedings of the Seventh International Conference on Principles of Knowledge Representation and Reasoning, pp. 483–493 (2000)
60. Miled, Z., Webster, Y., Liu, Y., Li, N.: An ontology for semantic integration of life science web databases. International Journal of Cooperative Information Systems 12(2), 275–294 (2003)
61. Nagy, M., Vargas-Vera, M., Motta, E.: DSSim - managing uncertainty on the semantic web. In: Proceedings of the Second International Workshop on Ontology Matching, pp. 160–169 (2007)
62. Nagy, M., Vargas-Vera, M., Stolarski, P., Motta, E.: DSSim results for OAEI 2008. In: Proceedings of the Third International Workshop on Ontology Matching, pp. 147–159 (2008)
63. Neches, R., Fikes, R., Finin, T., Gruber, T., Swartout, W.: Enabling technology for knowledge engineering. AI Magazine 12(3), 26–56 (1991)

64. Le Novère, N., Bornstein, B., Broicher, A., Courtot, M., Donizelli, M., Dharuri, H., Li, L., Sauro, H., Schilstra, M., Shapiro, B., Snoep, J.L., Hucka, M.: BioModels database: A free, centralized database of curated, published, quantitative kinetic models of biochemical and cellular systems. Nucleic Acids Research 34, D689–D691 (2006)

65. Le Novère, N., Finney, A., Hucka, M., Bhalla, U.S., Campagne, F., Collado-Vides, J., Crampin, E.J., Halstead, M., Klipp, E., Mendes, P., Nielsen, P., Sauro, H., Shapiro, B., Snoep, J.L., Spence, H.D., Wanne, B.L.: Minimum information requested in the annotation of biochemical models (MIRIAM). Nature Biotechnology 23, 1509–1515 (2005)

66. Noy, N.F.: Semantic integration: A survey of ontology-based approaches. SIGMOD Record 33(4), 65–70 (2004)

67. Noy, N.F., Griffith, N., Musen, M.: Collecting community-based mappings in an ontology repository. In: Sheth, A.P., Staab, S., Dean, M., Paolucci, M., Maynard, D., Finin, T., Thirunarayan, K. (eds.) ISWC 2008. LNCS, vol. 5318, pp. 371–386. Springer, Heidelberg (2008)

68. Noy, N.F., Musen, M.: PROMPT: Algorithm and tool for automated ontology merging and alignment. In: Proceedings of Seventeenth National Conference on Artificial Intelligence, pp. 450–455 (2000)

69. Noy, N.F., Musen, M.: Anchor-PROMPT: Using non-local context for semantic matching. In: Proceedings of the IJCAI Workshop on Ontologies and Information Sharing, pp. 63–70 (2001)

70. OBO. Open biomedical ontologies, http://obofoundry.org/

71. EU FP6 Network of Excellence on Reasoning on the Web with Rules and Semantics REWERSE. Deliverables of the A2 working group, http://www.rewerse.net/

72. Ossewaarde, R.: Simple library thesaurus alignment with SILAS. In: Proceedings of the Second International Workshop on Ontology Matching, pp. 255–260 (2007)

73. Oster, S., Langella, S., Hastings, S., Ervin, D., Madduri, R., Phillips, J., Kurc, T., Siebenlist, F., Covitz, P., Shanbhag, K., Foster, I., Saltz, J.: caGRID 1.0: An enterprise grid infrastructure for biomedical research. Journal of the American Medical Informatics Association 15, 138–149 (2008)

74. Mitra, P.P., Wiederhold, G.: Resolving terminological heterogeneity in ontologies. In: Proceedings of the ECAI Workshop on Ontologies and Semantic Interoperability (2002)

75. Prasad, S., Peng, Y., Finin, T.: Using explicit information to map between two ontologies. In: Proceedings of the AAMAS Workshop on Ontologies in Agent Systems (2002)

76. PubMed, http://www.ncbi.nlm.nih.gov/pubmed/

77. Quix, C., Geisler, S., Kensche, D., Li, X.: Results of GeRoMeSuite for OAEI 2008. In: Proceedings of the Third International Workshop on Ontology Matching, pp. 160–166 (2008)

78. Robbins, R.: Bioinformatics: Essential infrastructure for global biology. Journal of Computational Biology 3(3), 465–478 (1996)

79. Ruebenacker, O., Moraru II, Schaff, J.C., Blinov, M.L.: Kinetic modeling using BioPAX ontology. In: IEEE International Conference on Bioinformatics and Biomedicine, pp. 339–348 (2007)

80. Sabou, M., Gracia, J.: Spider: bringing non-equivalence mappings to OAEI. In: Proceedings of the Third International Workshop on Ontology Matching, pp. 199–205 (2008)

81. Sahoo, S., Bodenreider, O., Rutter, J.L., Skinner, K.J., Sheth, A.P.: An ontology-driven semantic mashup of gene and biological pathway information: application to the domain of nicotine dependence. Journal of Biomedical Informatics 41, 752–765 (2008)

82. Seddiqui, M.H., Aono, M.: Alignment results of Anchor-Flood algorithm for OAEI 2008. In: Proceedings of the Third International Workshop on Ontology Matching, pp. 120–127 (2008)

83. Shaker, R., Mork, P., Brockenbrough, J., Donelson, L., Tarczy-Hornoch, P.P.: The biomediator system as a tool for integrating biologic databases on the web. In: Proceedings of the VLDB Workshop on Information Integration on the Web (2004)

84. Shvaiko, P., Euzenat, J.: A survey of schema-based matching approaches. In: Spaccapietra, S. (ed.) Journal on Data Semantics IV. LNCS, vol. 3730, pp. 146–171. Springer, Heidelberg (2005)

85. Shvaiko, P., Euzenat, J.: Ten challenges for ontology matching. In: Proceedings of the 7th International Conference on Ontologies, Databases, and Applications of Semantics (2008)

86. Smith, B., Ceusters, W., Klagges, B., Köhler, J., Kumar, A., Lomax, J., Mungall, C., Neuhaus, F., Rector, A., Rosse, C.: Relations in biomedical ontologies. Genome Biology 46, R46 (2005)

87. Spiliopoulos, V., Valarakos, A.G., Vouros, G.A., Karkaletsis, V.: SEMA: results for the ontology alignment contest OAEI 2007. In: Proceedings of the Second International Workshop on Ontology Matching, pp. 244–254 (2007)

88. Stevens, R., Goble, C., Bechhofer, S.: Ontology-based knowledge representation for bioinformatics. Briefings in Bioinformatics 1(4), 398–414 (2000)

89. Stevens, R., Goble, C., Bechhofer, S., Ng, G., Baker, P.: Complex query formulation over diverse information sources in TAMBIS. In: Lacroix, Critchlow (eds.) Bioinformatics: Managing Scientific data, pp. 189–223. Kaufmann, San Francisco (2003)

90. Strömbäck, L., Hall, D., Lambrix, P.: A review of standards for data exchange within systems biology. Proteomics 7(6), 857–867 (2007) (invited contribution)

91. Strömbäck, L., Jakonienė, V., Tan, H., Lambrix, P.: Representing, storing and accessing molecular interaction data: a review of models and tools. Briefings in Bioinformatics 7(4), 331–338 (2006) (invited contribution)

92. Strömbäck, L., Lambrix, P.: Representations of molecular pathways: An evaluation of SBML, PSI MI and BioPAX. Bioinformatics 21(24), 4401–4407 (2005)

93. Stumme, G., Mädche, A.: FCA-Merge: Bottom-up merging of ontologies. In: Proceedings of the 17th International Joint Conference on Artificial Intelligence, pp. 225–230 (2001)

94. Su, X.M., Hakkarainen, S., Brasethvik, T.: Semantic enrichment for improving systems interoperability. In: Proceedings of the ACM Symposium on Applied Computing, pp. 1634–1641 (2004)

95. Sunna, W., Cruz, I.: Using the AgreementMaker to align ontologies for the OAEI campaign 2007. In: Proceedings of the Second International Workshop on Ontology Matching, pp. 133–138 (2007)

96. Tan, H., Jakonienė, V., Lambrix, P., Aberg, J., Shahmehri, N.: Alignment of biomedical ontologies using life science literature. In: Bremer, E.G., Hakenberg, J., Han, E.-H(S.), Berrar, D., Dubitzky, W. (eds.) KDLL 2006. LNCS (LNBI), vol. 3886, pp. 1–17. Springer, Heidelberg (2006)

97. Tan, H., Lambrix, P.: A method for recommending ontology alignment strategies. In: Aberer, K., Choi, K.-S., Noy, N., Allemang, D., Lee, K.-I., Nixon, L.J.B., Golbeck, J., Mika, P., Maynard, D., Mizoguchi, R., Schreiber, G., Cudré-Mauroux, P. (eds.) ISWC 2007. LNCS, vol. 4825, pp. 494–507. Springer, Heidelberg (2007)

98. Tan, H., Lambrix, P.: SAMBO results for the ontology alignment evaluation initiative 2007. In: Proceedings of the Second International Workshop on Ontology Matching, pp. 236–243 (2007)

99. Tannen, V., Davidson, S., Harker, S.: The information integration system K2. In: Lacroix, Critchlow (eds.) Bioinformatics: Managing Scientific data, pp. 225–248. Kaufmann, San Francisco (2003)

100. Taylor, C.F., Paton, N.W., Lilley, K.S., Binz, P.-A., Julian Jr., R.K., Jones, A.R., Zhu, W., Apweiler, R., Aebersold, R., Deutsch, E.W., Dunn, M.J., Heck, A.J.R., Leitner, A., Macht, M., Mann, M., Martens, L., Neubert, T.A., Patterson, S.D., Ping, P., Seymour, S.L., Souda, P., Tsugita, A., Vandekerckhove, J., Vondriska, T.M., Whitelegge, J.P., Wilkins, M.R., Xenarios, I., Yates III, J.R., Hermjakob, H.: The minimum information about a proteomics experiment (MIAPE). Nature Biotechnology 25, 887–893 (2007)

101. Tyson, J.J.: Modeling the cell division cycle: cdc2 and cyclin interactions. Proceedings National Academy of Sciences U.S.A. 88(16), 7328–7332 (1991)

102. Uhrmacher, A., Rolfs, A., Frahm, J.: DFG research training group 1387/1: dIEM oSiRiS - integrative development of modelling and simulation methods for regenerative systems. it - Information Technology 49(6), 388–395 (2007)

103. UMLS. Unified medical language system, http://www.nlm.nih.gov/research/umls/about_umls.html

104. Wächter, T., Tan, H., Wobst, A., Lambrix, P., Schroeder, M.: A corpus-driven approach for design, evolution and alignment of ontologies. In: Proceedings of the Winter Simulation Conference, pp. 1595–1602 (2006) (invited contribution)

105. Wang, P., Xu, B.: LILY: the results for the ontology alignment contest OAEI 2007. In: Proceedings of the Second International Workshop on Ontology Matching, pp. 179–187 (2007)

106. Wang, P., Xu, B.: Lily: ontology alignment results for OAEI 2008. In: Proceedings of the Third International Workshop on Ontology Matching, pp. 167–175 (2008)

107. WordNet, http://wordnet.princeton.edu/

108. BioPAX working Group. BioPAX - biological pathways exchange language. level 1, version 1.0 documentation (2004), http://www.biopax.org

109. Ben Yaghlane, B., Laamari, N.: OWL-CM: OWL combining matcher based on belief functions theory. In: Proceedings of the Second International Workshop on Ontology Matching, pp. 206–218 (2007)

110. Zargayouna, H., Safar, B., Reynaud, C.: TaxoMap in the OAEI 2007 alignment contest. In: Proceedings of the Second International Workshop on Ontology Matching, pp. 268–275 (2007)

111. Zghal, S., Ben Yahia, S., Nguifo, E.M., Slimani, Y.: SODA: an OWL-DL based ontology matching system. In: Proceedings of the Second International Workshop on Ontology Matching, pp. 261–267 (2007)

112. Zhang, S., Bodenreider, O.: Hybrid alignment strategy for anatomical ontologies: results of the 2007 ontology alignment contest. In: Proceedings of the Second International Workshop on Ontology Matching, pp. 139–149 (2007)

113. Zhang, X., Zhong, Q., Li, J., Tang, J.: RiMOM results for OAEI 2008. In: Proceedings of the Third International Workshop on Ontology Matching, pp. 182–189 (2008)

Author Index